MOBILE MIDDLEWARE

MOBILE MIDDLEWARE
ARCHITECTURE, PATTERNS AND PRACTICE

Sasu Tarkoma
Helsinki University of Technology, Finland

A John Wiley and Sons, Ltd., Publication

This edition first published 2009
© 2009 John Wiley & Sons Ltd

Registered office
John Wiley & Sons Ltd, The Atrium, Southern Gate, Chichester, West Sussex, PO19 8SQ, United Kingdom

For details of our global editorial offices, for customer services and for information about how to apply for permission to reuse the copyright material in this book please see our website at www.wiley.com.

Library of Congress Cataloging-in-Publication Data

Tarkoma, Sasu.
  Mobile middleware architecture, patterns, and practice / Sasu Tarkoma.
      p. cm.
  Includes bibliographical references and index.
  ISBN 978-0-470-74073-6 (cloth)
  1. Mobile computing. 2. Middleware. 3. Wireless communication systems. I. Title.
  QA76.59.T385 2009
  005.3–dc22
                                                                            2008053909

A catalogue record for this book is available from the British Library.

ISBN: 978-0-470-74073-6 (H/B)

Typeset in 10/12 Times by Laserwords Private Limited, Chennai, India
Printed and bound in Great Britain by CPI Antony Rowe, Chippenham, Wiltshire

This book is dedicated to the memory of Professor Kimmo Raatikainen.

Contents

Preface

Mobile computing has become truly one of the breakthrough technologies of today with over three billion mobile phones in use. As the computing power and capabilities of the devices are rapidly improving, software has become a crucial issue in the mobile marketplace. Indeed, the current trend is towards converged communication where web resources integrate seamlessly with mobile systems.

Middleware plays a vital role in current distributed systems. This wide technology domain consists of infrastructure and support services for applications and application developers. Middleware typically provides support for various interoperable service deployment and execution related functions.

This book provides a comprehensive overview of mobile middleware technology. The focus is on understanding the key design and architectural patterns, middleware layering, data presentation, specific technological solutions, and standardization.

The book is aimed at software professionals wanting to learn about mobile middleware as well as undergraduate and graduate level university students. The aim of this book is not only to give a comprehensive summary of relevant mobile middleware technologies, but also to give the reader insight into middleware architecture design and the well-known and useful design patterns. The role of middleware in the protocol stack will be extensively discussed in the book. The more theoretical notions and architectural solutions will be motivated and demonstrated by presenting how they are applied in practice.

About the Author

Sasu Tarkoma received his M.Sc. and Ph.D. degrees in Computer Science from the University of Helsinki, Department of Computer Science. He is currently professor at Helsinki University of Technology, Department of Computer Science and Engineering. He is Docent at the Faculty of Science, University of Helsinki. He has managed and participated in national and international research projects at the University of Helsinki, Helsinki University of Technology (TKK), and Helsinki Institute for Information Technology (HIIT). He has worked in the IT industry as a consultant and chief system architect, and he is principal member of research staff at Nokia Research Center. He has over 100 publications including 60 refereed scientific articles, and he has also contributed to several books on mobile middleware.

List of Contributors

Dr. Jaakko Kangasharju contributed most of the content for Chapters 5, 6, and 8 of the book. He received his Ph.D. degree in Computer Science from the University of Helsinki in 2008. He is currently working as a senior researcher at the Department of Computer Science and Engineering of the Helsinki University of Technology, managing projects on middleware in sensor networks and mobile computing. His previous work includes research on XML and web services in mobile and wireless environments. He has contributed to mobile middleware standardization at the OMG and in XML-related standardization at the W3C.

Ms. Nelli Tarkoma produced most of the diagrams used in this book. She is a professional graphic artist and illustrator.

The book has been inspired by the long running mobile middleware research project, Fuego Core, that was active at the Helsinki Institute for Information Technology HIIT from 2002 to 2007.

1

Introduction

1.1 Mobile Middleware

Mobile devices are increasingly dependent on good software and ultimately good user experience. In order to be able to rapidly develop good software in this operating environment that is very different from desktop computing, a lot of support services are needed. Typically, these are provided in the *middleware layer*, which is lower than applications, but above the operating system and basic TCP/IP protocol stack. Thus middleware provides a level of indirection and transparency for applications. By providing commonly used services using standardized or well-known interfaces, application developers save time, and ultimately cost, when developing their products. To consider some examples, websites and mashups, industrial systems, banking systems, and stock market systems rely extensively on middleware.

This development time and cost has traditionally been high for mobile applications and services, because the environment is more challenging than the typical fixed network environment. Namely, the wireless and mobile environment is not reliable, has long latency, small bandwidth, and there are many different terminal types. Thus one specific implementation is not necessarily usable on all mobile terminals and phones on the market. This motivates the development of mobile middleware solutions that abstract many of the issues pertaining to the operating environment for the developers and that supports adaptability to the current devices and operating environment.

Mobile middleware has evolved by leaps and bounds during recent years. From the client system viewpoint, developments such as the *Java 2 Micro Edition (Java ME)*, have enabled the creation of custom software for phones and other small devices that can be deployed at runtime. In addition to Java technologies, browser-based technologies have also evolved, and both thin browsers and rich browsers are available. A more recent advance has been the introduction of mobile web servers, thus allowing people to provide resources on their mobile devices. Today's systems are not very flexible in terms of the operating environment and usage context. Indeed, it is expected that forthcoming mobile applications are more asynchronous in nature, *context-aware*, and cope better with the different usage environments available.

Mobile Middleware   Sasu Tarkoma
© 2009 John Wiley & Sons, Ltd

1.2   Mobile Applications and Services

We characterize mobile applications and services by dividing them into four generations. This allows us to inspect the evolution of the mobile application landscape and outline some of the significant trends.

Early mobile phones in the 1980s did not support applications at all, but provided only the basic voice services. The first generation mobile applications and services, introduced around 1991, were restricted by technology. The two key enablers for applications were mobile data and the *Short Message Service (SMS)*. Indeed, the role of SMS has been paramount in the path towards mobile services. The iMode has also been instrumental in Japan for mobile applications, combining low-cost data and messaging.

The second generation of mobile applications has been supported by built-in browsers, such as the *Wireless Application Protocol (WAP)* browsers, and more recently light-weight web browsers, such as the ones on available on Nokia series 60 devices and the Apple iPhone. The second generation introduced a new messaging service around 2001 that supports multimedia content messages, namely the *Multimedia Messaging Service (MMS)*. MMS allows users to compose messages that include images, audio, and video content.

The third generation applications are supported by a more sophisticated environment than just basic messaging and data services, and limited browsing, as in the case of the previous generations. The third generation applications and services are built on top of a platform that offers various services, such as location support, content adaptation, storage, and caching, to name a few well-known support services. We are now witnessing the emergence of such third generation applications and example platforms include Series 60, J2ME, Android, and iPhone. Indeed, these new devices are able to support middleware and more complex applications as *smartphones*.

A web browser that has been adapted or created for mobile devices is typically called a *microbrowser* if the offered feature set is considerably limited due to device constraints. The feature set needs to be limited due to the processing power and display capabilities of various devices. Since 2006, an increasing number of mobile devices have appeared on the market that are capable of supporting mobile web browsers with advanced features, such as CSS 2.1, JavaScript, and *Asynchronous Javascript (AJAX)*. Indeed, the previously two separate worlds of wireless telecommunications and the Internet can be seen to be on a converging evolutionary paths.

The fourth generation of applications has not yet arrived and we briefly sketch the expected properties of these applications in the light of recent proposals in research and standardization communities. The fourth generation is expected to be adaptive not only in terms of application behaviour and content, but also regarding the networking stack and air interface. Always-on connectivity, multi-mode communications, mesh networking, and adaptive use network interfaces and physical communication medium can be said to be important parts of future mobile computing devices [1].

The following list summarizes the evolution of mobile applications and services with approximate dates for the generations:

- *1st (1990–1999).* Text messages (SMS) and mobile data. Speeds up to tens of Kbps.
- *2nd (1999–2003).* Limited browsers, WAP, iMode, and MMS. Speeds up to 144 Kbps.

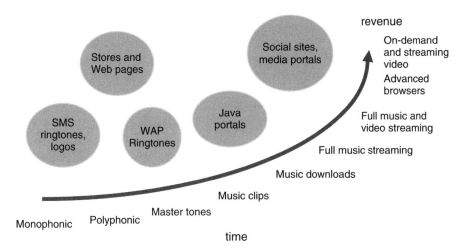

Figure 1.1 Evolution of mobile content

- *3rd (2003–2008)*. Mobile platforms, middleware services. Series 60, J2ME, Android, iPhone. Speeds up to several Mbps.
- *4th (2008–)*. Adaptive services, user interfaces, and protocols. Context-awareness, always-on connectivity. Speeds up to hundreds of Mbps.

Figure 1.1 illustrates the evolution of mobile content. Early mobile content was based on short text messages, namely SMS, ringtones, and logos. The early ringtones were monophonic and purchasable using SMS messages. Later monophonic ringtones were replaced with polyphonic tones and then music clips. More recent developments include storing more and more audio and video on the mobile devices and sharing it with other devices. This is typically supported using portals and other web services. The current trend is towards full music and video streaming over the Internet, and Video-on-Demand with the emergence of massively popular sites such as YouTube.

1.3 Middleware Services

Figure 1.2 illustrates the layered nature of the communications stack of today's computing devices. The network layer, namely the *Internet Protocol (IP)*, acts as the common technology for interoperability. Indeed, the TCP/IP protocol is said to have an hourglass shape due to this reliance on IP. In this book, we make the assumption that middleware and applications are built on top of the TCP/IP stack. Given the current dominant role of IP in networking and its emerging role in telecommunications systems, this appears to be a very reasonable assumption.

The four-layer TCP/IP has gained a somewhat fuzzy layer called middleware, which consists of various support services and APIs for higher level applications. Middleware typically includes services such as messaging and *Remote Procedure Call (RPC)* facilities, resource discovery, transactions, security, directory, and storage services.

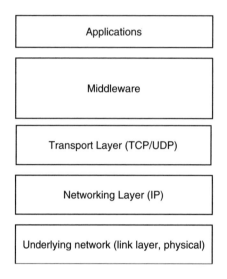

Figure 1.2 Layered network architecture

The basic communications middleware typically includes the following four services:

- *Messaging service*, which allows entities to send and receive one-way messages. The messaging service is a crucial part of a subset of middleware, called *message-oriented middleware*. This service is used to realize more complex interactions, for example the other three services. UDP can be seen as the primitive message service provided by the TCP/IP stack.
- (RPC), which allows the invocation of a remote procedure as if it were local. This requires the marshalling and unmarshalling of the request parameters and return values.
- *Remote Method Invocation (RMI)*, which allows the call of a remote method as if it were local. RMI is the Java way of doing RPC and requires the use of distributed garbage collection to remove unnecessary state in the distributed system.
- *Event service*, which allows entities to receive asynchronous notifications. A notification is the result of some observable event, and results in a possible state change in the listener. Thus, events are a basic building block for adaptive distributed systems.

We observe that the mobile environment, due to its intermittent connectivity and reachability challenges, presents significant challenges for data synchronization. Given that mobile devices have increasing amounts of volatile and non-volatile storage, it becomes challenging to keep data items synchronized over the networks. Therefore, data synchronization service can be seen as a basic service that must be included in a mobile middleware solution.

From the server-side viewpoint, we are currently witnessing the convergence of telecommunications technologies, such as the *Session Initiation Protocol (SIP)*, used in *Beyond 3G networks (B3G)*, and websites and services. Middleware plays a crucial role in this converged communications environment. This server-side part of mobile middleware facilitates the development, deployment, and execution of services for mobile devices.

1.4 Transparencies

Middleware provides various transparencies for higher layers that are not typically supported by the lower layers. These transparencies include location, transport, operating system, programming language, and failure transparencies. In the following, we briefly summarize the transparencies typically provided by middleware.

Location transparency abstracts network attachment points of communicating entities from applications. This allows applications to communicate with each other without knowing the locations of the destinations. A typical example of this is the host name, which abstracts the IP address, the topological location of an entity in the Internet, from the application.

Transport protocol transparency means that a given interaction can utilize any transport protocol suitable for the task. *Transparency of the operating system and programming language* means that external data representation is used to describe interactions and any data associated with them. Language and operating system independent description of both procedures and data types is needed. External data representation is crucial for supporting software for multiple operating environments and ensuring communications between them.

For example, web servers have been implemented in many different environments ranging from small embedded devices and mobile phones to large-scale clusters. The *Hypertext Transfer Protocol (HTTP)* protocol standardized by the *Internet Engineering Task Force (IETF)* ensures ubiquitous access to web resources. Hence, standards play a very important role in data communications. HTTP has become the de facto protocol for middleware and application communications due to its central role in web access, simple specification, and firewall friendliness. In addition to HTTP, the *Extensible Markup Language (XML)* standardized by the *World Wide Web Consortium (W3C)* has become the universally accepted data interchange format.

1.5 Mobile Environment

The mobile environment is radically different from the traditional fixed-network environment, which typically has low latency, high bandwidth, and reliable communications. The mobile environment is subject to disconnections and reachability problems. Moreover, mobile devices typically have multiple interfaces for communication, limited battery, limited processing capability, and limited memory.

The capabilities and limitations of the device dictates the constraints on how the user is able to access the services and what kind of content is provided for the user. The capabilities may be divided into two categories: *computational capabilities* and *interface capabilities*. The computational characteristics define what kind of resources the client device has in terms of providing services. Some devices are only capable of acting as an interface to the services running in the network node (e.g. web browsers), whereas more capable devices may run services or parts of the services by themselves. The interface capabilities dictate the characteristics of how the service is displayed to the end-user. For instance, a mobile phone has limited ways of showing high-end images and video, whereas a tablet computer is less limited in this respect.

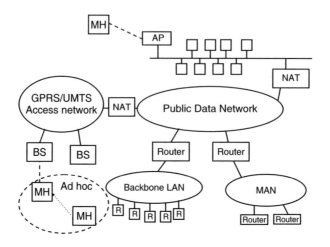

Figure 1.3   The mobile and wireless environment

Figure 1.3 presents an overview of the mobile and wireless communication environment. The environment consists of a number of distinct communications scenarios, including:

- fixed-network communications;
- communications with a wireless access point; and
- wireless ad hoc communications.

We define the central elements of the environment as follows: *MH* denotes the Mobile Host, *BS* is a wireless Base Station that provides connectivity to MHs, an Access Network is a network that provides connectivity for MHs and connects a set of BSes. *AP* denotes an Access Point, which provides connectivity for MHs. We refer to BSes as part of a mobile access network, such as *Global System for Mobile (GSM)*, *General Packet Radio Service (GPRS)*, or *Universal Mobile Telecommunications System (UMTS)*, and an AP as part of a wireless *Local Area Network (LAN)*. Different access networks and other networks are reachable over the Internet backbone. In the figure, *MAN* denotes Metropolitan Area Network and *R* a single router.

We briefly summarize the state of the art wireless technologies that have been deployed today. The phenomenally successful 2G digital cellular networks have been extended with 2.5G data services such as GPRS and EDGE, which are now being widely used in Europe and Asia. After a slow start, 3G networks and services are now available with the data rates from hundreds of kilobytes to several megabits per second. A significant trend has been toward all IP access networks. Indeed, it is expected that Beyond-3G systems will be based on IP technologies.

In addition to cellular technologies, 802.11 *Wireless LAN (WLAN)* has proved to be hugely popular in the residential market and in the enterprise sector, despite security concerns. WLAN offers typically 11Mbit/s and 54Mbit/s data rates. *Worldwide Interoperability for Microwave Access (WiMax)* is another example of wireless LAN technology. WiMax is an industry forum that develops point-to-point wireless access standards for home and mobile users. WiMax offers much larger bandwidth than 802.11 of up to 70 Mbit/s.

Now, given the popularity of WLAN access points and the cost of building a cellular data access, some companies are building WLAN access networks on a peer-to-peer basis. These mesh networks are currently an active topic in research and development, and may well be part of future wireless access solutions.

In addition to long-range technologies, *radio frequency identification (RFID)* and *Wireless Personal Area Network (WPAN)* are examples of short-range systems. An RFID tag is a small integrated circuit that can respond to a signal with some data. RFIDs are extensively used to track goods and people. RFIDs can be passive and only react to external interrogating signals, or they can be active and send beacons. Passive RFIDs do not require a battery whereas active ones do. WPAN is another example of short-range wireless communications. WPAN is a collection of short-range, low-power technologies, including infrared, Bluetooth, and ultra-wideband (UWB) technologies.

1.6 Context-awareness

Context information is important for applications that need to adapt to different situations [2]. To be context-aware, a system must gather contextual information from the surrounding environment. This information can then be used to support adaptability and ultimately better user experience.

There are many ways to define context and context-awareness. A. Dey states context is any information that can be used to characterize the situation of an entity. An entity is a person, place or object that is considered relevant to the interaction between a use and an application, including the user and applications themselves. Primary context types include location, identity, time and activity that characterize the situation of a particular entity.

1.7 Mobility

Wireless devices are not stationary and may move from one network to another. This kind of *terminal mobility* requires special protocol support both on the mobile device, and in the infrastructure. Another form of mobility, called *user mobility*, happens when a user relocates and switches devices. The user should be able to use the same services irrespective of the time, location, and device. This form of user mobility also requires support from the distributed system. The notion of mobility can be taken further and considered as *logical mobility*. Logical mobility pertains to changes in user interests, context, and other attributes defined in a multi-dimensional space that does not necessarily have a one-to-one mapping with the physical world. We shall return to these different notions of mobility later in this book.

Typically networks are separated by firewalls and *Network Address Translation (NAT)* devices. NAT devices support the separation of network address spaces by allowing private IP addresses which are then mapped to one or more public IP addresses. Only public IP addresses are globally routable. This means that if a communication path has one or more NAT devices, the end points of a communication may not necessarily be able to communicate with each other. Indeed, *NAT traversal* has become an active area of standardization at IETF. The goal of NAT traversal is to detect the presence of NATs and then utilize different techniques depending on the environment to ensure end-to-end connectivity. We shall return to the issue of reachability in the following chapter.

1.8 Example Use Case

The Information Society and Technology Advisory Group's (ISTAG)[1] future ambient intelligent scenarios illustrate future ubiquitous environments.

To give an example, we consider the first scenario, namely 'Maria – The Road Warrior'. In this scenario the user, Maria, is on a business trip and arrives at a foreign country. Maria is carrying a personalized communications device, the *P-Com*, which automates various tasks during her travels. For example, from the airport to the hotel: her visa is automatically checked at the immigration, her car rental is arranged beforehand, the P-Com recognizes and personalizes the rental car for Maria. Real-time traffic instructions are provided by the car and P-Com during the drive to the hotel. Finally, the hotel room facilities are customized for Maria by interactions between the P-Com and the hotel room computing infrastructure.

All of these above tasks require the knowledge of the user preferences and the computing and networking infrastructure at a given time and in a particular location. The networking environment is heterogeneous and may consist of various technologies, such as Wireless LAN, infrared, Bluetooth, 2G and 3G, Wibree, and WiMax. WLAN, 2G, 3G, and WiMax are examples of long-range technologies, whereas Bluetooth and infrared are short-range technologies.

For example, at the airport there may be WLAN hotspots available, and at the immigration there could be local Bluetooth coverage. During the walk from the airport arrival hall and at the car garage a 3G network is available, and once Maria enters the car, there could be a Bluetooth network specific to the car. During the drive to the hotel, traffic information is delivered using a 3G network. At the hotel, the hotel room may contain a WLAN hotspot to control room facilities and to access the Internet. In order to make all the different network infrastructures transparent to Maria, the P-Com can make use of the wireless network and location information, and based on these, change seamlessly from one network to another.

The scenario points out several notable characteristics of the ubiquitous environment including the presence of multiple and possibly overlapping wireless networks, seamless roaming between the access networks, different kinds of client terminals and network elements, location awareness, context awareness, data and content centric communications, and personalization.

We also observe that the interactions are user-centric and the goal of the distributed system is to support the user and anticipate user requirements. The interactions are also inherently context-dependent and content-centric. Maria is not interested in knowing which Web server is providing certain information, she as a user is only concerned that she receives the relevant content, for example driving instructions.

Compelling usage scenarios can also be found in the Wireless World Research Forum's Book of Visions [3, 4].

---

[1] ISTAG is the advisory body to the European Commission in the area of ICT.

1.9   Requirements for Mobile Computing

Mobile computing consists of many facets. As we have discussed already, the wireless and dynamic nature of the environment presents many challenges for application and service developers [5].

Mobile computing can be seen as overlapping with several related areas of computing, namely pervasive and ubiquitous computing. They are based on Mark Weiser's vision of ubiquitous computing [6]. The basic setting of pervasive and ubiquitous computing revolves around interactions between distributed sensors and other devices with computational capability. The devices are embedded and seamlessly integrated with the physical environment. One way to distinguish these areas is to consider the goal of mobile computing to offer content to users 'anytime, anywhere', and the goal of pervasive and ubiquitous computing to assist users 'all the time, everywhere'. Therefore the former is a more reactive and the latter a more proactive approach.

We briefly outline the different users of middleware:

- *End user.* The goal of middleware is not to directly interact with the end users, but rather support the applications and services that are visible to the users. This means that middleware should provide sufficient APIs and mechanisms to cope with different kinds of failures and faults, and in general support enhanced usage experience.
- *Device Manufacturers.* Device manufacturers use middleware in order to provide extended features that interface with device drivers.
- *Internet Service Providers.* Internet service providers utilize middleware to monitor and administer the network.
- *Platform Providers.* Platform providers develop middleware platforms that integrate with different operating systems.
- *Application Service Providers.* Application service providers utilize middleware in order to facilitate application development and deployment in a scalable and secure manner.

In this section, we briefly consider the important facets of mobile computing and highlight the central requirements for mobile middleware. We consider the following five non-functional requirements as key elements of mobile computing, namely:

- accessibility;
- reachability;
- adaptability;
- trustworthiness; and
- universality.

These non-functional requirements do not stand alone, but rather they are intertwined. Accessibility means that resources are available and accessible for end users irrespective of the current location and where the resource is located. This has many implications for the protocol stack, middleware, and applications. To be able to ensure accessibility, external data representation is needed. Furthermore, the desired resources must be discovered and located in the network before any access can happen.

Reachability is needed to ensure that resources are available in any location. In order for resources to be available for global access on mobile devices, they need to accessible and reachable. Reachability cannot be taken for granted in today's heterogeneous and dynamic environment depicted in Figure 1.3. As mentioned in this chapter, firewalls and NAT devices pose grave connectivity challenges. Moreover, mobility complicates reachability and requires special solutions in order to ensure the reachability of mobile devices.

Adaptability is a crucial requirement for this environment, because the environment is subject to changes, for example due to physical, logical mobility, or some other change in the environment. The nature of data access is inherently related to adaptability. A mobile device may not have the processing capability to handle all incoming data, or the data may not be presentable to a user without first adapting it to the current display and mode of user interaction. Adaptability implies that the mobile entity, either alone or assisted by some other entities, can monitor these changes in the environment, and then react accordingly and reconfigure the system. Therefore, context-awareness is an essential part of mobile computing. Adaptability also implies reflectivity, namely the ability of the applications to query and modify different parameters and context attributes at runtime.

From the security point of view, the mobile environment poses a number of challenges that must be addressed by the applications, middleware, and the protocol stack. We are faced with questions concerning the authenticity and integrity of data being accessed, and the authenticity of a network entity. There are many ways that malicious entities can disrupt, eavesdrop, and hinder data communications. Well-known examples of attacks include *man-in-the-middle attack*, *impersonation*, *data injection*, and various *Denial of Service (DoS)* attacks. Protocols that are engineered for the mobile environment to support the special requirements of the environment, for example mobility and multi-homing, must not introduce new security problems.

Trust is a challenging notion for distributed systems. An entity trusts another entity if it believes that the other entity follows a contract which is shared by the parties. Before any trust in external entities can be established, the applications must be able to trust the middleware and protocol stack. The notion of a contract is evident in the typical communications setting in which a user pays for a network operator for the services rendered, for example Internet data access. The notion of trust becomes complicated when there are no explicit contracts between the entities, which is typical of decentralized communication environments. Therefore, various security techniques are needed to build and maintain trust.

Security features can be realized on different layers of the protocol stack. Solutions exist for the link layer, network layer, transport layer, middleware, and applications. One current challenge in security is that these different security solutions are independent of each other, and applications do not have mechanisms to determine whether or not a lower layer security solution is being used. In practice, this means that multiple security solutions are used at the same time. The goal of mobile middleware is to provide security services for applications that enable end-to-end and end-to-middle security. These security solutions can then be used by applications to determine the trustworthiness of entities and data items.

Universal data access is one of the key reasons for the success of the Internet. Resources on the Internet are accessible from anywhere, at anytime, using standard protocols and formats. Therefore, universality is also a crucial requirement for mobile computing. This implies that standardized protocols and formats are used to discover and access resources.

Convergent communications, in which telecommunication solutions, such as SIP, and web protocols are used seamlessly, is an example of this requirement in action. Indeed, although a number of dedicated protocols have been developed for mobile data access, the current trend is to utilize web protocols also in the mobile environment. This allows mobile entities to be joined directly as part of the Internet ecosystem rather than a new ecosystem for mobile users being created.

Scalability is implied by the notion of universality. A network system needs to attain critical mass in order to sustain the service ecosystem. Therefore massive scalability is needed before universality can be achieved. Scalability can be perceived in many ways, for example the number of nodes that are supported by the network and the maximum number of messages that can go through a system. Currently there are over three billion mobile phones in the market and the expectation is that the five billion mark will be passed in recent years. The network therefore must be able to support these five billion, or more, devices and able to provide the requested services according to agreements with users. This ensures a crucial role for middleware to be able to support this requirement for scalability both in the device systems and on the server-side.

1.10   Mobile Platforms

Mobile middleware aims to support the development, deployment, and execution of distributed applications in the heterogeneous and dynamic mobile environment. The goals for mobile middleware include adaptability support, fault-tolerance, heterogeneity, scalability, and context-awareness.

The presented non-functional requirements pose significant challenges for software. The industry solution to these challenges has been to create middleware *platforms*. A platform collects frequently used services and APIs under a coherent unified framework. In this book, we consider state of the art mobile platforms including Java ME, *Open Mobile Alliance (OMA)*, Symbian, .NET, Android, and Apple's iPhone. These platforms have many common elements and they follow similar design principles and patterns. In the following chapter, we focus on these platforms and then analyze their commonalities and differences.

1.11   Organization of the Book

In this chapter, we examined the motivation for mobile middleware and considered some important properties of the wireless and mobile environment. The need for supporting easy service development, deployment, and execution has resulted in a fuzzy layer of middleware solutions that are above the operating system, but below the end-user applications. This book aims to present a synthesis of mobile middleware and how it is motivated and positioned with the current Internet protocol stack. We also investigate mobile applications and services and consider their requirements towards the middleware and the network.

Chapter 2 presents key mobile middleware architectures and platforms. We start with a brief overview of the current TCP/IP protocol suite and some of the current networking challenges. Then the key components of middleware architectures and platforms are covered, including objects, components, services and communication mechanisms. We take the Java platform and a recently proposed SPICE platform as examples of large platforms.

The mobile platforms that are examined include Java ME, iPhone, Symbian and Series 60, BREW, WAP, .NET Compact Framework, NoTA, Maemo, and Android. We highlight some of the similarities and differences between the platforms.

Chapter 3 presents a set of important support technologies that are the building block for more complex middleware solutions. We start by giving a brief summary of the SIP architecture and protocol, which is a fundamental protocol for current IP-based voice services, and it is becoming an enabler for other types of services as well. After examining SIP and the IMS system, we continue with web services. Web services and issues pertaining to web service integration are crucial for current and forthcoming mobile services. We consider the different facets of web services and compare them with a more lightweight approach called REST. Other considered technologies include SQLite and OpenGL. Then we focus on service discovery techniques, which are needed to realize information accessibility. We consider UPnP, Jini, SLP, and ZeroConf. One of the aims of mobile middleware is to support mobile terminals and users. The SIP architecture has intrinsic support for different kinds of mobility. In addition to SIP, we consider Mobile IP and Host Identity Protocol for host mobility and Wireless CORBA for object mobility. The advanced topics discussed in this chapter include overlay networks, context-awareness, service composition, security and trust, and charging and billing. Finally, we draw some of the discussed topics together by presenting an example of middleware platfom.

Chapter 4 presents a number of principles and patterns pertinent for middleware and mobile middleware. We start by discussing different distributed system design principles, including Internet, Web, SOA, Security, and mobile principles. After this, a number of crucial architectural patterns are examined. The patterns are presented briefly and the goals and liabilities are summarized. After the architectural patterns, some general patterns are considered before presenting a family of patterns for mobile computing. We focus on patterns pertaining to mobile communications and synchronization.

Chapter 5 presents an overview of relevant standards and motivates interoperability between specifications and implementations. We discuss standardization process in general and examine the key standards organizations from the mobile middleware and computing viewpoint. Key wireless communication standards and middleware standards are briefly mentioned. Some interesting emerging standards are also considered.

In Chapter 6 we consider mobile messaging, which is one of the core middleware services in today's systems. The examined solutions include the web services protocol stack, namely the SOAP protocol, REST, and *Java Message Service (JMS)*. We discuss techniques to optimize messaging for the mobile environment and pay particular attention to push technologies. Message push solutions are needed in order to realize universal data access.

We continue the discussion on asynchronous communication in Chapter 7, which focuses on the *publish/subscribe (pub/sub)* paradigm. Pub/sub is an emerging solution for the mobile and pervasive environment, in which the communication is based on current demand and supply of data. We examine the commonly used information dissemination topologies and consider how middleware can support information delivery based on supply and demand of data. The relevant standards include SIP event package, JMS, and the *Data Distribution Service (DDS)*. We also discuss issues pertaining to mobile data subscribers and publishers, and consider some advanced topics in this area.

Chapter 8 examines a number of data synchronization techniques especially for the mobile environment. We start with an overview of data synchronization, and then proceed to the SyncML standard, which is implemented on most current high-end mobile phones. We also consider other systems including Coda and Unison.

Chapter 9 focuses on security issues and we start with an overview of basic security principles and then consider cryptographic operations and public key infrastructures. This chapter considers security solutions needed on multiple layers of the protocol stack, including link, network, transport, and application layers. We briefly introduce the core security solutions for 3G and Beyond-3G networks, namely AAA, RADIUS, and Diameter. And then move on to the use of security tokens in mobile service access, and how security tokens can be used to realize single-sign on solutions. Finally, towards the end of the chapter, we consider web services security, and the security implications of downloaded active objects and code.

Chapter 10 presents a number of applications and service case studies. We start this chapter with a general overview of mobile applications and services, and compare the Internet and mobile application development processes. We then continue to examine several important use cases and application domains, including Mobile Web Server, mobile advertisement, push email, video delivery, mobile widgets, and airline services. We discuss how the mobile patterns described in Chapter 4 are used in these examples.

Finally, we conclude the book in Chapter 11 with a discussion and outlook for mobile middleware.

Bibliography

[1] Raatikainen, K., Christensen, H. B. and Nakajima, T. 2002 Application requirements for middleware for mobile and pervasive systems. *ACM SIGMOBILE Mobile Computing and Communications Review* **6**(4), 16–24.

[2] Dey, A. K. 2001 Understanding and using context. *Personal Ubiquitous Comput* **5**(1), 4–7.

[3] (ed. David, K.) 2008 *Technologies for the Wireless Future: Wireless World Research Forum (WWRF)* vol. **3**. Wiley.

[4] (ed. Tafazolli, R.) 2006 *Technologies for the Wireless Future: Wireless World Research Forum (WWRF)* vol. **2**. John Wiley & Sons Ltd.

[5] Satyanarayanan, M. 1996 Fundamental challenges in mobile computing *PODC '96: Proceedings of the fifteenth annual ACM symposium on Principles of distributed computing*, pp. 1–7. ACM, New York, NY, USA.

[6] Weiser, M. 1993 Ubiquitous computing. *Computer* **26**(10), 71–72.

2

Architectures and Platforms

There are many ways to approach the design and structure of current mobile architectures and platforms. Middleware, and mobile middleware more recently, have been developed from the need to find ever faster software development processes and easier software deployment. Typically, the networking protocol stack, such as TCP/IP, provides only basic interfaces for communication. In the case of TCP/IP, programmers have the *Sockets API* for sending and receiving packets and streams. Thus middleware has emerged as an intermediate layer between the protocol stack and applications with the goal of providing re-usable functions for software developers.

This chapter presents the key middleware architectures and platforms, and identifies the core mobile middleware services available today for software developers. Towards the end of the chapter, we discuss current trends and extrapolate on the nature of future mobile middleware.

2.1 Overview

Today's wireless and mobile services are typically monolithic and centralized in nature, and service access is tied to a single device or delivery channel. There is a growing need for supporting heterogeneous service access and sharing service usage experience. Envisaged new sources of revenue for Internet Service Providers include tailored, personalized, and dynamically composed services. The key requirements for service platforms include achieving faster time to market, cost efficiency, and compelling user experience.

This mobile service environment can be categorized into two parts: *service access* and *service provision*. Service access pertains to a consumer's device and environment, and the possible interactions with a service. Service provision focuses on platform side operation and includes mechanisms for service creation, life-cycle management, reconfiguration, and composition.

Key requirements and scenarios for future wireless networks have been extensively covered by the *Wireless World Research Forum (WWRF)* [1, 2]. The requirements for future wireless services include: heterogeneous networks, mobility, composition, security, privacy, dependability, backward compatibility and migration, network robustness and fault tolerance, quality of service, multi-domain support, accountability, personalization,

Mobile Middleware   Sasu Tarkoma
© 2009 John Wiley & Sons, Ltd

context communications, extensibility of network services provided, and application innovation.

In order to address these requirements, many different middleware components are needed. The role of middleware is crucial in the development of modern distributed applications and services, in which middleware typically provides various transparencies and collects frequently used functions under standardized APIs. New technologies and emerging standards such as SIP, *IP Multimedia Subsystem (IMS)* [3, 4]), *Open Mobile Alliance (OMA)*, *Open Service Access (OSA)* and Parlay, and web services are crucial building blocks for mobile service architectures that support converged communications.

2.2 Networking

Mobile middleware builds on the networking stack, typically the TCP/IP stack or the *Wireless Application Protocol (WAP)* stack. Before considering current middleware solutions, we briefly outline the key features of the protocol stack. The three well-known protocol stack examples are the TCP/IP, *Open Systems Interconnect (OSI)*, and WAP stacks. In this section, we focus on the four-layer TCP/IP stack since it is the dominant communications suite available today and plays a fundamental role in converged telecommunications environments.

2.2.1 TCP/IP

Figure 2.1 presents the TCP/IP communications suite, which is typically presented using four layers, namely the physical and link layer, network layer, transport layer, and application layer. TCP/IP is based on the exchange of *Internet Protocol (IP)* packets over interconnected networks. The *connectionless* hop-by-hop nature is a flexible and versatile form of communication and typically contrasted with *connection-oriented* networking techniques such as circuit-switching. Although hop-by-hop packet-based routing cannot provide the same level of *Quality of Service (QoS)* as circuit-switched technologies, the versatility has proven to be a winning strategy for TCP/IP. The current trend in telecommunications is towards all IP wireless access networks.

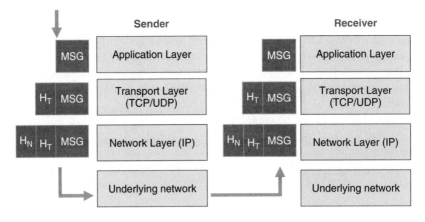

Figure 2.1   Protocol headers and layering

In the protocol stack, each layer provides functions for layers above. The layers below the network layer are responsible for delivering bits over a link, for example a wireless transmission link or an Ethernet network. The two important layers here are the physical layer and link layer. The former is responsible for the actual delivery of bits, and the latter is responsible for framing, flow control, error detection on the link.

The network layer is responsible for internetworking and uses the IP to deliver data from upper layers between end hosts. IP is a best effort protocol and the delivery of packets to destinations are not guaranteed. Whereas the lower layers only pertain to a link, the network layer and higher layers are end-to-end. IP packets are short sequences of bytes that contain a header and a body. The header describes the packet's destination, which is used by routers to forward the packet towards the destination. The higher order bits of the IP address determine the network on which the host resides and the remaining low-order bits determine the host number.

A *routing algorithm* is responsible for building and maintaining routing tables, and a *forwarding algorithm* is responsible for determining the proper next hop link given a destination address. Essentially, routing is the process of selecting paths in a network and it directs forwarding, which is responsible for the transmission of data packets between network nodes.

Packet routing involves use of protocols such as *Border Gateway Protocol (BGP)*, *Open Shortest Path First (OSPF)* in order to decide the route that each packet has to travel. The route is decided using a routing table consisting of *<destination address,next hop>* tuples. Routing tables also include additional information, for example metrics pertaining to the entries.

Forwarding models can be categorized into three cases, namely:

- *Unicasting*, in which a packet traverses a chain of links from a source to a destination. The majority of traffic on the Internet is unicast; however, different forwarding strategies can be implemented on the application layer. For example, application layer multicast is used by most data dissemination systems such as *peer-to-peer (P2P)* networks.
- *Multicasting*, in which a packet selectively traverses multiple chains of links from typically one source to a multiple destination.
- *Broadcasting*, in which a packet is sent on multiple links to every device on the network. In practice, broadcast is applied only within a specific broadcast domain, such as a local area network. Broadcast has an important function in supporting discovery of devices and services in a network.
- *Anycasting*, which is about selecting a suitable chain of links from a number of possible candidates. Packets are therefore sent to the nearest or best destination.

In the prior classful network architecture, IP address allocations were based on octet (8-bit) boundary segments of the 32-bit IP address, requiring either 8 (class C), 16 (class B), or 24-bit (class A) address prefixes. This resulted in significant scalability problems when the C-class addresses were running out. As a solution, *Classless Inter-Domain Routing (CIDR)* was introduced to *IP version 4 (IPv4)*. CIDR is based on *variable-length subnet masking (VLSM)* that allows the allocation of arbitrary length network prefixes. The length of the network prefix of an IP address is indicated using a suffix, typically using the following convention, *IP-adresss/suffix*, where the suffix indicates the number of high order bits that determine the network prefix. CIDR effectively allows better utilization than the original three network classes.

2.2.2   IPv4 and IPv6

IPv4 is the currently dominant IP protocol version. Due to several limitations in this protocol, including address exhaustion and lack of security support, a new version, IP version 6 (IPv6), has been defined by IETF and is now being deployed, albeit slowly. IPv6 features 128 bit addresses and therefore can support 2^{128} distinct addresses. IPv6 uses the CIDR convention of indicating prefix length with a suffix in the IP address, but the increased addressing space has made it less urgent to allocate minimum amount of address space to an organization. Although technically superior, there has not been sufficient economic incentive for operators to swiftly adopt the new protocol version. The significant new features of IPv6 include security, *IP Security (IPSec)*, autoconfiguration support, and optional mobility support using the *Mobile IP (MIP)* protocol.

The two protocol versions are supported by modern mobile operating systems. The typical solution is to have a dual networking stack for IPv4, and IPv6, respectively. Indeed, the vast numbers of mobile devices are expected to benefit from the increased address space in IPv6, giving the possibility for a unique globally addressable IP address for each device.

IPv6 offers significant improvements for mobile devices due the expanded addressing space and mobility and security support. A dual networking stack can be found in Symbian OS, and Windows Mobile; however, not all devices support the new protocol version, form example iPhone 2.0 does not support IPv6. The protocol is supported in OS X, which is used in iPhone, so it is expected that future versions of iPhone will support also IPv6.

2.2.3   TCP and UDP

The transport layer consists of two protocols, namely *Transmission Control Protocol (TCP)* and *User Datagram Protocol (UDP)*. The former provides reliable end-to-end delivery for data streams with flow control, congestion control, and ordered error free transfer. The latter provides unreliable packet delivery. Since the majority of the Internet's traffic is TCP traffic, such as WWW, email, and File Transfer Protocol, it has been the subject of intense research. TCP's reliability has a drawback in terms of latency. This is why many real-time streaming applications use UDP. *Real-time Transport Protocol (RTP)* uses UDP to deliver data streams over the Internet.

TCP's congestion control mechanism is based on the *additive increase multiplicative decrease*, which is a feedback control algorithm that uses linear increments to the congestion window, and an exponential reduction when congestion takes place. The congestion control mechanism is fundamental to today's Internet. In order to guarantee reasonable fairness for different network flows, the flows of one transport protocol implementation must be fair to the flows of other transport protocols. Moreover, the transmission rate for each flow should be roughly similar to that of TCP. Components of TCP-friendly congestion control are slow-start, additive increase multiplicative decrease, and retransmission timers relative to round-trip time.

Selective Acknowledgments (SACK) specified in *Request for Comments (RFC)* 3517 is a commonly used TCP improvement that allows additional information to be sent from receiver to sender about missing packets. SACK is a TCP option that reports discontinuous blocks of received data. This allows more efficient retransmissions.

TCP was originally developed for wired and reliable networks. The TCP congestion control algorithm assumes that packet losses are due to congestion in the network and slows down the send rate to alleviate congestion. With wireless networks, packet drops are typically due to the conditions of the wireless environment, such as handoffs, temporary losses due to fading, handoffs, and not congestion. As a consequence, TCP does not perform well in wireless environments. After a backoff of the congestion window size, the congestion control algorithm increases the window size conservatively.

A number of solutions to wireless TCP have been proposed, which can be classified into link layer solutions, proxy-based solutions, and end-to-end solutions [5]. Link layer protocols can be divided into two classes, the first class includes protocols that use error correction and those that use retransmission and decide to look at how the link layer protocol interacts with the transport layer.

A proxy-based solution, or split connection, involves the splitting of the TCP connection into two parts at the network edge using a specialized transport protocol over the wireless hop. Example systems include Mowgli and the Snoop protocol. The Snoop protocol sits at the base station and monitors TCP traffic. When a loss is sensed, the packet is retransmitted and any duplicate acknowledgments are suppressed.

The end-to-end solutions are implemented in the communicating end hosts, namely senders and receivers. The cumulative acknowledgments used by TCP do not allow efficient recovery from multiple packet losses. SACK can be used to inform the sender what packets are still missing and it thus alleviates some of the efficiency problems with the cumulative acknowledgments.

Stream Control Transmission Protocol (SCTP) specified in RFC 2960 is a reliable transport protocol similar to TCP. SCTP supports SACK, preserves message boundaries, supports multiple streams in a single connection, and supports multi-homing. From the mobile computing viewpoint, especially the multi-homing feature is useful as well as the message-oriented nature. SCTP guarantees that within a single stream, messages are delivered in order; however, between streams they are delivered in arbitrary order. SCTP offers separate congestion control for each address pair with multi-homing. This is not possible with TCP and network layer mobility protocols.

2.2.4 *MANETs and Wireless Mesh Networks*

In addition to the standard Internet protocols, such as OSPF and BGP, a number of protocols have been developed for ad hoc networks, which are very different in nature from traditional fixed networks [6]. A *Mobile Ad Hoc Network (MANET)* is a wireless ad-hoc network, consisting of self-configuring mobile devices, namely hosts and routers. The devices are free to move arbitrarily and the network topology is subject to dynamic and unpredictable changes. A MANET can be a standalone network or connected to the Internet using a gateway or a number of gateway devices.

The IETF MANET Working Group is developing a number of MANET protocols, which can be classified under two types, namely *reactive* and *proactive* protocols. Standardized protocols include *Ad Hoc On Demand Distance Vector (AODV)* specified in RFC 3561, and the *Dynamic Source Routing Protocol (DSR)* specified in RFC 5148. The AODV algorithm enables dynamic, self-starting, multihop routing between participating mobile nodes wishing to establish and maintain an ad hoc network. DSR allows the network to be completely self-organizing and self-configuring, without the need for any existing network infrastructure or administration. The protocol is composed of the two main mechanisms of *Route Discovery* and *Route Maintenance*, which work together to allow nodes to discover and maintain routes to arbitrary destinations in the ad hoc network.

MANETs have several characteristics that have been identified in RFC 2501:

- *Dynamic topologies.* Nodes are free to move arbitrarily; thus, the network topology, which is typically multihop, may change randomly and rapidly at unpredictable times, and may consist of both bidirectional and unidirectional links.
- *Bandwidth-constrained, variable capacity links.* Wireless links will continue to have significantly lower capacity than their hardwired counterparts.
- *Energy-constrained operation.* Some or all of the nodes in a MANET may rely on batteries or other exhaustible means for their energy. For these nodes, the most important system design criteria for optimization may be energy conservation.
- *Limited physical security.* Mobile wireless networks are generally more prone to physical security threats than are fixed-cable nets. The increased possibility of eavesdropping, spoofing, and denial-of-service attacks should be carefully considered.

Wireless mesh networks (WMNs) consist of mesh routers and mesh clients, where the clients can move arbitrarily and the mesh routers have minimal mobility and form the backbone of WMNs [7]. The mesh routers provide network access for both mesh and conventional clients. They may also provide connectivity with other networks, such as the Internet or wireless access networks. WMNs are anticipated to improve the performance of ad hoc communications environments and support the integration of personal area networks with other types of networks, such as local, campus, city area networks. WMNs are currently an active topic in network research.

2.3   Naming and Addressing

The way in which entities are named and addressed is a crucial element of a network architecture. As we will observe, naming and addressing can happen on many different layers. Indeed, the modern communications environment is essentially multi-layered, which on one hand provides flexibility but on the other hand introduces redundancy and overhead. Overlay networks can be seen as an example of a new naming and addressing system built on top of an existing underlying network.

2.3.1 Basic Definitions

A name identifies a network element in some scope, which may be local or global. A network address identifies how information should be delivered to the node. In the TCP/IP model, a host name is a label that identifies the node in question. A domain name is a label given to a collection of network devices that belong to a domain. Domain names define regions of administrative authority within the *Domain Name System (DNS)*. An IP address defines the topological location of a node or a subnetwork in the global network routing topology. IP uses prefix-based routing, in which variable bits of the IP-address (32 bits in IPv4 and 128 bits in IPv6) identify the network and the rest identify the host. This allows scalable routing tables using prefix aggregation.

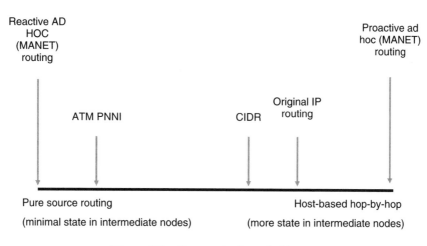

Figure 2.2   Characterization of addresses

Figure 2.2 illustrates the differences between different kinds of network addresses. An address may contain explicit information about how data should be forwarded over the network. This is called *source routing* in networking literature, and it can be strict or loose, depending how much flexibility routers have in determining the destination. Addressing is implicitly related with the notion of *routing state*. A network design can place routing state in the address, or it can maintain routing state in the routers. The *Private Network-Node Interface (PNNI)* is an ATM network-to-network hierarchical dynamic link-state routing protocol designed to support large-scale ATM networks. The original IP solution maintained more state in the routers and offered three network classes for routing decisions. The current classless approach (CIDR) allows more flexibility in defining the network prefix and thus requires less state in the network. Proactive MANET routing maintains paths in an ad hoc network by constantly probing the network. At the other extreme, we have reactive MANET routing, which discovers paths to destination on demand.

The *FARA (Forwarding directive, Association, and Rendezvous Architecture)* network architecture proposal defines an abstract naming model that decouples end-systems from their network addresses. They focus on abstract definition of the architecture and do not rely on a global namespace, but rather a rendezvous function that routes the first packet from a source to a destination [8]. Later in this book, we will revisit rendezvous as a central design pattern for supporting mobile computing.

2.3.2   Challenges

The current dominant networking stack, the Internet Protocol suite, suffers from a number of limitations, but works reasonably well for the current demands. These include the failure of global end-to-end reachability that was one of the requirements of the original Internet architecture [9, 10]. Firewalls and *Network Address Translation (NAT)* middleboxes make end-to-end data delivery difficult. Moreover, the current network is vulnerable to spam and *Denial-of-Service (DoS)* attacks. In addition, flexible global multicast has proved to be difficult to deploy.

The current TCP/IP network faces several challenges when wireless and mobile hosts are introduced to the global network. Many of these mobility and security related problems stem from the fact that the IP address specifies both the identity and location of a host. This means that a location management scheme is needed to update any changes in a mobile node's IP address to its peers. In addition, various intermediaries, such as NAT-boxes may filter and modify packets. Private networks created using NAT create yet another problem for the architecture, because typically only outbound connections are allowed.

The Internet architecture was not originally designed for mobility and multi-homing and they present grave challenges for mobile users. Current areas of research include security, location management, naming of entities, and the integration of various intermediary entities on the routing path.

2.3.3 Mobility and Multi-homing

Multi-homing is a technique to increase the reliability of the Internet connection for an IP network. The two well-known forms of multi-homing are host and site multi-homing. In the former, a host can have multiple IP addresses and multiple interfaces towards the Internet. In the latter, a network site advertises multiple prefixes in order to improve service availability. From the mobile computing viewpoint, multi-homing is a desirable feature, because it allows multiple interfaces and access networks to be utilized at the same time, thus offering flexibility and service availability.

Figure 2.3 illustrates the many faces of Multi-homing. Multi-homing is inherently related with mobility, because multi-homed devices can move between networks. For a single multi-homed device, any changes to its IP addresses need to be signalled to the corresponding hosts. A number of protocols can be used to support host mobility with or without multi-homing, for example Mobile IP or SCTP. The situation becomes more complex when whole subnetworks are moved.

The *Network Mobility (NEMO)* Working Group at IETF is concerned with managing the mobility of an entire network. Due to mobility, the point of attachment to the Internet and thus its reachability in the topology changes. The NEMO solution is to use one

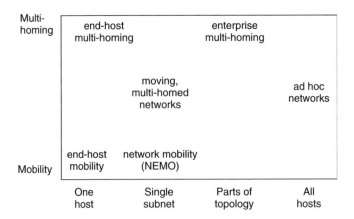

Figure 2.3 Many faces of multi-homing

or more mobile router that connects the subnetworks to the global Internet. A mobile network is assumed to be a leaf network, i.e. it will not carry transit traffic. However, it can be multi-homed, either with a single mobile router that has multiple attachments to the Internet, or by using multiple mobile routers that attach the mobile network to the Internet.

2.3.4 Network Address Translation

Network Address Translation (NAT) is the process of modifying network address information in packet headers. In essence, NATs transform addresses between address spaces, namely between private and public address spaces. The motivation for address translation is the increased addressing space gained through private addressing spaces. In addition, address translation typically hides the internal network which improves security. This hiding of the internal address space is called *network masquerading* (RFC 1918). This feature is implemented in a NAT device that uses stateful translation tables to map the hidden address space into a single public address. Outgoing IP packets are then rewritten so that they originate from the NAT device. In the reverse path, the NAT device maps the incoming packet to a private address using state stored in the translation tables. The established tables are flushed in a short period if they are not refreshed by new traffic. *Network Address Port Translation (NAPT)* discussed in RFC 2662 is a variation of NAT, which uses the TCP and UDP port numbers to map a number of internal addresses to a single public address.

Figure 2.4 illustrates the key NAPT functionalities, namely packet rewriting and state maintenance for outgoing flows. In the figure, the host sends a request to a server. This request creates an associated stateful entry in the NAT, which is used to forward the reply from the server to the host. The entry includes information pertaining to mapping the internal address to the public address.

We have already mentioned that NAT devices break the end-to-end connectivity of the Internet. This follows from the fact that a NAT device with network masquerading hides

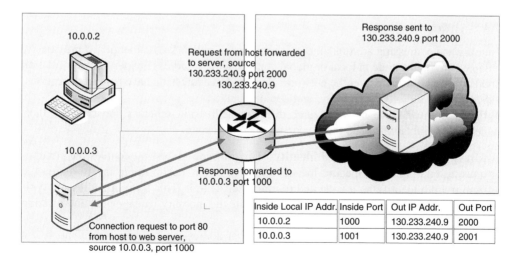

Figure 2.4   Example of Network Address Translation

the internal address space. Thus the devices in this private network cannot be contacted from outside without them first initiating contact and establishing state in the NAT device. Since the current practice is to use NATs between most networks, including mobile access networks, this creates challenges for both traditional and mobile software. The worst case happens when there are multiple NAT devices on the communicating path. In this case, a special relay server must be used to be able to support communications between the hosts.

Several NAT traversal solutions have been developed for UDP and TCP based applications, namely:

- *Simple Traversal of UDP through Network Address Translators (STUN)* defined in RFC 3489 that allows applications to discover the presence and specific type of NAT, and obtain the mapped IP address and port number that the NAT has allocated for the application's UDP connections to remote hosts. The protocol relies on the assistance of a STUN server located on the opposing public site of the NAT, typically the public Internet.
- *Interactive Connectivity Establishment (ICE)* is a protocol for gathering data about the communication path and then finding the best configuration for the communicating end points that guarantees reachability. ICE is developed by IETF's MMUSIC working group and the main application is to allow SIP-based VoIP clients to traverse firewalls and NATs.
- UPnP and Bonjour, which require specific support from a NAT device.

NATs also complicate tunneling protocols such as IPsec, because NAT devices modify values in packet headers. The modification of packet interferes with the integrity checks performed by IPsec and other tunneling protocols. As a solution to some of the problems with NATs, *an Application Layer Gateway (ALG)* can be used to update any packet payload data invalidated by address translation. ALG is a module running in the NAT device. The challenge is that ALGs must understand the higher-level protocol in order to be able to update the packet payload.

2.3.5 A Taxonomy

Naming and addressing are fundamental properties of networks. As mentioned previously, a name identified a node in local or global context, and an address defines how information should be sent to a node over the network. Naming and addressing have many implications, especially for security, scalability, and deployment.

In the following, we consider three different addressing schemes and compare their benefits and limitations. The three addressing schemes are:

Address with both location and identity This form of addressing couples the communicating end-points to specific locations in a network. For example the IP address is used in both identifying a node and routing packets to it. This form of addressing typically uses a mediating stationary node to handle the mobility management and location updates for the mobile nodes.

Address with locator/identity split This way of addressing separates the identity of a node and the location of the node. This allows more flexible mobility support since the identity may be used to look up the physical location of a node. For example the *Internet Indirection Architecture (i3)* and the *Host Identity Protocol (HIP)* [11] are based on this form of addressing.

Data-centric and content-based addressing Data-centric and content-based addressing goes beyond locator/identity split, because it decouples the destination from both identity and location. The destination is no longer defined by a single identity, such as the IP address or a cryptographic public key, but rather it is defined by logical rules set by applications running on the destination host. The rules are applied on messages or packets in order to make forwarding decisions. This means that using content-based addressing we have decoupled many-to-many communication. On the other hand, the realization of content-based communication is more complex and costly. The cost of mobility in content-based routing is high when compared with the other forms of addressing. Research systems such as SIENA [12] and Rebeca [13] use content-based addressing.

Identity-based mechanisms may be extended to support anonymous communication and multicast. For example, i3 supports multicast using triggers and anonymity by chaining private and public triggers. Content-based routing may, on the other hand, be extended to support identity-based communication by subscribing public keys, for example.

The addressing models have differing notions of the addressing space, in which addresses are defined. These differing notions can be used to characterize the difference between identity-based addressing and content-based addressing. The identity vector (public key) is a point in the flat one-dimensional addressing space of an overlay system. The content-based address, which is defined using a logical rule, is a subspace of a multi-dimensional addressing space. This illustrates the main difference, which is the expressiveness of the communication. In essence, for IP mobility there is a single fixed indirection point, for locator/identity split there is a single indirection point, and with content-based there are multiple indirection points.

2.4 Middleware and Platforms

2.4.1 Overview

Middleware is a loosely defined set of software components and services that support the development and deployment of applications. Typically these enabling services allow multiple processes running on one or more machines to communicate over a network. Examples of middleware services include web servers, application servers, XML processing, SOAP processing, web services, and *Service-Oriented Architecture (SOA)*.

As briefly discussed in Chapter 1, middleware typically includes services such as messaging and *Remote Procedure Call (RPC)* facilities, resource discovery, transactions, security, directory, and storage services. Middleware provides various transparencies for higher layers that are not typically supported by the lower layers. These transparencies include location, transport, operating system, programming language, and failure transparencies.

Mobile middleware can be categorized into client-side and server-side functionality. A mobile middleware architecture is a collection of these functional elements that defines how different parts can interact and communicate. Architecture defines the elements, their connections, and possible configurations in a loose way. A middleware platform is a concrete realization of an architecture, following the principles and constraints set by architecture. A middleware platform therefore implements an architecture or a subset of it.

Data communications and data synchronization are key services provided by mobile middleware solutions. The ability to reliably and securely communicate over mobile and wireless access networks is central.

2.4.2 Objects, Components, and Services

Decomposing software into flexible units has been the trend in software development for a long time. The many benefits of being able to compose software using units are obvious, for example the benefits include separation of concerns and unit re-use. Over the past decades this notion of building complex applications from smaller units has evolved significantly.

The first solutions to decomposition were about functional decomposition with functions or subroutines as the individual units of software. Later, the notion of a class has evolved to describe the behaviour of objects from a problem domain. A class embodies certain semantics. In object-oriented software engineering, classes and interfaces, and the notions of polymorphism and inheritance have become powerful software development features for programmers. *Polymorphism* allows increased flexibility and *inheritance* supports code reuse.

These object-oriented notions can be applied in a distributed environment. Indeed, with the advent of Java and web services, distributed objects and communication between various distributed software platforms has become an integral part of today's Internet services.

Figure 2.5 illustrates the relationship of key components of distributed software development, namely objects, network, security, and directories. We observe that objects rely on network features, namely TCP/IP stack, a number of security features, and directories. Directories are needed for the system to be able to locate and find objects and other information.

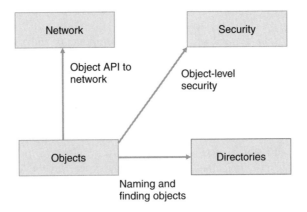

Figure 2.5   Objects and their relation to components

Local and distributed objects differ in many respects [14]. Important differences include the following:

- *Life cycle*. Creation, migration, and deletion of distributed objects is different from local objects.
- *Reference*. Remote references to distributed objects are more complex than simple pointers to memory addresses.
- *Request Latency*. A distributed object request is in order of magnitude slower than local method invocation.
- *Object Activation*. Distributed objects may not always be available to serve an object request at any point in time.
- *Parallelism*. Distributed objects may be executed in parallel.
- *Communication*. There are different communication primitives available for distributed object requests.
- *Failure*. Distributed objects have more points of failure than typical local objects.
- *Security*. Distribution makes the objects vulnerable to attack.

As such, objects and classes do not describe the process of deployment. The notion of a component has been developed to separate application logic, namely objects and classes, from deployment issues. A component is typically associated with a deployment descriptor that defines the behaviour of the component with respect to security, state management, persistency, and transactions.

Services can be seen as an evolutionary step of components. They can be used to abstract the particulars of a component model and thus unify the access and invocation of distributed components from the programmer. The developer need only think in terms of web services and the machinery in the platform takes care of interfacing with various components and the specifics of their component models. The motivation for this kind of service is interoperability. Through this interoperability, it becomes possible to compose services in a manner not possible before.

Although the terms application and service are used to mean similar functionalities provided by an entity to other entities, the two differ in some of the implicit assumptions. A service can be expected to provide a set of interoperable interfaces to other entities, whereas an application does not necessarily provide such interfaces. Therefore the term service implies that it is more general and interoperable than an application. In the end, however, both are software that is executed by a node or a set of nodes in the environment, and applications and services can utilize the same libraries.

It is often said that a service is logically centralized when it is assumed that interaction is required with some fixed entity, but that this entity can be distributed over a set of nodes. This is contrasted by decentralized operation, in which no such centralized entity is required or expected. This makes software development for decentralized environments challenging. Indeed, most current mobile applications and services have a logically centralized nature, and the current trend is to utilize servers in the fixed network for computationally intensive operations. This is called *cloud computing* and the proponents include Google, Microsoft, and Amazon who develop data centers and have massive computation and storage requirements.

The notions of objects, classes, and components are central in today's distributed platforms. From the mobile client perspective, accessing distributed resources has become crucial, and HTTP and web services support are basic building blocks of mobile applications. One challenge facing universal data access is allowing clients to invoke arbitrary service interfaces available in a network. This challenge has many facets, and we will briefly decompose key challenges, which include the following:

- Establishing basic connectivity including authentication and authorization for network access. Security decisions typically utilize *Authentication, Authorization, and Accounting (AAA)* systems such as RADIUS and Diameter.
- Discovering local resources and services, for example Universal Plug and Play (UPnP) and Jini can be used to realize this. These technologies will be examined later in this chapter.
- Discovering global resources and services available on the Internet if Internet connectivity can be established. The process of service discovery requires the use of one or more service brokers. Typically, *Universal Description, Discovery, and Integration (UDDI)* and search engines can be used to do this.
- Given that a resource or service has been located, the client may send a request or invocation to the entity responsible for the resource or service. Before any request can be sent, the client needs to be able to understand what kind of arguments are required by the recipient, and also what kind of values can be returned. Typically, in order to help programmers cope with mapping data between various formats and ensure interoperability in distributed systems, a number of interface specification languages have been developed, namely IDL used in *Common Object Request Broker Architecture (CORBA)* from *Object Management Group (OMG)*, and *Web Services Description Language (WSDL)* from W3C. The process of mapping an arbitrary language specific structure into an interoperable representation, such as specified by *Interface Definition Language (IDL)* or XML schemas and WSDL, is called data binding.
- Finally, given that the client and the destination are able to understand each other, requests sent by the client and responses from the destination can be properly processed.

We observe that the above example highlights two very different environments, namely local and Internet. The former is an emerging communications environment for mobile phones, whereas the latter is about bridging the wireless and mobile, and fixed Internet worlds aiming at Mobile Internet. A crucial challenge is that many of the standard Internet protocols do not work well in wireless environments. This has been addressed by a number of dedicated standardized wireless protocols, such as the WAP stack, however the industry has been adopting TCP/IP as the primary communications stack for several years now for applications despite some of the limitations of the stack pertaining to wireless operation.

With the advent of web services and XML, the XML processing capability of a device becomes important. Unfortunately, the verbose human readable XML 1.0 requires quite a lot of processing from mobile devices, and it is difficult to manipulate large XML files on such systems. We will return to XML processing on small devices in the Mobile Messaging chapter.

2.4.3 Message Passing and RPC/RMI

Modern distributed applications are typically object oriented. Distributed objects are software modules that are used to create a service or an application, but they reside on different computers connected via a network. The objects communicate by sending messages using the underlying network and protocol stack. *Inter-Process Communication (IPC)* is a technique for exchanging data between one or more processes. IPC techniques can be local or distributed. IPC techniques are typically divided into methods for message passing, synchronization, shared memory, and RPC. IPC is central to middleware systems.

Distributed object and remote method invocation systems including CORBA, Java *Remote Method Invocation (RMI)*, DCOM, SOAP, and .NET Remoting are message passing systems. They follow the message passing model in which a copy of a data item is sent to a communication endpoint over the network. The endpoint has differing names and definitions, but the underlying principles are the same and exemplified by SOAP and REST in today's web. The interaction has typically a number of features, such as reliability, security, and transaction related features.

Serialization of objects is vital for distributed object systems. This is the process of creating a representation of an object suitable for storing it into a storage system or transferring the data across a network. Similarly, *deserialization* is the process of extracting the data structure from a given sequence of bytes. Alternative terms for serialization are marshalling and deflating, and unmarshalling and inflating, correspondingly.

XML serialization has become popular with the advent of SOAP and web services. XML is heavily used by Ajax web application to exchange structured data between clients and a server. *JavaScript Object Notation (JSON)* is a lightweight text-based alternative for XML serialization that uses JavaScript syntax. JSON is supported in many programming languages.

XML markup has the benefit of being a human readable text-based format, which is useful for persistent and interoperable representation of objects. The limitation of the text-based serialization of XML are processing requirements and size overhead. For high performance environments, a more compact and byte stream based solution is desirable. Indeed, a number of bit efficient and streamable XML processing systems have been developed. We will consider XML processing in more detail in the chapter on Mobile Messaging.

As an example of a distributed object oriented system, we consider the Java RMI API presented by Figure 2.6. RMI is a Java application programming interface for performing the object equivalent of remote procedure calls. There are two common implementations of the API. The original implementation depends on Java Virtual Machine (JVM) class representation mechanisms and it thus only supports making calls from one JVM to another. The protocol underlying this Java-only implementation is known as *Java Remote Method Protocol (JRMP)*. In order to support code running in a non-JVM context, a CORBA version was later developed (RMI over IIOP).

2.4.4 Mediation and Delegation

Mediation is a mechanism that can be used to enhance or enable an existing service. The main purpose is to separate the added service functions from core platform functions in

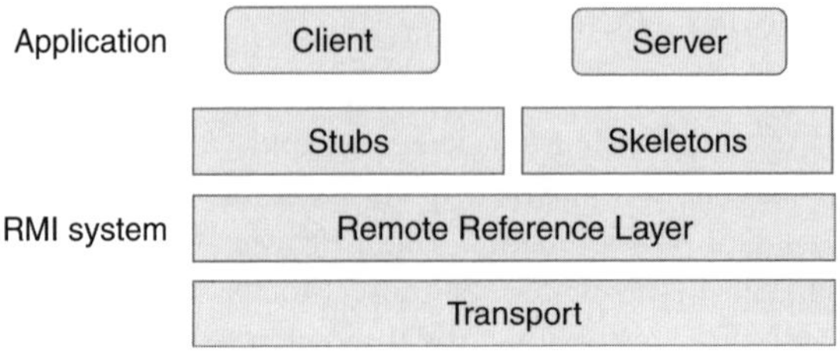

Application   Client   Server

                   Stubs   Skeletons

RMI system   Remote Reference Layer

                      Transport

Figure 2.6   Overview of RMI

order to simplify service development and maintain a coherent architecture. The extended functionality offered by the mediation framework can be transparent to the service user, e.g., for authentication or logging.

This separation of concerns offered by mediation allows service providers to focus on the content and leverage the signaling and mediation capabilities of the core platform. Mediators add specific functionality to a service and are a specific form of service composition and selection. Mediators sit on the message signaling path and are able to intercept messages and act on the message flow according to platform policies. In many cases, mediation will modify the message flow in order to adapt a service for the user. For instance, it may forward a message to the AAA that requests additional credentials or proof of identity from a user. The concept of mediation also creates independence from a specific service implementation.

For instance, let us consider a video-on-demand service. First, the basic service components, such as the video search and video playback components, are introduced in the form of public interfaces and usage policies. End-user systems may then discover the public entry-point to the service or it may be advertised in a service catalog. Upon requesting the service, it is then composed with mediators that provide authentication, authorization, charging, and load balancing. Furthermore, users may access the services via different access networks, for example WLAN or cellular technologies, which typically have different authentication technologies. In this case, the mediators have to take care of the interworking of different security standards. Hence the service developer and operator do not have to know about technology specific details and the service can be used in a more flexible way in different settings.

In the current *Enterprise Service Bus (ESB)* systems, the concept of mediation is used in a very similar way. The ESB provides a messaging system between different applications and enhances the interactions by intercepting and modifying messages. Typical usage patterns include monitoring, transcoding, routing, and aggregation.

In a web services based service environment, mediation can be realized using SOAP and a number of SOAP intermediaries on the message forwarding path. A message path comprises of the nodes that process the message starting from the sender to the ultimate receiver. The SOAP header system is a flexible basis for building mediation systems; however, an intermediary may also need to inspect the content of a message realizing content-based routing.

In content-based routing the destination is determined by an active intermediary based on the content of the message and subscriber interests. In addition to maintaining security policies and tagging messages with security tokens, mediators may also be used for various transformations, augmentation, and handling special situations, such as detecting spam and denial-of-service attacks.

2.4.5 Mobility

Mobility is an important requirement for many application domains, where entities change their physical or logical location. Physical location denotes the real-world location of a device, whereas logical location is not necessarily dependent on the physical environment. Mobility support may be divided into several technical layers and also categories depending on the nature of mobility.

Middleware support for mobility is required in order to provide location transparency for objects, agents, and other components, support efficient and reliable communication in wireless environments, and buffer messages and other data for disconnected operation. In addition, the middleware may support scalability and availability of resources and services.

Mobility is inherently tied with the way nodes are addressed in a distributed network. We have examined three different ways to address mobile nodes and components: addresses with both location and identity, locator/identity split, and data and content based addressing. The first addressing model is used by the IP protocol. The second model is an extension of the first and used, for example, in the Host Identity Protocol (HIP) and the i3 overlay. The third model has been proposed for expressive communication in mobile environments.

2.5 Overview of Platforms

A service platform is the realization of a service architecture following its principles and patterns. A *Service Delivery Platform (SDP)* is a set of components that are used to implement a service delivery architecture that includes service creation, life-cycle, session control, and security support. A service platform revolves around three main actors: service providers, service requesters and a service registry. Service providers publish service descriptions, and service requesters discover services and bind to the service providers. Publication and discovery are based on service descriptions. Current telecom SDP's utilize SIP, IMS, web, and IPTV technologies in delivering services to mobile users. As mentioned in Chapter 1, the current trend is towards converged communications, in which TCP/IP and web technologies seamlessly integrate with telecom technologies such as SIP and IMS.

As an important part of a SDP, the *Service Creation Environment (SCE)* is responsible for supporting end-user or developer driven software development. The main motivation for a SCE is to support easier and more flexible service creation and deployment. Typically, the SCE is used by a developer experienced with the supported scripting language and other tools; however, it is also envisaged that end users use SCEs to compose services. For end users, scripting and programming are expected to be replaced with adapting pre-generated service composition templates for current needs and requirements.

We consider two concrete examples of distributed software platforms. First, we examine Java 2 Platform which has become massively popular for service development and deployment. Later, we will investigate the mobile and wireless part of Java Platform, namely the Java Micro Edition. Second, we examine the SPICE platform which is an example of a next generation service platform that features converged communications and SCE. Finally, we discuss service ecosystems and their value chains.

2.5.1   Java Platform

The Java platform is a collection of programs from Sun which allow the development and execution of programs created using the Java programming language. A notable feature of the platform is that it is not dependent on a specific hardware or operating system. Central parts of the platform are the programming language, execution engine (e.g. virtual machine), and a compiler with a set of standard libraries and APIs.

The Java programming language is syntactically similar to C++, but it is executed in a different manner. While C++ uses unsafe pointers and programmers are responsible for allocating and freeing memory, the Java programming language uses type-safe object references. A process called *garbage collection* is used to reclaim any unused memory automatically. The virtual machine forms the foundation of the platform. The virtual machine can be implemented to run on top of a variety of operating systems and hardware and enables the safe execution of untrusted code.

The Java platform includes several different editions of the platform tailored for different environments and ends (Figure 2.7):

- Java Card, which allows small Java-based applications to be executed on smart cards.
- Java ME (Micro Edition, formerly J2ME), which specifies several different profiles, collections of libraries, for devices that are sufficiently limited that it is not feasible to support the full Java platform on them.
- Java SE (Standard Edition), which is the platform for general purpose desktop PCs.
- Java EE (Enterprise Edition), which includes the Java SE and a number of additional APIs for multi-tier client-server enterprise applications.

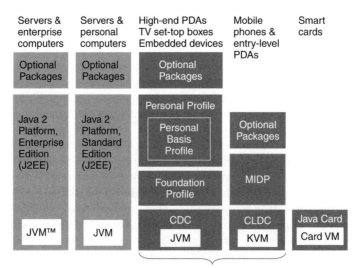

Figure 2.7   Java platform

The important components of the Java platform are the Java language compiler, the libraries, and the runtime environment. The runtime environment is responsible for executing the intermediate bytecode according to the rules specified in the virtual machine specification.

Java is based on the write once, run anywhere concept, which has proven to be tremendously succesful. Indeed, the more recent .NET platform from Microsoft incorporates a good number of the good features of Java. One major difference between these platforms is that Java is available on many different operating systems, .NET has been built for the Windows family of operating systems.

Java Enterprise Edition is specified by its specification derived using the Java Community Process. The platform includes a number of API specifications including RMI, *Java Message Service (JMS)*, Web services, XML, *Java Database Connection (JDBC)*, and email. More unique components of the platform include *Enterprise JavaBeans (EJB)*, servlets, portlets, *JavaServer Pages (JSP)*, and a number of web service technologies. The aim of these technologies is to allow developers to create portable and scalable enterprise applications. A Java EE application server is responsible for managing transactions, scalability, security, and concurrency of the components that are deployed on the server. The motivation is that developers can focus on the business logic rather than integration and deployment issues.

A web container implements the web component contract of the J2EE architecture, which specifies a runtime environment for web components that includes security, concurrency, life-cycle management, transaction, deployment, and other services. The Java Servlet API allows a software developer to add dynamic content to a web server. Servlets are the Java solution to dynamic web content technologies. Servlets may maintain state across many server transactions by using HTTP cookies, session variables or URL rewriting.

Figure 2.8 illustrates the Java EE platform and the multi-tier nature of modern service platforms. The figure presents an Enterprise JavaBeans container, which is typically used to host the business logic. This is the third tier. The web container hosts JSP pages and Servlets and is the middle tier responsible for data presentation and interaction with users. Finally, enduser applications are deployed in the first tier, namely Applet containers and Application client containers. The figure also presents the commonly used middleware technologies, such as JMS, Java Mail, and Connectors.

2.5.2 SPICE

The EU IST *SPICE (Service Platform for Innovative Communication Environment)* project addressed issues in mobile service provisioning by developing a framework for the rapid development of new mobile services. The aim of the framework is to hide the complexities of the converged communications environment [15, 16].

The SPICE middleware layer has four sublayers focusing on the core aspects of the distributed service execution environment, namely the *Capabilities and Enablers Layer*, the *Component Service Layer*, the *Knowledge Layer* AQ, and the *Value Added Service (VAS) Layer*. In addition, the platform features a *Mediation and Exposure* layer that is used to interface external platforms and resources. Figure 2.9 gives an overview of the SPICE platform.

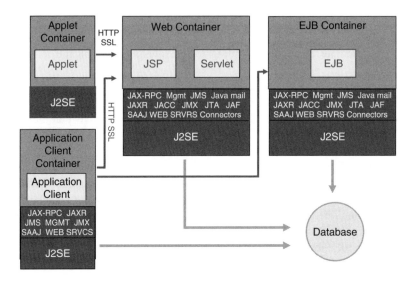

Figure 2.8   Java 2 Enterprise Architecture

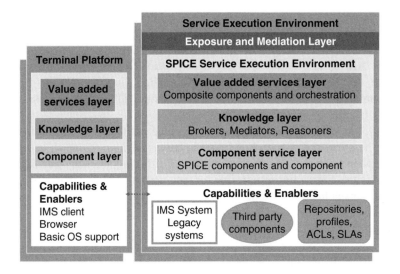

Figure 2.9   Overview of the SPICE platform

The SPICE project has a platform-centric architecture approach. It targets fixed-mobile convergence and manages heterogeneous access technologies and devices. The architecture developed within this project is based on components, which are discovered, federated, combined, and executed in a distributed environment. The platform consists of four different kinds of components. The basic components are generic building blocks that take advantage of the SPICE project platform features. Resource adapters are special basic components, which act as proxies to components in legacy systems. Components that support interfaces from the knowledge framework are called intelligent components.

Both may further be used to create VAS components and composite VAS components. The composite components may be created at runtime by using real-time information provided by various sources and processed by the knowledge layer components.

The capabilities and enablers layer is responsible for providing the various support functions that are vital for the SPICE platform. Support functions play a major part in enabling the core functions of the *Service Execution Environment (SEE)*. The support functions are external to the platform and used to realize functions such as profile management and storage, and session control. This layer provides support for the publishing of profiles and other information to the component service layer. The component service layer has corresponding functionality that exposes previously published information to interested components.

The component service layer provides support for component-based development and deployment. This layer includes services such as the service broker, controllers, various managers, and resource adapters. Resource adapters are used to integrate legacy components with the SPICE-project architecture components. The component model and methodology support cross-system service deployment and provisioning. The components are seen as a set of modelling artifacts that are constructed using technology-neutral notations such as the *Unified Modeling Language (UML)* and XML.

The use of technology-neutral notations makes it easier to manage service life cycle. In order to invoke component services seamlessly, component metadata and interface semantics must be developed and published, and made available. SPICE applies the best-practice architecture of SOA to support IP-based multimedia services on networking architecture such as IMS.

A number of components are exposed to the outside world using the exposure and mediator layer. These components can then be used and combined in a multi-platform environment to create more complex composite services. A third party service execution environment can use the publishing capability of the platform to publish and advertise services or components in the SPICE system. Several support functions are needed, such as the IMS system or a subset of it. These support functions are not part of the core SPICE platform.

The knowledge layer supports the delivery, discovery, and transformation of information, such as context and presence attributes. This layer introduces several building blocks for intelligent components, e.g., knowledge brokers, reasoners, and recommenders. The layer is realized as a collection of distributed knowledge sources and brokers. Sources produce information, which is delivered to sinks. Information is either delivered directly or by employing the publish/subscribe pattern. For example, the profile manager is one source of knowledge.

Enablers on this layer include knowledge brokers and recommendation engines. Brokers are responsible for mediating and transforming information, and providing interfaces for knowledge discovery. Discovery is performed either using direct query to a broker or by using the publish/subscribe pattern. The learning and prediction algorithms are pluggable and new techniques can be added to the system.

The VAS layer facilitates the creation of compound components and services from the basic components. A compound component is useful in realizing personalized and context-aware services. We can take a personalized ticket booking service as an example. In this case a compound component would integrate several logically separate functions, such as finding interesting films and other events for the user, prioritizing them, and then reserving seats after the user has selected the preferred option. These interactions require that the services and components of several different stakeholders need to be contacted. Run-time meta-data and context-based discovery of components can be used to find suitable components for composition. An orchestration engine is responsible for ensuring that the components of a composite component are properly synchronized and the interactions follow predefined rules.

2.5.3 Service Ecosystem

Middleware also serves an important function in enabling the service ecosystem by supporting service discovery, access, security, and composition. Website mashups have become popular in recent years and a number of interoperable specifications have been developed based on best practices, such as OpenID and OAuth specifications for security, and a large number of microformats for content exchange and sharing.

There are different value network actors in the service ecosystem, for example security solution providers (namely AAA maintainers), network providers, core platform providers, and various service providers. Given that users and developers can create custom mashups, or service compositions, the whole ecosystem needs to support dynamic and flexible operation.

To be able to sustain the value chain, we consider the following architectural features:

- Loose coupling and late binding.
- Asynchronous communication model.
- Semantic and context-aware matching of messages.
- Evolvability and extensibility.
- Interoperability through standards and well-defined interfaces.
- Intermediary-based security model.

The first feature is a typical feature of current SOA, and a common design pattern for distributed systems. Components communicate using asynchronous message passing, and coupling is performed by the core service platform at run-time, and as late as possible. This allows great flexibility in finding suitable service components that meet the current and foreseeable requirements of the whole value chain.

This flexibility is improved when the binding between component implementations is done based on component interfaces, metadata, and current context. In order for the system to be able to do this kind of matching, it needs to have semantic and context-aware matching capabilities.

Moreover, the architecture and platform needs to be evolvable and extensible to the expected and unexpected usage scenarios of tomorrow. This further motivates the loosely coupled component-based model. To be able to sustain the value network and the service ecosystem, the component and platform interfaces as well as communication protocols must follow established standards, such as those from W3C, IETF, and *3rd Generation Partnership Project (3GPP)*.

The value chain must also be trustworthy to enable a prosperous service ecosystem, and profit to be made between different stakeholders. Therefore, security and trust are important ingredients that must be supported by the platform and any components introduced by entities in the value chain.

2.6 Mobile Platforms

In this section, we present a number of well-known concrete mobile platforms and consider their middleware aspects. We consider the following platforms: Java 2 ME, iPhone, Symbian, Brew, WAP, Windows Mobile and .NET, Compact Framework, NoTA, Linux Maemo, Android, OSGi, Python, Flash LiteAdobe.

2.6.1 Java ME

Java Platform, Micro Edition (Java ME or previously J2ME) is a specification of a subset of the Java platform aimed at providing a standardized collection of Java APIs for the development of software for small and resource-constrained devices. Target devices are from many industries including consumer devices, home appliances, security, defense, automotive, industrial, industrial control, and multimedia. From December 2006, the Java ME source code has been licensed under the *GNU General Public License*.

The platform has a layered structure illustrated by Figure 2.10. First, the configuration specifies the virtual machine and the core libraries. There are two well-known configurations, namely *Connected Device Configuration (CDC)* and *Connected Limited Device Configuration (CLDC)*. The former is for high end PDAs and the latter is intended for mobile phones and other small devices.

The configurations are then augmented by profiles, which define additional APIs for applications. The most common profile is the *Mobile Information Device Profile (MIDP)* aimed at mobile phones. Another well-known profile is the *Personal Profile* which is aimed at consumer products and embedded devices such as set-top boxes and PDAs.

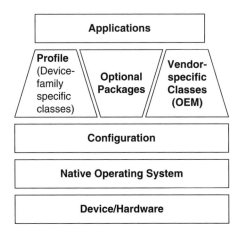

Figure 2.10   Overview of J2ME

The current Java ME platform consists of CLDC version 1.x (JSR 30) and CLDC version 1.1 (JSR 139). The current MIDP version is 2.x (JSR 118) with MIDP 3.0 being under development (JSR 271).

2.6.1.1   Connected Device Configuration

The *Connected Device Configuration (CDC)* is a subset of Java SE and it includes almost all the libraries that are not related with the *graphical user interface (GUI)*. Comparing CDC to CLDC, the latter has a much reduced set of Java SE libraries available. The three frequently used profiles for CDC are the Foundation Profile, Personal Basis Profile, and Personal Profile.

The Foundation Profile is a CDC profile that is intended to be used by devices requiring a complete implementation of the Java virtual machine up to and including the entire Java Platform, Standard Edition API. This specification describes the facilities that the Foundation Profile provides to the device and other profiles that use it. The Personal Basis Profile extends the Foundation Profile to include lightweight GUI support. The Personal Profile is an extension of the Personal Basis Profile, and includes a more comprehensive GUI subset based on AWT and includes applet support.

2.6.1.2   Connected Limited Device Configuration

The *Connected Limited Device Configuration (CLDC)* contains a subset of the Java class libraries. It is specifically designed to meet the needs for a Java platform to be executed on devices with limited memory, processing power, and graphical capabilities. CLDC defines a virtual machine for resource-constrained devices such as mobile phones, pagers, and mainstream personal digital assistants.

The Mobile Information Device Profile was designed for mobile phones and it includes the relevant APIs. MIDP applications are called MIDlets and it is widely supported in modern mobile phones. The *Information Module Profile (IMP)* is a profile for embedded devices that do not have a display or have a very simple display.

The first handsets using MIDP 1.0 arrived on the market in 2001. MIDP allowed phone users to download MIDlets from websites and interact with a richer custom interface rather than the default handset browser. However, MIDP 1.0 had a number of major drawbacks, which included restrictive access to system features such as storage, audio, and rendering features, and being only limited to HTTP interactions. The latter meant that pushing content to mobile devices required the polling of servers. Some of these limitations of MIPD 1.0 have been addressed in the form of optional JSR specifications; however, there are no guarantees that these are supported by handset makers, and if supported, they vary from device to device. The MIDP version 2.0 alleviated some of the technical limitations of MIDP 1.0, especially from the viewpoint of game developers. Indeed, MIDP 2.0 features gaming and multimedia APIs and some optional packages.

The MIDP specification contains two APIs for two different ways of programming applications:

- *High-level API.* This API is designed portability in mind. To ensure portability, some API functions have been limited, for example applications cannot change fonts.

- *Low-level API.* This API provides applications with direct control of the user experience by allowing them to change how the presentation and interaction is done. Device specific API is also part of low-level API (such as Nokia UI API). Applications using low-level API should not use device specific APIs. They should also cope with different screen sizes.

The main object of the UI API is the Displayable abstract class. An object that implements Displayable can be placed on the Display and thus it can become visible to the user. Figure 2.12 presents an example MIDlet conforming to MIDP 1.0. The code illustrates how a textbox is added to the Display.

MIDlets follow a specific lifecycle (illustrated by Figure 2.11), namely they can be in the following states:

- paused;
- active; and
- destroyed.

While creating the MIDlet, the runtime system calls the constructor of the MIDlet class. MIDlet's constructor cannot access the Display. When the application receives focus (at the start or when returning from Paused state), startApp() is executed. Application then should set the desired display by calling Display.getDisplay(this).setCurrent() method. When the application loses the focus, the method pauseApp() is called. When paused, the display is not active and it is not shown to the user. While paused, all timers and threads are still running. The application can call setCurrent() to indicate the screen that should be presented when returning from the Paused state.

The method destroyApp() is called when the application is being destroyed. At this time, there is no access to the Display; however, the application can save its state to persistent storage. The application can request to not enter the Destroyed state by throwing MIDletStateChangeException, but this is only possible if the unconditional flag associated with the destroyApp method is set to false.

The MIDP 2.0 specification extends the features of the much older 1.0 version. The key MIDP 2.0 packages are the following:

- javax.microedition.rms (from MIDP 1.0). Record Management System provides a form of persistent storage for Java ME.

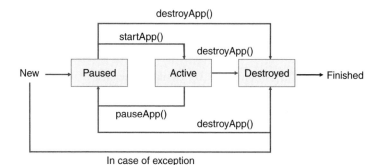

Figure 2.11   MIDlet lifecycle

```
import javax.microedition.midlet.*;
import javax.microedition.lcdui.*;

public class HelloWorld extends MIDlet implements CommandListener
{
    private Command exitCommand;
    public HelloWorld() {}
    public void startApp()
    {
        Displayable cur = Display.getDisplay(this).getCurrent();
        if(cur == null)
        {
            TextBox helloScreen = new TextBox("Hello!", "Hello!",
              256, 0);
            exitCommand = new Command("Exit", Command.EXIT, 1);
            helloScreen.addCommand(exitCommand);
            helloScreen.setCommandListener(this);
            Display.getDisplay(this).setCurrent(helloScreen);
        }
    }

    public void pauseApp() {}
    public void destroyApp(boolean b) {}

    public void commandAction(Command c, Displayable d)
    {
     if (c == exitCommand) {
     destroyApp(false);
     notifyDestroyed();
     }
    }
}
```

Figure 2.12   An example MIDlet

- javax.microedition.io (from MIDP 1.0). Contains the classes used for I/O operations.
- javax.microedition.lcdui (from MIDP 1.0). Contains the Java ME-specific classes used for the GUI.
- javax.microedition.midlet(from MIDP 1.0). Contains the base classes for Java ME applications.
- javax.microedition.media (from MIDP 2.0). Contains the base classes of the multimedia playback. These are approximately a subset of the JSR 135 Java Mobile Media API.
- javax.microedition.lcdui.game (from MIDP 2.0). A gaming API aimed at simple 2D sprite based games.
- javax.microedition.pki (from MIDP 2.0). Authentication APIs for secure connections.

MIDP 3 specified in JSR 271 will specify the third generation mobile APIs. A key design goal of MIDP3 will be backward compatibility with MIDP2 content. This revision will come with several new features, Notably MIDP3 will be supported by both CLDC and CDC virtual machines. In addition, MIDlet concurrency will bring multi-tasking to applications, and an InterMIDlet Communication facility is included as part of the javax.microedition.io package. The new package javax.microedition.event will provide application-to-application communication and notification of system events.

2.6.1.3 Extensions

Java ME has been extended with optional JSR specifications that add new APIs for developers. A mobile phone JSR support is specific to the phone model and version and thus JSR support varies with phone type and manufacturer. Table 2.1 presents a number of frequently supported JSR specifications, including File Connection and PIM, Bluetooth, web services, and Mobile Media API.

The Java ME platform's *Mobile Service Architecture (MSA)* specification (JSR 248) defines a standard set of application functionality for mobile devices. It also examines interactions between various technologies associated with the MIDP and CLDC specifications. An MSA version 2 device can use either CLDC 1.1 or CDC 1.1 as its configuration. The MIDlet execution environment is extended to the Connected Device Configuration.

Table 2.1 Important JSRs

JSR	Description
75	File Connection and PIM
82	Bluetooth
120	Wireless Messaging API (WMA)
135	Mobile Media API (MMAPI) Audio, video, multimedia
172	Web Services
177	Security and Trust Services
179	Location API
180	SIP API
184	Mobile 3D Graphics
185	Java Technology for the Wireless Industry (JTWI) General
205	Wireless Messaging 2.0 (WMA)
211	Content Handler API
226	SVG 1.0
229	Payment API
234	Advanced Multimedia Supplements (AMMS) MMAPI extensions
238	Mobile Internationalization API
239	Java Bindings for the OpenGL ES API
248	Mobile Service Architecture General
256	Mobile Sensor API
287	SVG 2.0

The interesting JSRs of MSA version 2 include the following:

- JSR 271: MIDP 3.0.
- JSR 256: Mobile Sensor API, which allows on-device sensors such as accelerometers to be accessed.
- JSR 257: Contactless Communication API, which supports technologies such as *Think Radio Frequency Identification (RFID), Near Field Communication (NFC)*, and barcodes.
- JSR 258: Mobile User Interface Customization API that allows interface customization, e.g., different 'skins'.
- JSR 272: Mobile Broadcast Service API for Handheld Terminals, which supports streaming of multimedia to mobile phones. The JSR includes an electronic program guide, purchasing and DRM, connection management, and presentation.
- JSR 280: XML API for Java ME. The XML API for Java ME will enable widespread use of XML in Java ME applications for data storage and exchange.
- JSR 281: IMS Services API.
- JSR 287: Scalable 2D Vector Graphics API 2.0 for Java ME.
- JSR 293: Location API 2.0, which adds support for geocoding, maps incorporating landmarks, and navigation.

As a summary, we observe that the Java ME is becoming a versatile platform for mobile application development. The early MIDP applications were restricted to a very small set of APIs; however, with the introduction of various JSRs and the MSA version 2 this has gradually changed as more and more vendors support newer specifications. Moreover, software portability challenges between CLDC and CDC are being addressed in MIDP3, which allows CDC to run MIDlets. From the mobile computing viewpoint, Mobile Sensor API, Contactless Communication API, and Location API support the creation of applications that are aware of their surroundings and context, and can adapt and tailor content accordingly. Mobile Broadcast Service API supports the delivery of streaming multimedia content to mobile phones. Converged communications support is provided by the XML API and IMS Services API. The latter allows MIDlets to utilize SIP and IMS platform features. Usability is crucial in mobile devices, as amply demonstrated by the iPhone's usability aspects which have become a selling point for the product, and this is addressed by the Mobile User Interface Customization API and Scalable 2D Vector Graphics API. The former allows users to the customize MIDlets, and the latter provides APIs for processing SVG vector graphics.

2.6.2 *iPhone*

The iPhone OS is a mobile operating system developed by Apple Inc. for their iPhone and iPod touch products. The OS is derived from Max OS X and uses the Darwin foundation. Darwin is built around XNU, a hybrid kernel that combines the Mach 3 microkernel, various elements of Berkeley Software Distribution (BSD) Unix, and an object-oriented device driver API (I/O Kit).

The iPhone OS is based on four abstraction layers, namely the *Core OS layer*, the *Core Services layer*, the *Media layer*, and the *Cocoa Touch layer*.

The Core OS layer includes the OS X Kernel, TCP/IP networking stack and the Sockets interface, power management, file system, and security features. The Core Services layer includes Mac OS X application programming interfaces that are below the Media and Cocoa Touch layers. These include APIs to OS services such as networking and threads, and web services APIs. In addition, this layer provides embedded SQLite database and support for geolocation. The Media layer pertains to various functions regarding the input and output interfaces of the device, namely audio mixing and recording, video playback, OpenGL, and animation support. Finally, the Cocoa touch layer supports multi-touch events and controls, and has interface to use accelerometer input. The layer also has support for localization (i18n), and camera support.

The iPhone OS's user interface is based on *multi-touch gestures*. Interface control elements consist of sliders, switches, and buttons. Interaction with the OS includes gestures such as swiping, tapping, pinching, and reverse pinching. Additionally, using internal accelerometers, rotating the device on its y-axis alters the screen orientation in some applications.

Apple provides the SDK as a download without fee, but in order to release software for the iPhone platform, developers must obtain Apple's approval. In order to upload applications, payment and registration are needed.

The App Store is an application for the iPhone and iPod touch created by Apple Inc., which allows users to browse and download applications that were developed with the iPhone SDK. They are available to purchase or are free, depending on the application. The applications are downloaded directly to iPhone or iPod touch.

Figure 2.13 presents an overview of the MacOS X architecture, which has been adapted in the iPhone architecture. The iPhone system is built on an ARM processor and the Core

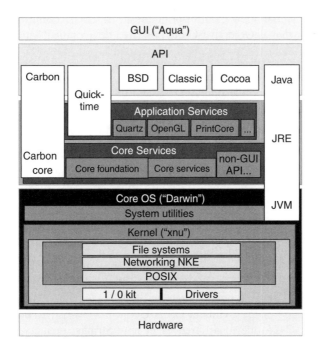

Figure 2.13   The MacOS X architecture

OS (Darwin) includes the XNU kernel and system utilities. The XNU kernel includes POSIX support, networking, and file system support, as well as device drivers. Above the kernel, we have the system utilities. Above the operating system, we have the middleware, which has a layered structure, namely core services, application services, API layer, and finally the GUI (Aqua).

The five important APIs available for the developer are illustrated in the API layer in the diagram, namely Carbon, Quicktime, BSD/Posix, Classic, Cocoa. Java support is shown in the figure; however, the current iPhone systems do not support Java. Sun Microsystems have announced plans to release a Java Virtual Machine (JVM) for iPhone OS based on the Java ME.

Carbon is a procedural API that consists of separate manager entities that realize part of the API. Each Manager offers an API related to some functionality, thus defining the necessary data structures and functions. Managers are often interdependent or layered. Managers in Carbon include the file manager, resource manager, font manager, and event manager.

The POSIX specifications define crucial operating system software interfaces. POSIX also defines a standard threading library API which is supported by most modern operating systems. The POSIX standard has three important parts, Kernel APIs, Commands and Utilities, and Conformance Testing. The kernel APIs include real-time services, threads, security interface, network file access, and network process-to-process communications.

Classic, or Classic Environment, is a backwards compatible hardware and software abstraction layer. Classic is no longer supported in the current Mac OS version.

Cocoa Touch provides an abstraction layer of the iPhone OS. Cocoa Touch is based on the native object-oriented application program environment for the Mac OS X. Cocoa's design is based on Model-View-Control principles. The Cocoa frameworks are written in Objective-C, which makes Objective-C the preferred development language.

2.6.3 Symbian and Series 60

Symbian OS is an open mobile operating system developed by Symbian Ltd. for ARM processors. The system includes a microkernel OS, associated libraries, user interface, and reference implementation of common tools. The OS is structured in the fashion of many desktop operating systems with pre-emptive multitasking and memory protection. The multitasking model features server-based asynchronous access based on event passing. The choice of servers, a microkernel design, and event passing were motivated by the three design rules, namely minimize response times to users, maximize integrity and security, and utilize scarce resources efficiently.

The three default user interfaces for Symbian OS are Series 60, UIQ, and MOAP. Symbian Ltd. was formed in 1998 as a partnership between the mobile vendors, namely Nokia, Ericsson, Motorola, and Psion. Recently, Nokia has acquired the ownership of Symbian (2008) and has established Symbian Foundation that aims to provide royalty-free software for the mobile environment. They have stated that the Symbian OS and S60, UIQ, and MOAP will become open source in 2010.

The native language of the Symbian OS is C++; however, the language is not compatible with ANSI C++ due to differences in how Strings, memory management, and exceptions are realized. The OS and applications are based on the Model-View-Control (MVC) design pattern, which support the separation of different functions. All Symbian applications are built up from three classes defined by the application architecture: an application class, a document class, and an application user interface class. These classes create the fundamental application behavior. The remaining required functions, the application view, data model, and data interface, are created independently and interact solely through their APIs with the other classes.

Symbian OS emphasizes resource recovery using several programming features, such as a cleanup stack and descriptors. The event-based nature of the OS allows the minimization of thread switching using a technique called active objects.

The Symbian OS System Model contains the following layers that are illustrated by Figure 2.14:

- UI Framework Layer.
- Application Services Layer.
- Java ME.
- OS Services Layer: generic OS services, communications services, multimedia and graphics services, connectivity services.
- Base Services Layer.
- Kernel Services and Hardware Interface Layer.

The Base Services Layer is the lowest level reachable by user-side operations; it includes the File Server and User Library, the Plug-In Framework which manages all plug-ins, Store, Central Repository, DBMS, and cryptographic services. Symbian OS has a micro-kernel architecture, which includes a scheduler, memory management, and device drivers, but other services like networking, telephony, or filesystem support are placed in the OS Services Layer or Base Services. Symbian OS applications are distributed using SIS files, which may be installed over-the-air, using a cable, or using Bluetooth or memory cards.

Figure 2.15 presents a more detailed overview of the communications support Symbian architecture. The Base Services layer is responsible for basic connectivity and serial communications as well as telephony. The communications infrastructure has been developed on this layer and two prominent networking stacks are the TCP/IP and WAP stacks. In addition, a number of narrow band protocols, infrared, and Bluetooth are supported. The

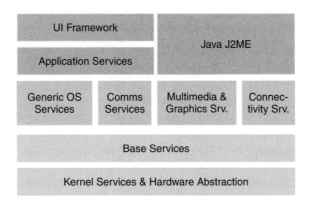

Figure 2.14   The Symbian architecture

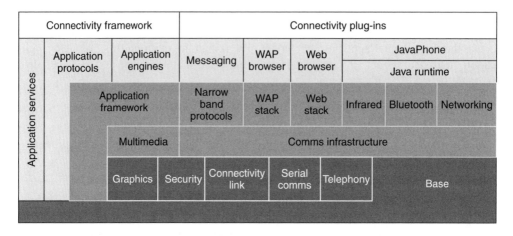

Figure 2.15 Communications support in the Symbian architecture

Web and WAP browsers are available for the respective protocol stacks. Moreover, the Java runtime and JavaPhone are available for applications. On top of the stack, the connectivity framework and plug-ins offer extension possibilities for communications. The figure does not include SyncML support, which is also a core feature.

Recently, in Symbian OS 9.x, most applications must be signed using a centralized process provided by Symbian Ltd. in order to be installable and executable on a mobile phone. An unsigned application has very limited features and at least in theory cannot perform harmful actions. One motivation for application signature process, called Symbian Signed, is to improve mobile phone security by preventing the installation and execution of unknown and possibly hazardous programs. Several viruses and trojan horse programs have been developed for the OS, for example Cabir, which have caused some concerns for the trustworthiness of mobile software and prompted a number of anti-virus products for mobile phones.

The most recent version of the Series 60 platform is the 5th Edition, which was released in October 2008. This version includes widescreen support, advanced camcorder features, multimedia support for MP3, ACC, H.264, Windows Media, Flash Video, and a built-in Internet browser and Flash Lite support. Given the recent trend towards tactile user interfaces, the Touch UI has a tactile feedback and sensor framework.

The Nokia Web Browser is based on the S60WebKit (illustrated in Figure 2.16), which is a port of the open source WebKit project to the S60 platform and Symbian OS. WebKit contains the WebCore and JavaScriptCore components that Apple uses in its Safari browser. Based on KHTML and KJS from KDE's Konqueror open source project, this software offers improvements in website usability on smartphones through the re-use of an existing desktop rendering engine that has been developed and optimized by a large open source community over many years. The Nokia Web Browser supports Dynamic HTML, AJAX applications, and W3C's XHTML 1.0, DOM, CSS and SVG-Tiny. Other supported web standards include SSL and ECMAScript; and Netscape style plug-ins such as Flash Lite and audio.

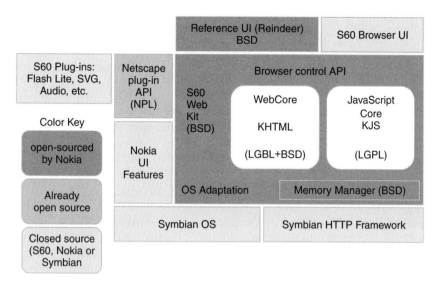

Figure 2.16 Overview of the S60WebKit

2.6.4 BREW

Binary Runtime Environment for Wireless (BREW) is a mobile application development platform created by Qualcomm in 2001. BREW was originally developed for *Code Division Multiple Access (CDMA)* handsets, but has later been ported to support other types of handsets including GSM and GPRS handsets. BREW offers similar functionality as Java ME and the platform is capable of downloading small programs and then executing them. Typical programs include games, messaging, and photo sharing.

From the software developers' perspective, BREW offers a set of APIs and a SDK for development in C, C++, and Java. The SDK includes a BREW Simulator that allows developers to test their software. In order for a handset to execute a BREW application, the application must first be signed. Signing is needed because BREW gives control of the handset hardware to the application, and thus security measures are needed to prevent malicious software from being downloaded. Content providers or authenticated BREW developers have the tools necessary to sign applications. There is a centralized application testing process that is used to ensure that a BREW application can be offered to a mobile operator for delivery to end users.

In order to help developers test software, a BREW Simulator is included in the SDK. This tool does not emulate a mobile phone's hardware, but rather compiles the BREW application to native code linked to a x86 compatible BREW runtime library. The simulator does not detect all possible problems with memory management and possible incompatibilities with hardware versions. Therefore, testing on real BREW devices is necessary to ensure that an application works properly.

BREW application deployment involves transferring the necessary components of the application to a mobile handset. The necessary components include a *name.mif* file that describes the application and its features and permissions, a *name.mod* file that is the compiled application, a *name.bar* that contains the string and image resources if they are needed, and a *name.sig* file which is the application signature. Should the signature check fail, the application is deleted by the BREW platform.

2.6.5 WAP

Wireless Application Protocol (WAP) is an open international standard for application layer network communications in a wireless communication environment developed by the WAP Forum standards organization. WAP Forum is now part of the *Open Mobile Alliance (OMA)* organization that is developing core wireless and mobile standards. The WAP Forum proposed a protocol suite that would allow the interoperability of WAP equipment and software with many different network technologies. The motivation was to build a single platform for competing network technologies such as GSM and CDMA. The main usage scenario of WAP is to enable access to the Internet (HTTP) from a mobile phone or PDA [17]. A WAP browser provides all of the basic services of a computer based web browser but simplified to operate within the restrictions of a mobile phone, such as its smaller view screen. WAP sites are websites written in, or dynamically converted to, *Wireless Markup Language (WML)* and accessed via the WAP browser.

The WAP protocol suite consists of a layered protocol stack, which is outlined in Figure 2.17. In the following, we briefly outline each of the layers.

- The lowest protocol in the suite is the *Wireless Datagram Protocol (WDP)*. WDP is an adaptation layer that maps the underlying data network to UDP like API by offering unreliable transport of data. WDP is considered by all the upper layers as one and the same protocol with different technical implementations. On native IP bearers such as GPRS and UMTS, WDP is the same as UDP.
- *Wireless Transport Layer Security (WTLS)* provides a public-key cryptography-based security mechanism similar to TLS. This is an optional protocol.
- *Wireless Transaction Protocol (WTP)* provides transaction support (reliable request/response) for wireless networks. WTP is more efficient than TCP with the problem of packet loss and burst errors.

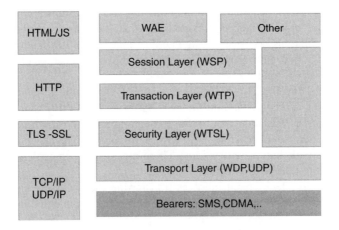

Figure 2.17   WAP architecture

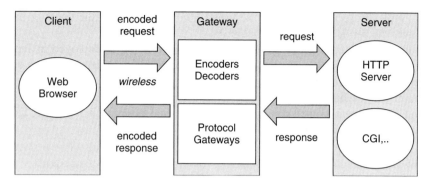

Figure 2.18   WAP Gateway

- *Wireless Session Protocol (WSP)* is a compressed version of HTTP.
- *Wireless Application Environment (WAE)* defines the WML for the definition of WAP web pages, and an associated scripting language.

The prominent feature of the original WAP architecture was the WAP gateway, which was used to connect mobile phones using the WAP protocol suite to the global Internet. The WAP gateway performed protocol conversion, content adaptation and transcoding, and caching. Figure 2.18 presents an overview of the WAP gateway. A client runs a WAP compatible web browser and the WAP protocol stack encodes requests and decodes responses. The gateway is responsible for transforming the encoded requests into HTTP requests and then transforming the HTTP responses to encoded responses.

The latest version of the protocol suite is WAP 2.0 which is based on a subset of XHTML and end-to-end HTTP. This standard does not require a gateway and custom protocol suite, which makes it more compatible with the current Web 2.0 protocols. A gateway can be used with WAP 2.0; however, the gateway is in this case simply a proxy server. The new version also supports mobile push. WAP Push allows WAP content to be pushed to mobile devices with minimum user intervention. This is achieved by using a specially encoded message which includes a link to a WAP address. The message is sent using WDP, typically GPRS or SMS.

Although the WAP protocol suite addressed some of the problems of the wireless and mobile environment, namely TCP performance over wireless and content adaptation, WAP did not meet the high expectations for the technology. Several concerns have been voiced over WAP technology and the business models. From the technological viewpoint, WML was very constrained and simple, and not directly compatible with HTML used in the web. This required the use of the gateway and transcoding as a means of displaying web contents to users. This meant that for best usability, content had to be hand crafted for WAP.

From the business model viewpoint, WAP services required that service developers add WAP support to their services and then negotiate with wireless providers to make the service usable on that provider's network. With WAP, the service developers did not have convincing incentives to develop WAP interfaces. On one hand this was due to the uncompelling business opportunities with wireless operators and on the other hand the WAP service ecosystem was a collection of closed walled gardens. The default WAP page was typically served by the mobile user's home operator and the operator had tight control over the services available through WAP.

WAP is often compared with the i-mode architecture that was deployed in Japan and became a very successful technology. In contrast to WAP, i-mode was based on more open services and offered better incentives for service developers. From the technical perspective, i-mode builds on top of the fixed Internet data formats, such as C-HTML based on HTML; however, in a similar vein to the original WAP stack, it features DoCoMo proprietary protocols ALP (HTTP) and TLP (TCP, UDP).

2.6.6   Windows Mobile and .NET Compact Framework

Windows Mobile 6 was released by Microsoft at the 3GSM World Congress 2007 and it comes in three flavours, namely standard version for smartphones, a version for PDAs with phone functionality, and a classic version for PDAs without phone features. Windows Mobile 6 is based on the Windows CE 5.0 operating system and has been designed to integrate with Windows Live and Exchange products. Software development for the platform is typically done using Visual C++ or .NET Compact Framework. When native client-side functionality is not needed, server-side code can be used that is deployed on a mobile browser, such as the Internet Explorer Mobile bundled with Windows Mobile.

The .NET Compact Framework is a subset of the .NET Framework and shares many components with the desktop software development environment. The framework includes an optimized *Common Language Runtime (CLR)* and a subset of the .NET Framework class library, which supports features such as *Windows Communication Foundation (WCF)* and Windows Forms. It also contains classes that are designed exclusively for the .NET Compact Framework. A platform adaptation layer exists between Windows CE and the common language runtime to map the services and device interfaces required by the CLR and Framework onto Windows CE services and interfaces. The expectation is that managed components developed using .NET languages, such as C#, are used to create the applications. It is also possible to use the Win32 API with .NET CF, which allows direct access OS features.

The advantages of a managed environment such as CLR include more trustworthy and platform independent software. The disadvantage of managed code is the performance penalty in real-time environments. Garbage collection and *Just in Time (JIT)* compilation can introduce unexpected delays to program execution and thus introduce non-determinism.

Figure 2.19 illustrates the .NET Compact Framework architecture and its components. The lowest layer of the architecture consists of the actual hardware of the mobile device. On top of this, there is the operating system that provides access to the features of the hardware. Directly on top of the operating system, we have the runtime environment, which executes instructions of a .NET application and maps them to the instructions of the operating system and the underlying hardware.

The .NET Compact Framework CLR is made up of the following three components:

- class libraries;
- execution engine; and
- platform adaptation layer.

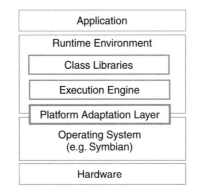

Figure 2.19   The .NET Compact Framework architecture

The purpose of the class libraries is to provide a basic set of classes, interfaces, and value types that constitute the foundation for developing applications in .NET. The execution engine is the core component of the CLR. It provides the fundamental services needed for executing managed code. The execution engine includes components such as a JIT compiler, a class and module loader, and a garbage collector. Portability between different operating systems is achieved using the *Platform Adaptation Layer (PAL)*. The PAL layer maps calls from the execution engine to the functions of the underlying operating system. Consequently, the PAL layer is instrumental in supporting different operating systems.

The .NET Compact Framework advocates the creation of smart clients, which are able to better harness the processing power of the device. In contrast to micro-browser-based mobile thin clients, the smart clients are envisaged to be more reliable and efficient by utilizing aggressive caching and local processing to minimize network round trips. Smart clients are motivated by the unreliable and slow wireless networks and offline capability. Moreover, one major benefit is the instant user interaction that they can provide thus improving the usability of mobile software.

The .NET Compact Framework supports WCF, which is Microsoft's unified programming model for building service-oriented applications. Support for a new WCF transport, the Microsoft Exchange Server mail transport, has been introduced to both .NET Compact Framework applications and desktop applications. *Language-Integrated Query (LINQ)* introduces general-purpose query facilities to the .NET Compact Framework pertaining to various sources of information such as relational databases, XML data, and in-memory objects.

The .NET Compact Framework supports different operating systems, most importantly Windows CE. The framework has been also ported successfully to work with Symbian [18]. The main differences between Windows CE and Symbian for the porting were related with multitasking, error handling, file access, and networking. Figure 2.20 illustrates the Symbian port of .NET Compact Framework. The main differences observed in the porting process were the following:

- A C++ dialect that redefines basic language structures.
- No writable global and writable static variables allowed in DLLs.

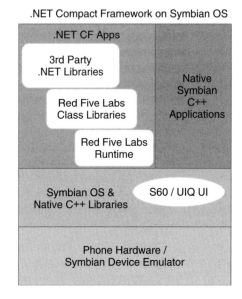

Figure 2.20   The Symbian port of .NET Compact Framework

- Extensively used client/server model that, for example, implies constraints for accessing file and networking functions.
- Event-driven programming model with a focus on non-preemptive multitasking.
- Symbians error handling and cleanup model.
- Concepts from the Unix/Windows world such as environment variables as well as several file and networking functions are missing.

2.6.7   NoTA

Network on Terminal Architecture (NoTA) developed by Nokia is a modular service-based architecture framework for embedded devices. The architecture is interconnect-centric, and emphasizes flexible communication between different system components. The basic paradigm has been inspired by the web service paradigm. The interconnection model is based on the NoTA *Device Interconnect Protocol (DIP)*. DIP is a device level protocol that can be implemented for a number of different interfaces, including Bluetooth and WLAN. The NoTA core technology has been opened and it is available on the Internet.

The motivation for the NoTA architecture has been the need to reduce time-to-market of mobile phone products, which are very complex. The product development challenges become highlighted at the system integration phase, where the interactions of different parts of the system need to be understood and managed. From the software developer viewpoint, the situation is also complicated because the available features and APIs differ between devices, operating systems, and software platforms. Therefore open interfaces to subsystem services are needed and they need to be discoverable and accessible to higher level software.

The idea behind NoTA is to move from tightly coupled architectures used today to flexible decoupled modular systems. Thus a mobile phone would no longer be centered around a single application engine, but user a network of subsystems, in which each subsystem performs certain tasks. These tasks can include application processing, storage, connectivity, and multimedia support.

The aim of NoTA is to open the hardware for platform innovations. NoTA will also be open for open source community. Traditionally mobile devices have been a very closed and monolithic system, so this is going to be a revolutionary change. The migration of new ideas into mobile devices has been very controlled and in many ways hard process. Since NoTA is not dependent on implementation technologies, it can also extend beyond product boundaries. The path to radical innovations in this traditionally closed business is therefore open.

A NoTA platform consists of loosely connected services (service nodes, SN) running on top of heterogeneous subsystems. The applications (application nodes, AN) in the final product exploit these services. In NoTA based systems all service and data communication is routed via the interconnect network as illustrated in Figure 2.21. The architecture is a distributed architecture enabling for example direct communication channels between service nodes needed in efficient data streaming. Service Nodes and Application Nodes communicate through logical *Interconnect (IN)*. IN provides two basic means of communication, namely message based and streaming type. The former is bi-directional and used for Service Messages. The latter one is uni-directional and used for large amounts of data like media content. Service Nodes have unique *Service Identifier (SID)*.

The Interconnect is divided into two layers, namely *High Interconnect (H_IN)* and *Low Interconnect (L_IN)*. The former provides means for service activation and deactivation as well as service and stream accesses. Low Interconnect provides network socket interface with uniform addressing mechanism. L_IN internally can be divided into transport network independent and dependent parts.

The NoTA design offers separation of concerns between the following three domains:

- end-user requirements;
- platform functionality that includes the services provided by the platform;
- platform architecture that includes the definition of subsystems and communication infrastructure, and implementation of subsystems and interconnect protocols.

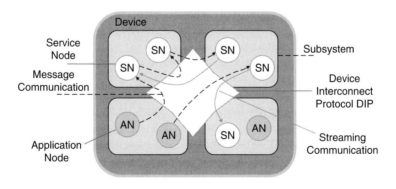

Figure 2.21   NoTA architecture

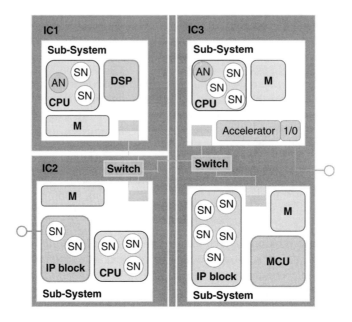

Figure 2.22   Four NoTA subsystems

NoTA emphasizes automated design steps and the reuse of models and services. The driving idea is to be able to postpone the implementation decisions and allow easy modification of the design without compromising the quality of the resulting product. A NoTA based product consists of subsystems connected by an interconnect. NoTA defines two main level of protocols for the interconnect. The high level protocol stack provides communication functionality for platform services and applications. The low level protocol provides the physical connection between subsystems. A NoTA subsystem implements a set of services. It is an architectural concept that does not necessarily align with chip boundaries. There may be several subsystems on a chip and a subsystem may extend outside the boundaries of a chip. Figure 2.22 shows an example of a NoTA system consisting of four subsystems.

In a product all the subsystems can be developed independently by subsystem providers which supports the separation of concerns. The providers can include various technologies and optimised functionality. In addition, the same subsystems can also be used in other products and in completely different type of products.

NoTA services are defined using a *Service Interface Specification (SIS)*. Services register with the Interconnect using a service ID associated with the interface specification. Other services may then find the service and request a connection conforming to the interface specification by using the ID. The Interconnect is responsible for setting up the communication channel.

A subsystem specification comprises two parts:

- A list of services (and their SIS) that the subsystem should implement.
- A set of usage scenarios (use cases) that the subsystem should support.

Table 2.2 Comparison of NoTA, Web services, and Sun RPC

	NoTA	Web services	Sun RPC
Service Access	Invocation, streaming	Invocation	Invocation
Invocation style	Many	RPC,SOA	RPC
Addressing	Indirect, NoTA service type	Indirect, various	Direct, node
Wire protocol	Standard, application specific	XML based	RPC specific
Byte transport	NoTA specific	TCP	TCP

Table 2.2 presents a comparison of NoTA with Web services and Sun RPC.[1] Although NoTA focuses on the interconnection of device subsystems, it borrows features from web services and SOA. NoTA is a hybrid architecture that involves both hardware and software. Applications have interfaces modelled after Web services, which makes it palatable for the modern programmer. The comparison highlights some notable differences, namely NoTA supports various invocation styles.

Figure 2.23 presents the NoTA service STUB decoding and encoding when an AN sends a request to a SN via the Interconnect. The Interconnect is responsible for connecting the two components. A NoTA stub generator is available for generating communication related code to ANs and SNs. Stub generator can be used to generate both AN and SN side stubs. The figure illustrates the stubs generated from SIS descriptions at SN and AN. Part of the interfaces may come from standards bodies or be based on agreements between vendors.

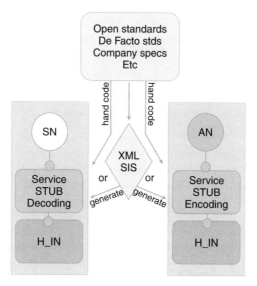

Figure 2.23 NoTA interoperable interconnection

---

[1] Service Concepts and NoTA. Presentation at the 1st NoTA Conference, 11 June 2008 by Tuomo Vehkomäki.

2.6.8 Linux Maemo

The Maemo platform from Nokia includes the Internet Tablet OS, which is based on Debian GNU/Linux and draws much of its GUI, frameworks, and libraries from the GNOME project. Figure 2.24 presents an overview of the system architecture. Maemo is based on the Linux operating system kernel, which is a monolithic kernel that supports multiple hardware platforms. Linux is scalable from tiny embedded devices to server farms and clusters. It uses the Matchbox window manager, and like Ubuntu Mobile, it uses the GTK-based Hildon as its GUI and application framework. The Maemo platform is intended for Internet tablets, which are smaller than laptops, but larger and more versatile than PDAs. A tablet may have a small keyboard, and central characteristics include a stylus and a touch-sensitive screen. Graphical interfaces must be designed with the touch screen in mind.

The Maemo SDK features a sandboxed development environment on a GNU/Linux desktop system largely built using a tool called Scratchbox. This environment behaves in a similar manner than the actual OS on the Nokia Internet Tablet devices. Using this environment, the development of Maemo software is comparable to the development of normal Linux software. Cross-compilation to the actual device environment is handled by the Scratchbox tool. Currently C is the only official programming language for Maemo.

The Maemo user interface architecture is based on the GNOME framework, especially the GTK+ widget set. GNOME is an application framework for desktop Linux systems. Maemo extends GTK+/GNOME by providing the Hildon extensions in order to support a mobile desktop. The Hildon framework provides components on top of the GNOME components to support control panel, status bar, task navigator, and home applets. Hildon framework also provides backups/restore service, help framework, and an application installer.

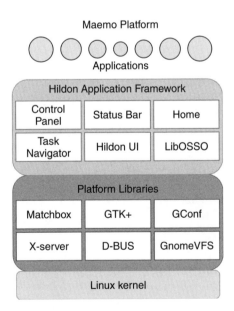

Figure 2.24   Maemo architecture

For hardware abstraction, Maemo provides HAL Hardware Abstraction Layer. It provides a shared library that has an API for device objects. HAL is capable of loading the right device driver, when a new device is detected, creating and maintaining /dev files, and tracking the status of devices.

The Maemo platform includes the normal networking protocols, such as TCP/IP stack, OpenSSL library for network security, and and libcurl that provides HTTP access for applications. The DBUS communication system is used as the primary channel between applications. Maemo also includes an SQL database, SQLite, that can be used to store user application data. SQLite database is accessed through a library interface and thus there is no centralized server process to connect into.

The applications are built on top of the Hildon framework. Simple applications use the Hildon libraries and GTK+ in order to use the graphical user interface elements.

The Maemo *Power Management (PM)* framework can be divided into two independent mechanisms, namely *OS idle* and *Dynamic Voltage-Frequency Scaling (DVFS)*. The first is based on the operating system scheduler. Whenever the scheduler has no tasks to perform, it calls the idle function. The idle function can then minimize power usage by choosing to shutdown all or parts of the hardware. The level of power savings depends on the clock and voltage resources in use. The second is used to scale down the System on a Chip (SoC) frequency and voltage at runtime to reduce leakage currents. The decision to scale the frequency of the ARM and DSP processors is based on the current load.

2.6.9 Android

Android is an operating system and software platform for mobile devices based on the Linux OS. Android has been developed by Google and the Open Handset Alliance. The platform allows developers to develop managed code using a language that is very similar to Java. The language follows the Java syntax, but does not provide the standard class libraries and APIs that come with Sun's Java Micro Edition or Standard Edition. The language utilizes libraries and APIs developed by Google.

Android and the founding of the Open Handset Alliance was announced in November 2007. The alliance consists of 34 hardware, software, and telecom companies. The Android platform is expected to be released under the Apache free-software open source license.

The notable features of the Android platform include the following:

- Dalvik virtual machine optimized for mobile devices.
- Integrated browser based on the open source WebKit engine.
- Optimized graphics powered by a custom 2D graphics library; 3D graphics based on the OpenGL ES 1.0 specification (hardware acceleration optional).
- SQLite for structured data storage.
- Media support for common audio, video, and still image formats (MPEG4, H.264, MP3, AAC, AMR, JPG, PNG, GIF).
- GSM and 3G Telephony (hardware dependent).
- Bluetooth, EDGE, 3G, and WiFi (hardware dependent).
- Camera, GPS, compass, and accelerometer (hardware dependent).
- Development environment including a device emulator, tools for debugging, memory and performance profiling, and a plugin for the Eclipse IDE.

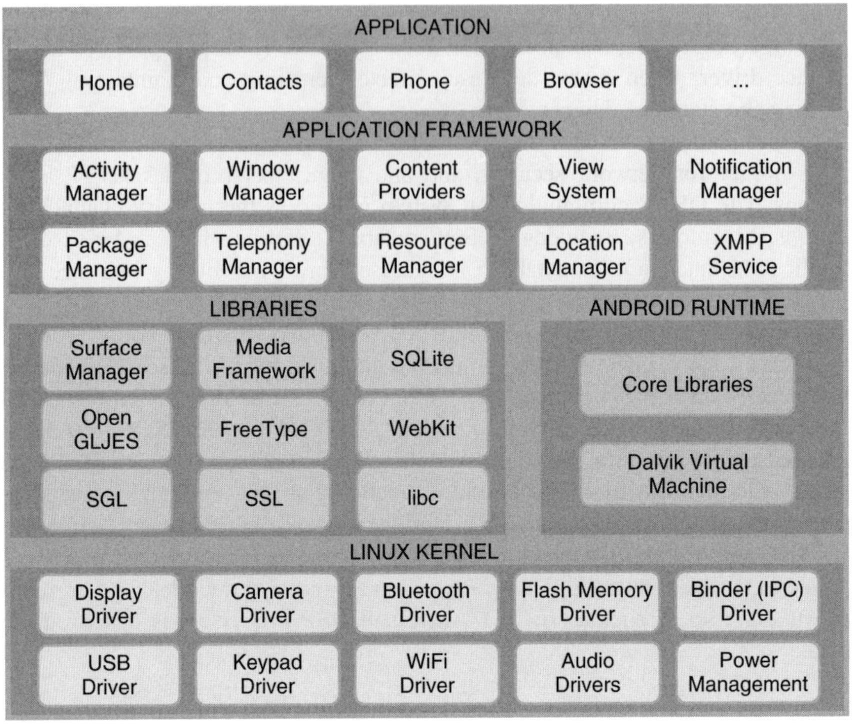

Figure 2.25   The Android architecture

Figure 2.25 presents the Android architecture. The architecture is based on the Linux kernel and a set of drivers for the various hardware components, such as display, keypad, audio, and connectivity.

Android includes a set of C/C++ libraries used by various components of the Android system. The capabilities of these libraries are exposed to developers through the Android application framework APIs. The core libraries include:

- System C library, a BSD-derived implementation of the standard C system library (libc), adapted for embedded Linux-based devices.
- Media Libraries based on PacketVideo's OpenCORE.
- Surface Manager that manages access to the display subsystem and seamlessly renders 2D and 3D graphic layers from multiple applications.
- LibWebCore, a web browser engine which powers both the Android browser and an embeddable web view.
- SGL, the underlying 2D graphics engine.
- 3D libraries, an implementation based on OpenGL ES 1.0 APIs.
- FreeType, bitmap and vector font rendering.
- SQLite, a lightweight relational database engine available to all applications through the framework API.

The Android runtime is responsible for the execution of the custom Java bytecode. The runtime includes the Core Libraries and the Dalvik Virtual Machine. On top of the libraries and the runtime, we have the application framework, which consists of various managers. Finally, the applications reside on top of the managers and include a number of bundled applications. Android-based mobile phones are expected to ship with a set of core applications that include an email client, SMS program, calendar, maps, browser, and contacts. All Android applications are written using the Java programming language. Developers utilize the same API that is also used by the built-in core applications. Android emphasizes component reuse and any component can publish its capabilities, which can then be utilized by other components given that security constraints do not prevent this.

In the following, we briefly outline the key parts of the Android API:

- *AndroidManifest.xml.* This XML document contains the configuration that tells the system how the top-level components will be processed.
- *Activities.* An activity is an object that has a life cycle and performs some work. An activity can involve user interaction. Typically one of the activities associated with an application is the entry point for that application.
- *Views.* A view is an object that knows how to render itself to the screen.
- *Intents.* An intent is a message object that represents an intention to perform some action. As an example, we can consider the normal web browsing scenario, in which the user wants to open a web page. In Android terminology, the application has then intent to view the web page, and generates an Intent instance in order to view the web page using a URL. The Android system then decides how to implement the intent. In this case, a browser would be used to load and display the web page.
- *Services.* A service is code that runs in the background. The service exposes methods for components. Other components bind to a service and then invoke methods provided by using remote procedure calls.
- *Notifications.* A notification is a small icon that is visible in the status bar. Users can interact with this icon to receive information. The most well-known notifications are SMS messages, call history, and voicemail, but applications can create their own. To avoid interrupting the user, notification is the preferred mechanism for alerting the user.
- *ContentProviders.* A ContentProvider provides access to data on the device.

In Android, an activity is a single task that the user can perform. Most activities interact with the user and the Activity class supports this through the setContentView method, which allows the UI to be set. The activities are managed as an activity stack. When a new activity is created, it is placed at the top of the stack. The previous activity remains below the new one in the stack, and will not come to the foreground until the new activity exists. An activity has four main states (illustrated by Figure 2.26):

- Active. An activity is active when it is in the foreground of the screen and at the top of the activity stack.
- Paused. An activity is paused when it has lost focus, but is still visible. A paused activity is alive, but can be destroyed by the system if memory needs to be freed.
- Stopped. An activity is stopped when it is obscured by another activity. The stopped activity retains its state, but it is no longer visible and can be destroyed by the system when memory is needed.
- If an activity is paused or stopped, the system can remove the activity from memory. This can happen in two ways, the system can ask the application to finish or simply destroy the process.

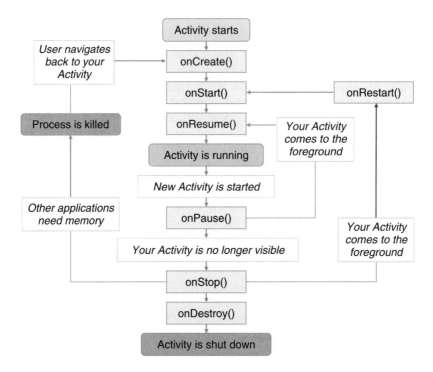

Figure 2.26   Activities in Android

Android applications are distributed as packages (.apk) that can be installed on a device. Each installed package is given its own unique Linux user ID. In addition, applications are executed in a sandbox environment to avoid security problems. A basic Android application does not have any security permissions so it cannot do anything that would have adverse effects on the system. In order to use protected features of the system, the AndroidManifest.xml file needs to have special *<uses – permission>* elements that declare the permissions that the application requires. Security enforcement happens at the process level. This means that code of any two packages cannot be normally run in the same process although there is a way to avoid this by making sure that the packages have the same user ID.

Since applications are written using Java and there is no support for native code, applications cannot access hardware features directly. Symbian and Linux-based mobile platforms offer the possibility of richer interactions with the hardware.

Android developer resources suggest the following programming techniques for improving application performance:

- Avoid Creating Objects. As a general rule for programming for small and resource constrained devices, unnecessary object creation should be avoided. This can be supported by using lazy instantiation of objects and by using object pools.
- Use Native Methods. For example, when processing strings, the String.indexOf() and other methods are native and can be considered to be efficient.

- Prefer Virtual Over Interface. Calling through an interface reference can take much longer than a virtual method call through a concrete reference.
- Prefer Static Over Virtual. If you do not need to access an object's fields, make your method static. It can be called faster, because it does not require a virtual method table indirection.
- Avoid Internal Getters/Setters. This is not a good idea when developing for Android. Virtual method calls are more expensive than instance field lookups.
- Cache Field Lookups. Accessing object fields is much slower than accessing local variables.
- Declare Constants Final. This makes their usage more efficient.
- Avoid Enums.
- Use package scope with inner classes.
- Avoid floating point math operations and use fixed point arithmetic instead.

2.6.10   OSGi

The OSGi Alliance is an open standards organization founded in March 1999. The Alliance and its members have specified a Java-based service platform that can be remotely managed and which supports dynamic discoverable and loadable modules. OSGi application areas include home network gateways, mobile phones, industrial automation, cars, and grid computing.

The core part of the specifications is a framework that defines an application life cycle management model, a service registry, an execution environment, and software modules. OSGi relies on Java to achieve portability across different platforms.

The OSGi Service Platform provides necessary functions for creating and changing compositions, called bundles, on a device. A service-oriented approach is used that enables components to dynamically discover each other. The OSGi Alliance has developed many standard component interfaces for functions including HTTP servers, configuration, logging, security, and user administration. Each bundle is a tightly-coupled, dynamically loadable collection of classes, jars, and configuration files that explicitly declare their external dependencies.

A service interface is described as a Java class or interface along with a variable number of name value pairs, the service properties. The service registry allows service providers to be discovered through queries formulated in a *Lightweight Directory Access Protocol (LDAP)* syntax. In addition, notification mechanisms allow service requesters to receive events signalling changes in the service registry. Service providers and requesters are part of a bundle. Service interfaces are implemented by objects created by the bundle. The bundle is responsible for run-time service dependency management activities, which include publication, discovery and binding as well as adapting to changes resulting from dynamic availability of services that are bound to the bundle.

Figure 2.27 highlights the important layers in the OSGi architecture. On top of the OS we have the Java virtual machine. The framework is conceptually divided into the following areas:

- Module, encapsulation and declaration of dependencies.
- Life Cycle, API for life cycle management.

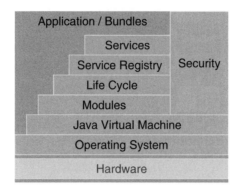

Figure 2.27 OSGi architecture

- Service Registry, providing functionality to other bundles.
- Security layer, restrict bundle functionality to pre-defined capabilities. Security is realized across the layers starting from the virtual machine to ensure that security requirements are not broken.

OSGi supports Java Configurations and Profiles, such as CDC, CLDC, MIDP are all valid execution environments. The OSGi platform has also standardized an execution environment based on Foundation Profile and a smaller version that specifies the minimum requirements. The class loading model is based on Java but introduces modularization with the possibility to add controlled linking between modules. The life cycle system introduces dynamics that are not normally part of an application. The service registry provides a cooperation model for bundles. Bundles can cooperate through traditional class sharing; however, this is not feasible in dynamic environments. The service registry addresses this by using a number of events that signify important state changes pertaining to services.

2.6.11 Python

Python is a general-purpose high-level programming language. Python's core syntax and semantics are minimalist, while the standard library is large and comprehensive. Python supports multiple programming paradigms, which primarily consist of object oriented, imperative, and functional. Python also features a fully dynamic type system and automatic memory management.

Python has become a succesful mobile language, especially for rapid prototyping of software. The favourable characteristic of Python is that it allows access to the underlying features of the device, which are hidden by Java ME. On the other hand, this opens up the possibility of creating havoc in the system. Python has been ported to Symbian Series 60 devices, Microsoft Mobile, Apple iPhone, Linux phones, Palm, and other systems including some MP3 players.

2.6.12 Flash Lite

Flash Lite is being currently introduced as the lightweight alternative to Adobe Flash player in mobile devices. Flash Lite aims to bring the same kind of experience to these resource constrained devices that we currently have on websites, namely interactive audio-visual user interfaces. Flash Lite is implemented at the client-side, namely in the user interface layer and thus it competes with other technologies including Sun Microsystem's JavaFX script. Flash Lite can be also seen as an alternative technology to Java ME and BREW.

Flash Lite 1.1 supports Flash 4 ActionScript. Flash Lite 2.0, which is based on Flash Player 7, supports the newer ActionScript 2.0. Both versions also support the W3C's Standard *Scalable Vector Graphics (SVG)* Tiny, a mobile profile of the consortium's SVG recommendation. Flash Lite can support audio and interactive elements without relying on JavaScript. XML content is supported natively. Flash Lite 3 is based on Flash 8, which supports the H.264 video standard and *Flash Video (FLV)* content. Flash Video is the name of a file format used to deliver video over the Internet using Adobe Flash Player. Well-known users of the Flash Video format include YouTube, Google Video, and Yahoo! Video.

2.6.13 Opera Mini

MIDlets and .NET CF applications are examples of smart clients, which can include complex processing at the client side. An alternative solution for mobile computing is to use so called thin clients, which require only minimal processing on the client side and place most processing on the server side. Opera Mini is an example of the latter solution. It is a web browser specifically designed for mobile phones and PDAs. The system is based on Java ME platform and thus requires that the phone is capable of running ME applications. Opera Mini is distributed free of charge and it is supported through a partnership between Opera Software and Google.

Opera Mini has been derived from the Opera web browser. Opera Mini requests web pages through servers maintained by the Opera Software company. The servers process and compress the pages before relaying them back to the mobile phone. Thus the architecture is very similar to the original WAP architecture, in which the WAP gateway performed content adaptation. The Opera Mini servers reformat web pages into a format suitable for small screens. The system uses a language called *Opera Binary Markup Language (OBML)*. In the default mode of operation, Opera Mini opens only one connection to the proxy servers and then keeps this connection open. Multiple requests and responses can then be multiplexed over this single connection.

Opera Mini supports two rendering modes, namely a full page view and a small-screen rendering. The former displays a zoomed out view to the page as it would appear on a normal desktop computer. The latter mode reformats pages into a single vertical column so that the page can be scrolled only up and down.

The browser has limited support for JavaScript due to the server centric system design. Since a server formats pages for easy viewing on the device, any JavaScript must be first processed by the server. Before a page is sent to a mobile device, any onLoad events on the page are fired, and all scripts are allowed a maximum of two seconds to execute. Scripts designed to wait for a certain time before executing are not supported. After the scripts have been executed, they are stopped, and the page is compressed and delivered to the mobile device. Once on the device, only a small set of events are allowed to trigger scripts, namely:

- *onUnload.* Fired when the user navigates away from a page.

- *onSubmit.* Fired when a form is submitted.
- *onChange.* Fired when the value of an input control is changed.
- *onClick.* Fired when an element is clicked.

When an event is fired on the client-side, the Opera Mini generates an associated request to the proxy server to process the event. The proxy server then executes the JavaScript and returns the updated page. This way of rendering allows Opera Mini to support almost all of the current web standards supported by the desktop version of Opera. Current limitations include missing support for SVG, Web Forms 2.0, and limited support for frames.

The connection between the mobile device and the server is encrypted for privacy and security. The encryption is created when the browser is first started by requesting the user to provide input for randomization. The problem with the used security strategy is that it only secures connection from the mobile device to the server in the middle and not from the mobile to the ultimate content web server. This means that websites that require HTTPS require special solutions. The employed solution involves doing the secure connection from the Opera Mini server; however, this breaks the end-to-end security.

2.6.14 Summary

Now, we briefly summarize the salient features of the presented mobile platforms.

- Java ME is a portable solution for creating various mobile applications that can be downloaded to mobile devices. Java ME includes a number of configurations and profiles for different environments. Currently, Java ME has an impressive selection of optional modules. The most important of these are currently being collected under the Mobile Service Architecture (MSA). Newer JSR specifications include support for pervasive communications, converged communications, as well as 2D/3D rendering.
- Symbian is a versatile general purpose platform. Nokia's Symbian based Series 60 is the most popular and well known mobile platform available today. In order to face challenges in the competitive mobile market, Symbian has been announced to become fully Open Source.
- .NET Compact Framework is the mobile development environment for Windows Mobile devices.
- BREW is a mobile development environment and platform developed by Qualcomm for CDMA-based networks.
- Android is a Linux based platform for mobile devices. All Android applications are created using the Java programming language.
- Python is a portable language for rapid prototyping of mobile concepts. Python offers the possibility to interface the OS and hardware. This contrasts Java ME, which hides the underlying system and only exposes system features through standard APIs.
- Flash Lite is suitable for graphics intensive applications with or without user interaction.
- Opera Mini and related microbrowsers. These solutions are good when lightweight functionality is needed and latency variations are allowable.

Bibliography

[1] (ed. David K) 2008 *Technologies for the Wireless Future: Wireless World Research Forum (WWRF)* vol. **3**. Wiley.

[2] (ed. Tafazolli R) 2006 *Technologies for the Wireless Future: Wireless World Research Forum (WWRF)* vol. **2**. Wiley.

[3] Camarillo G and Garcia-Martin M 2004 *The 3G IP Multimedia Subsystem (IMS): Merging the Internet and the Cellular Worlds*. Wiley.

[4] Cuevas A, Moreno JI, VIdales P and Einsiedler H 2006 The IMS Service Platform: A Solution for Next Generation Network Operators to Be More Than Bit Pipes. *IEEE Communications Magazine*.

[5] Balakrishnan H, Padmanabhan VN, Seshan S and Katz RH 1997 A comparison of mechanisms for improving TCP performance over wireless links. *IEEE/ACM Transactions on Networking* **5**(6), 756–769.

[6] Toh CK 2001 *Ad Hoc Wireless Networks: Protocols and Systems*. Prentice Hall PTR, Upper Saddle River, NJ, USA.

[7] Akyildiz IF, Wang X and Wang W 2005 Wireless mesh networks: a survey. *Computer Networks* **47**(4), 445–487.

[8] Clark D, Braden R, Falk A and Pingali V 2003 Fara: reorganizing the addressing architecture. *SIGCOMM Comput. Commun. Rev*. **33**(4), 313–321.

[9] Carpenter B 1996 *Architectural Principles of the Internet Internet Engineering Task Force: RFC 1958*.

[10] Clark DD 1988 The design philosophy of the DARPA internet protocols *SIGCOMM*, pp. 106–114 ACM, Stanford, CA.

[11] Moskowitz R, Nikander P, Jokela P and Henderson T 2008 *RFC 5201: Host Identity Protocol* IETF.

[12] Carzaniga A, Rosenblum DS and Wolf AL 2001 Design and evaluation of a wide-area event notification service. *ACM Transactions on Computer Systems* **19**(3), 332–383.

[13] Mühl G, Ulbrich A, Herrmann K and Weis T 2004 Disseminating information to mobile clients using publish/subscribe. *IEEE Internet Computing* **8**(1), 46–53.

[14] Emmerich W 2000 *Engineering Distributed Objects*. John Wiley & Sons.

[15] SPI 2008 *SPICE (Service Platform for Innovative Communication Environment) homepage*.

[16] Tarkoma S, Bhushan B, Kovacs E, Kranenburg HV, Postmann E, Seidl R and Zhdanova AV 2007 Spice: A service platform for future mobile ims services *WOWMOM*, pp. 1–8. IEEE.

[17] WAP Forum 2002 WAP Push Proxy Gateway Service Specification. Technical report, Open Mobile Alliance.

[18] Siegemund F, Sugar R, Gefflaut A and van Megen F 2006 Porting the.net compact framework to symbian phones. *Journal of Object Technology* **5**(3), 83–106.

3

Support Technologies

A number of common technologies are used in many different platforms. This section collects well-known examples of these technologies, such as SIP, IMS, web services, and other technologies including SQLite and OpenGL. A particular emphasis is put on service discovery techniques and mobility solutions. Service discovery techniques allow mobile systems to locate interesting resources in the communications environment, and they are a crucial building block for mobile services that are aware of their environment. We consider UPnP, Jini, *Service Location Protocol (SLP)*, and ZeroConf. Mobility solutions allow devices, sessions, and other elements to change their location. Important technologies include Mobile IP, Host Identity Protocol, and Wireless CORBA. We present a summary of the differences between these standards. Towards the end of the chapter, we consider advanced topics including overlay networks, context-awareness, service composition, security and trust, and charging and billing. Finally, we present an example mobile middleware platform that combines some of the technologies to support mobile data communications and synchronization.

3.1 Session Initiation Protocol (SIP)

The Session Initiation Protocol (SIP) [1] is an ASCII-based, application-layer control protocol that can be used to establish, maintain, and terminate calls between two or more end points. The SIP protocol is a session layer protocol in the OSI model, and at the application layer in the TCP/IP model. SIP is designed to be independent of the underlying transport layer, for example it can be used with TCP, UDP, or SCTP.

The driving application for SIP has been telephony, e.g., the ability to be able to establish audio or video sessions between mobile devices. The SIP architecture has grown over the years and consists of a good number of IETF RFCs, Internet Drafts, and best practices. In November 2000, SIP was accepted as a 3GPP signaling protocol and thus an integral part of the IMS architecture for IP based multimedia services.

SIP can be used to implement many of the advanced call processing features found in the *Signaling System 7 (SS7)* used in traditional telecom systems. SS7 is a centralized protocol with a complex call processing network and dumb endpoints. SIP features are typically implemented in the communicating endpoints rather than in the network.

Mobile Middleware   Sasu Tarkoma
© 2009 John Wiley & Sons, Ltd

The SIP client is a network element that sends SIP requests and receives SIP responses. Clients interact directly with a human user or a terminal device. The SIP server is a network element that receives requests and sends back responses to requests. The SIP addressing model is based on URIs. Examples of URI usage in SIP include:

- a mailbox;
- a telephone number at a gateway service;
- a user of a service;
- a resource that needs to be tracked or manipulated;
- a group in an organization.

SIP servers include proxies, user agent servers, redirect servers, and registrars. A SIP proxy is similar to an HTTP proxy server. When a client sends requests to the proxy, the proxy either handles the requests or forwards them to other servers. A SIP redirect server accepts a SIP request and informs the originating client where to send the request. A SIP registrar server accepts registration requests and associates a client's address to a user's SIP URI. Typically, a registrar is combined with a proxy or redirect server. The proxy servers handle message routing, authentication, and authorization of SIP clients.

Figure 3.1 gives an overview of the distributed SIP signaling environment. In the figure, SIP phones are connected by a sequence of domains, namely a visited domain, home domain, and called domain. The AAA servers are responsible for authentication and authorization decisions, as well as gathering data for auditing. The home domain keeps track of the current location of SIP phones that belong to its administration. Since the SIP architecture has been designed with wireless and mobile phones and other devices in mind, mobility support plays a crucial role in the architecture. The SIP framework supports four different types of mobility [2]:

- *session mobility* allows a user to maintain a media session while changing terminals;
- *terminal mobility* allows a device to move between IP subnets while continuing to be reachable for incoming requests and maintaining sessions across subnet changes;

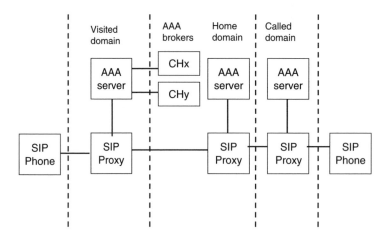

Figure 3.1   SIP proxies

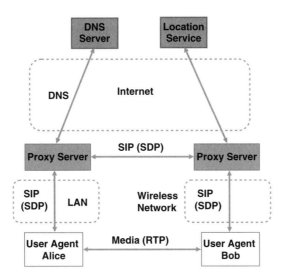

Figure 3.2   Overview of the SIP architecture

- *personal mobility* allows the addressing of a single user located at different terminals by using the same logical address; and
- *service mobility* allows users to maintain access to services while moving or changing devices and network service providers.

These different types of mobility support apply also for event delivery and subscribers receive notifications even when they are roaming. Mobility support is realized by updating any client address changes to respective servers. SIP supports personal mobility by using a technique called *forking*. With forking, a message is sent to all known addresses of a SIP phone or terminal.

Figure 3.2 presents an overview of the SIP architecture and Figure 3.3 illustrates how the user Alice establishes a new session with the user Bob. Alice's *User Agent Client (UAC)* is configured to communicate with a proxy server (the outbound server) in its domain. Alice's UAC begins by sending an INVITE message to the proxy server that indicates its desire to invite Bob's *User Agent Server (UAS)* into a session (1). The proxy server acknowledges the request (2). The outbound proxy server should forward the INVITE request to the proxy server that is responsible for the domain test.com. The outbound proxy consults a local DNS server to obtain the IP address of the test.com proxy server (3). This is done by requesting for the DNS SRV resource record that contains information on the proxy server for test.com.

The DNS server responds (4) with the IP address of the test.com proxy server (the inbound server). Alice's proxy server can now forward the INVITE message to the inbound proxy server (5), which acknowledges the message (6). The inbound proxy server now consults a location server to determine Bob's location (7). The location server responds with Bob's location (8).

The proxy server can now send the INVITE message on to Bob (9). A ringing response is sent from Bob back to Alice (10, 11, 12) while the UAS at Bob is alerting the local

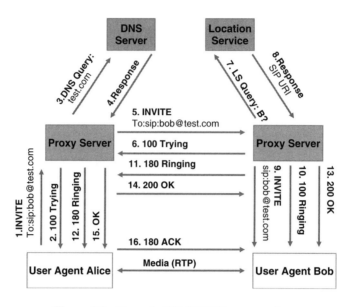

Figure 3.3 Example SIP INVITE message flow

media application. When the media application accepts the call, Bob's UAS sends back an OK response to Alice (13, 14, 15).

Alice's UAC sends an acknowledgment message to Bob's UAS to confirm the reception of the final response (16). In this example, the ACK is sent directly from Alice to Bob. This occurs because the endpoints have learned each other's address from the INVITE/200 (OK) exchange, which was not known when the initial INVITE was sent. Alice and Bob can exchange data over one or more connections.

There are six different types of request messages defined in the SIP specification that are distinguished by their method type. Additional methods are defined in other RFCs and they are extensions to the core SIP standard. The methods are as follows:

- REGISTER, which is used by a client to register an address with a SIP server.
- INVITE, which indicates that the user or service is being invited to participate in a session. The body of this message typically includes a description of the session.
- ACK, which confirms that the client has received a final response to an INVITE message.
- CANCEL, which is used to cancel a pending request.
- BYE, which is sent by a User Agent Client to indicate to the server that it wishes to terminate the call.
- OPTIONS, which is used to query a server about its capabilities.

The process of establishing a session begins with an INVITE message. This message is sent by a calling user to a called user and the INVITE message invites the called user to participate in a session. The message includes information about the session, for example the type of session and its properties. Typically, the *Session Description Protocol (SDP)* is used to describe sessions. We will return to SDP when discussing SLP later in this chapter.

For the INVITE message, the caller may receive a number of interim responses before the called user accepts the session. Typically the caller is informed that the user is being alerted, namely the phone is ringing. When the called user answers the call, an OK response is generated and sent to the calling client. The calling client then sends an ACK message and then the session has been setup and the actual data can be sent and received. The session teardown is started when one of the users hangs up, and a BYE message is generated and sent to the other client. The other clients confirm this by sending an acknowledgment, which then ends the call.

The SIP response messages contain Status Codes and Reason Phrases that indicate the current condition of a request. The status code values are divided into six general categories (RFC 3261):

- 1xx: Provisional. The request has been received and processing is continuing.
- 2xx: Success. An ACK, to indicate that the action was successfully received, understood, and accepted.
- 3xx: Redirection. Further action is required to process this request.
- 4xx: Client Error. The request contains bad syntax and cannot be fulfilled at this server.
- 5xx: Server Error. The server failed to fulfill an apparently valid request.
- 6xx: Global Failure. The request cannot be fulfilled at any server.

A number of extensions and enhancements have been made to the original SIP RFC 2543. These extensions include the introduction of the following new methods to SIP, which can be used for event notification, instant messaging and call control:

- *SUBSCRIBE.* The SUBSCRIBE method enables a user to subscribe to certain events. This means that the user should be informed when matching events occur.
- *NOTIFY.* The NOTIFY method is used to inform the user that a subscribed event has occurred. For example, Windows Messenger uses the SUBSCRIBE method to request contacts, groups, and allow and block lists from the server and to get the presence of contacts in a group.
- *MESSAGE.* SIP can also be used for Instant Messaging. A user sends an instant message to another user by sending a request that includes the MESSAGE method. This request carries the actual text in a body of a SIP packet.
- *INFO.* The INFO method is used for transferring information during a session, such as user activity. For example, Windows Messenger 5.0 uses the INFO method to inform the called user that a calling user is typing on the keyboard.
- *SERVICE.* The SERVICE method can carry a SOAP message as its payload. For example, Windows Messenger 5.0 uses the SERIVCE method to add contacts and groups on the server. This method is also used to search for contacts in the SIP domain.
- *NEGOTIATE.* The NEGOTIATE method is used to negotiate various kinds of parameters, such as security mechanisms and algorithms.
- *REFER.* A REFER request enables the sender of the request to instruct the receiver to contact a third party using the contact details provided in the request.

The SUBSCRIBE and NOTIFY methods are used by the SIP event package to enable a client to subscribe to the desired events and receive notifications when the expected event occurs (RFC 3680). The main application areas for the SIP event package are callback services, buddy lists (presence), and message waiting indications. The SIP event package requires that all notifications are subscribed beforehand, therefore unsolicited messaging is not allowed. However, this pertains only to the event package and does not affect call-control traffic and other signalling with SIP. As such SIP does not specify how subscription state is distributed or stored in the system.

3.2   IP Multimedia Subsystem (IMS)

The *IP Multimedia Subsystem (IMS)* is an architecture for delivering IP-based multimedia to mobile users. It was originally designed by 3GPP, and is a crucial part of the evolution towards beyond GSM. 3GPP2 has later updated the vision towards beyond 3G networks. In order to be compatible with current Internet technologies, IMS utilizes IETF protocols such as SIP.

The IMS standard defines the functional architecture for a managed IP-based network. It aims to provide a means for carriers to create an open, standards-based network that delivers integrated multimedia services to increase revenue, while also reducing network costs.

The IMS architecture has been designed to clearly separate the connectivity, control and service plane. IMS decomposes the networking infrastructure into separate functions with standardized interfaces between them. Each interface is specified as a reference point, which defines both the protocol over the interface and the functions between which it operates.

The 3GPP architecture is split into three main planes or layers, each of which is described by a number of equivalent names: Service or Application Plane, Control or Signalling Plane, and User or Transport Plane.

The service plane (application plane) provides an infrastructure for the provision and management of services, and defines standard interfaces to common functionality (e.g. configuration storage, identity management, billing, presence and location).

The control plane comprises network control servers for managing call session setup, modification and release. It is located between the application and transport plane and routes the call signaling, informs the transport plane what traffic to allow, and generates billing records for the use of the network. The core element of this plane is the *Call Session Control Function (CSCF)*, which includes several different functions.

The user plane consists of routers and switches, both for the backbone and the access network. Access into the core network is provided through Border Gateways that enforce policy provided by the IMS core and control traffic flows between the access and core networks.

Figure 3.4 illustrates the central components of the IMS system including several SIP servers or proxies called CSCF, the *Home Subscriber Server (HSS)*, the *Media Resource Function (MRF)*, *Breakout Gateway Control Function (BGCF)*, and *Application Servers (AS)*. Interfaces are defined between different entities, for example the Gw interface is used to exchange SIP messages between the user terminal and the CSCFs. The main protocols are SIP and Diameter. SIP supports a number of interaction styles with servers including *Common Gateway Interface (CGI)* and SIP servlets.

The following three SIP servers or proxies have a fundamental significance in the IMS design in the control plane:

• *Proxy-CSCF (P-CSCF)* is a SIP proxy that is the first point of contact for an IMS terminal. It can be located either in the visited network or in the home network. The

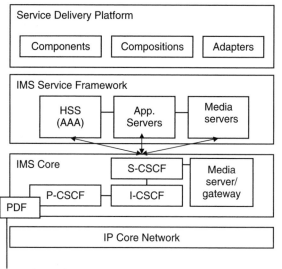

Figure 3.4   IP Multimedia Subsystem

terminal discovers its P-CSCF with either DHCP, or it is assigned in the *PDP Context*. The proxy is assigned to a terminal during registration and does not change for the duration. The proxy authenticates the client and establishes an IPSec association to ensure security of the registration. P-CSCF is in the signaling path and can inspect every message.

- A *Serving-CSCF (S-CSCF)* is the central node of the signaling plane and can inspect every SIP message. S-CSCF is a SIP server and it is located in the home network. It uses Diameter Cx and Dx interfaces to the HSS to access user profiles. The server handles SIP registrations, which allows it to bind the user location.
- An *Interrogating-CSCF (I-CSCF)* is a SIP function located at the edge of an administrative domain. I-CSCF's IP address is published in the DNS of the domain thus making it discoverable by remote servers and use it as a forwarding point for SIP messages to the domain.

3GPP has specified together with ETSI an *Open Service Access (OSA)* framework, which allows applications access to network related data and functions without the need to know all internal details of an operator network. The API for OSA is called Parlay (or OSA/Parlay) and it is developed jointly in a collaboration by 3GPP and the *European Telecommunications Standards Institute (ETSI)* and is distributed via the Parlay Group. The current version, Parlay 5.0, was developed in cooperation with a number of *Java APIs for Intelligent Networks (JAIN)* Community member companies. The aim of Parlay/OSA is to provide an API that is independent of the underlying networking technology and of the programming technology. The set of mappings to specific technologies include CORBA/IDL, WSDL, and Java.

An important role of the Parlay/OSA Framework is to provide a way for the network to authenticate applications. The framework allows applications to discover the capabilities of the network, and provides management functions for fault tolerance and overload. The Parlay service APIs allow applications to make telephone calls, query the location of an entity, and charge for provided services. An IMS *Open Service Access – Service Capability Server (OSA-SCS)* interfaces OSA Application Servers using Parlay. In addition, a subset of the Parlay API is provided through web service-based interfaces in Parlay X. This newer specification defines a set of simple telecom web services, including third party call control, SMS and messaging functions, charging, and location and user status.

The JAIN initiative has defined a set of Java APIs for the development communications products and services. The JAIN *Service Logic Execution Environment (JSLEE)* specifies the Java-based service platform for executing event oriented applications. SIP is supported through the JSIP API and *Java Call Control (JCC)* API.

3.3   Web Services

The web services paradigm introduces machine readable and accessible content on the web. In this model, applications use services by selecting and composing suitable service components. The key ingredients of the architecture are the web services provider, consumer or client, and the service registry. The aim of service composition is to allow one to create complex services from more basic parts, thus supporting software re-use. Typical applications involving composite services include personalized portals, electronic commerce service bundles, and mobile services.

The W3C web service definition is broad and allows many different realizations; however, typically the term refers to clients and servers that communicate using XML messages, more specifically according to the SOAP standard. A notable characteristic is that there is often a machine-readable description of the operations hosted by the service. In the W3C web services model, this description is written using the *Web Services Description Language (WSDL)*. There are many tools for WSDL and it can be used to generate client and server side code automatically. Recently, RESTful web services have become commonplace. *Representational State Transfer (REST)* is an architectural paradigm that is based on lightweight, decoupled, and stateless messages, typically implemented using HTTP.

Figure 3.5 presents the W3C web services architecture, which consists of the following four layers:

- Transport, which is responsible for transferring bytes and messages between network nodes. HTTP is used to deliver messages due to its universal nature and firewall friendliness. Other frequently used protocols are FTP, SMTP, and JMS.
- XML Messaging is responsible for ensuring that the sender and receiver understand the structure and format of the message. SOAP is the de facto solution for XML messaging in W3C's web services architecture. SOAP defines the header and body of the message and has various bindings to the transport protocols, such as HTTP. Other alternatives include XML-RPC and plain XML (in HTTP POST).
- Description layer is responsible for providing information regarding the service interfaces, namely what methods are available and what kind of input and output data types are supported. WSDL is the prominent standards based solution here.

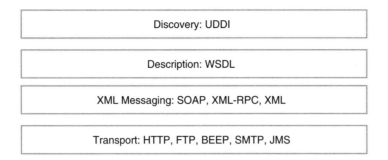

Figure 3.5 Web services stack

- Discovery layer is responsible for providing interfaces for the discovery of web services across networks. The premier technology is the *Universal Description, Discovery, and Integration (UDDI)*.

Web services have their history in RPC; however, the trend has been towards more decoupled communications, which is evident in the current flexible specification of SOAP and the emerging REST-based services. Moreover, web services are often used to implement SOA concepts, where the communications are based on messages rather than operations. The need for versatile solutions has been taken into account in the design of WSDL version 2.0, which offers support for binding to all the HTTP request methods.

Basic web services and XML security building blocks are the W3C's XML Encryption and Signature specifications. In addition, the *XML Access Control Markup Language (XACML)* standardized by OASIS provides a policy language which allows administrators to define the access control requirements for application resources. XACML model consists of *Policy Enforcement Points (PEPs)*, *Policy Decision Points (PDPs)*, *Policy Information Points (PIP)*, and *Security Assertion Markup Language (SAML)* assertions. Decisions made by PDPs, authentication assertions, and actions performed by PEP are represented using SAML. authorization, and auditing features of mobile service platforms.

Web services can be seen as an interoperable and easy way to deploy customized service elements also for mobile clients. The world of mobile platforms; however, is considerably different from the traditional web services execution environment within the fixed network. In the current and forthcoming mobile systems, users are able to access services through different heterogeneous access networks. The service access should be adapted by the service platform and customized based on the terminal, user preferences, and network capabilities.

Basically a mobile platform on a mobile client can vary between a thin client, i.e., a web browser with no additional functionalities and capabilities on the mobile client. At the other end would be a smart client. Both approaches have pros and cons and different variations between are possible. For example a web browser solution could be an interesting approach to harmonize heterogeneity of underlying platforms such as different operating systems, whereas for many advanced applications, like the ones relying on context information obtained from sensor and other information of the OS or for disconnected operation, a pure web browser is not sufficient.

One important solution for mobile platforms that is currently intensively investigated and developed is IMS, which can be extended to support web services. A number of key questions regarding the converged communications environment include the role of IMS, how web service access is realized, how security is implemented, and how component services are invoked seamlessly. From the security point of view, a key challenge is how to leverage existing authentication. The web services paradigm introduces machine readable and accessible content on the web. In this model, applications use services by selecting and composing suitable service components. The key ingredients of the architecture are the web services provider, consumer or client, and the service registry. The aim of service composition is to allow one to create complex services from more basic parts, thus supporting software re-use. Typical applications involving composite services include personalized portals, electronic commerce service bundles, and mobile services.

Figure 3.6 illustrates the components needed for a fully fledged web services platform. As mentioned above, a number of transport protocols are needed to support message exchange between nodes in the distributed system. Then in order to be able to send and receive XML message, an XML processing stack is needed. The XML processor is deeply connected with XML message routing subsystem and the security pipeline that ensures that security requirements are not broken. The security pipeline typically includes integrity checking, message validation, content checking to prevent injection attacks, and then authentication of the sender and parts of the message, and ultimately authorization.

In addition to XML and security relates per-message processing, a number of other components are needed as well. The figure highlights the management console, which is responsible for supporting the design and deployment of security policies. Identity management is typically provided in the form of managed credentials stored in LDAP or similar. ID management also involves protocols such as Kerberos and *Public Key Infrastructure (PKI)*.

A more advanced identity management component also supports *Single Sign-On (SSO)*, enabling the access of services across administrative boundaries. In addition, reporting and auditing are crucial for today's systems, especially when business transactions are

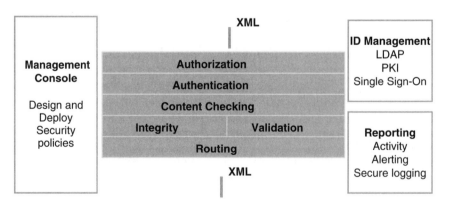

Figure 3.6   XML processing system

involved. Activity monitoring, alerting, and secure logging are crucial components in the design.

A logically centralized component called the *Enterprise Service Bus (ESB)* generally provides an abstraction layer that allows the integration of applications using messaging. The driving idea of ESB is to divide applications and services to their constituent parts and then facilitate flexible communication in a decoupled way between these parts. An ESB is a standards-based message based interconnect. Web services are a frequently used ESB implementation technology. ESB does not realize SOA but rather it is a building block for SOA.

Typical features of ESBs include:

- Invocation support for synchronous and asynchronous transport protocols.
- Routing addressability, static/deterministic routing, content-based routing, rules-based routing, policy-based routing.
- Mediator components that facilitate protocol transformation and service mapping.
- Message-oriented middleware that provides message processing and transformations.
- *Complex Event Processing (CEP)* responsible for event detection, correlation, and pattern matching.
- Choreography that realizes complex business processes.
- Orchestration that combines services that are exposed to a single aggregate service.

3.4 Other Technologies

3.4.1 IP Television (IPTV)

Internet Protocol Television (IPTV) is a loosely defined set of protocols for delivering digital television over the IP protocol. IPTV involves two popular forms of video delivery, namely live streaming and *Video on Demand (VoD)*. The former is analoguous to the current broadcast television, whereas in the latter form of media delivery movie files are delivered upon request. In general, IPTV playback requires a desktop or mobile phone with decoding software, or a dedicated set-top box. A broadband or mobile broadband connection is needed for the best viewing experience.

There is an active standardization effort pertaining to use of the 3GPP IMS as an architecture for supporting IPTV. The motivation for this is that carriers can utilize the same infrastructure for both voice and IPTV. This also enables convergent operation in which voice and TV features are combined.

The benefits of combining IMS and IPTV include Quality of Service (QoS) guarantees for media flows and support presence services or call forwarding. Moreover, IMS has a charging mechanism which allows pay-per-view business models. A number of different access networks can be supported such as 3G, DVB-H or WLAN.

Digital Video Broadcasting - Handheld (DVB-H) specification (ETSI standard EN 302 304) is a mobile television format. DVB-H enables mobile phones to playback digital television that is broadcast in a wireless DVB-H network. DVB-H is based on the successful digital terrestria television and addresses some of the requirements with mobile devices, such as energy awareness. Technically, DVB-H utilizes a time slicing technique and transmits IP datagrams as bursts in small time slots.

A set of specifications have been standardized by ETSI under the *IP Datacasting (DVB-IPDC)* label that address the key aspects of a commercial mobile TV system. The specifications pertain to electronic service guide, content delivery protocols, and service purchase and protection.

3.4.2 SQLite

Recently, lightweight database technologies have been introduced to mobile devices. This development has been motivated by the increased computing power and storage space on the devices, and also the need to store more and more data locally in the devices, and to synchronize the data across multiple systems. SQLite has become one of the most supported mobile database technologies, and in this section we will outline the main features of the system.

SQLite is a lightweight mostly ACID compliant relational database management system developed for small and embedded devices. SQLite has a relatively small size for a programming library, around 500kB. A transactional database is said to have the ACID property when changes and queries appear to be *Atomic, Consistent, Isolated, and Durable (ACID)*. SQLite implements most of the SQL-92 standard, including database transactions that are atomic, isolated, and durable. SQLite also supports triggers and most complex queries.

The favorable features of the SQLite system include:

- Zero-configuration. Written in ANSI-C. Simple programming API.
- Implements most of SQL-92.
- A complete database is stored in a single cross-platform disk file.
- Self-contained: no external dependencies.
- Cross-platform: Linux (unix), MacOSX, OS/2, Win32 and WinCE, and other systems.
- Sources are in the public domain.

SQLite differs from the traditional client-server database management systems in that it is not a standalone process that accepts incoming requests, but rather it is a library that is linked into the application programs. SQLite thus uses simple function calls which reduces the latency because client-server IPC processing is not needed. The entire database is stored as a single cross-platform portable file on the host machine. Several computer processes or threads may access the same database. When a transaction is performed to the file, the whole file is locked. This means that only one write access can be granted at a time.

SQLite can be used in many different environments. The envisaged uses for SQLite include:

- Database for small devices. SQLite is a popular option for the database engine in cellphones, PDAs, MP3 players, set-top boxes, and other electronic devices.
- Application file format. Rather than using the normal file system API to write structured data, an SQLite database can be used to store the data. This avoids the need for a parser and transactional updates can be guaranteed.
- Website database. Since SQLite does not require configuration it is a suitable database system for small or medium sized websites.

3.4.3 OpenGL ES

The current generation of mobile phones is equipped with advanced audio and video capabilities. The rendering capabilities have been extended to support 3D graphics, which has enabled the development of various mobile games. The de facto standard in mobile 3D graphics is the *OpenGL ES (OpenGL for Embedded Systems)*, which is a subset of the OpenGL 3D graphics API. OpenGL ES has been designed for embedded devices such as mobile phones, PDAs, and video game consoles. It is defined and promoted by the Khronos Group, a graphics hardware and software industry consortium.

A large part of the functionality in the original OpenGL API was removed in order to be able to make the system suitable for embedded devices. Two of the significant differences between OpenGL and OpenGL ES are the introduction of fixed point data types and the removal of the glBegin and glEnd calling semantics for primitive rendering. The fixed point types utilize integers instead of floating points in the 3D calculations in order to improve performance. This is motivated by the fact that many of the devices do not have FPUs making floating point operations cumbersome.

OpenGL ES has become a fundamental graphics technology for mobile devices. For example, Symbian OS, iPhone, and Android use it as the default 3D graphics API. The API is also used in some games consoles, namely Playstation 3.

3.4.4 PAMP

The acronym PAMP refers to a solution stack comprising of open source software that is used to run dynamic websites. The letters of the acronym are defined as follows:

- Apache, the Web server.
- MySQL, the database management system (or database server).
- PHP and sometimes Perl or Python, the programming languages.

When used with the Linux operating system, AMP becomes LAMP. PAMP is a scaled down version of AMP for Personal devices. Nokia has released PAMP as open source for the Symbian Series 60 platform. The most interesting feature of PAMP is that it includes a web server that is executed on the mobile device. This allows services not only to be implemented in the fixed network, but also in the mobile phones as well paving the way for new kinds of peer-to-peer usage scenarios.

3.5 Service Discovery

In order to develop applications and systems that cope with heterogeneous environments, there must be platform support for finding other entities in the environment. This process of locating entities based on some set criteria is called *service discovery*. Service discovery is an integral part of current wireless and mobile systems and current software platforms support it by providing various standards based interfaces. In this section, we consider three well-known service discovery mechanisms, namely the *Universal Plug and Play (UPnP)*, Jini, and *Service Location Protocol (SLP)*.

3.5.1   UPnP

Universal Plug and Play (UPnP) is a set of protocols developed by the UPnP Forum [3]. The aims of UPnP are to allow devices to discover each other and then communicate seamlessly. The technology has been designed with both home and corporate networking in mind. In order to foster interoperability, the UPnP device control protocols are built on open Internet standards. The architecture supports peer-to-peer networking of devices and other appliances, and builds on TCP/IP, UDP, HTTP, and XML.

One of the goals of UPnP has been support for zero-configuration networking. Irrespective of the vendor, a UPnP device can dynamically join a network, obtain an IP address, and convey its capabilities upon request and obtain information about the capabilities of other devices. Devices can leave a UPnP network without leaving any unwanted state information.

The architecture does not require DNS and *Dynamic Host Configuration Protocol (DHCP)* servers, but they are used if they are available. In managed network that has a DHCP server, a UPnP obtains an IP address from this server. If the network is unmanaged, that is there is no DHCP server, the device assigns itself an address.

3.5.1.1   Discovery

Discovery is an important part of UPnP and needed to bootstrap the communications. When a new UPnP device is added to a network, the UPnP discovery protocol advertises the device and its services to the control points in the network. When a control point is added, the same protocol is used by the point to find devices and their capabilities. The discovery protocol is based on a message that contains details about the device and its services. The UPnP discovery protocol is based on the *Simple Service Discovery Protocol (SSDP)*.

3.5.1.2   Description

After initial discovery has been done it is essential that devices are described in a detailed way to better support communications and device pairing based on capabilities. A control point needs to learn about the devices and their capabilities. There are two ways to obtain this information, it can be retrieved from the device by the control point, or a reference to the description can be provided in the discovery message. The UPnP description for a device is formatted in XML, and the description includes vendor-specific manufacturer information, a list of any embedded devices or services, and URLs for control, eventing, and presentation. For each service, the description includes a list of the commands supported by the service and their parameters.

3.5.1.3   Control

After a control point has a description of the device, it can send actions to the device's service. A control message is sent to the control URL of the service using SOAP, which then results in a return value.

3.5.1.4   Event Notification

Event notification or eventing is an important part of UPnP networking. A UPnP description for a service includes a list of actions the service supports and a list of variables that model the state of the service. The service publishes updates when these variables change. A control point may subscribe to receive this information.

A service publishes updates by sending event messages, which are then received by the subscribers. Event messages contain the names of one or more state variables and the current value of the variables. These messages are formatted using the XML-based *General Event Notification Architecture (GENA)*. A special initial event message is used to subscribe to a control point. This message defines the names and values of the subscribed variables and allows the subscriber to initialize its model of the state of the service it is monitoring. In order to support multiple control points, eventing is designed to keep all control points equally informed about the effects of any action.

3.5.1.5   Presentation

Finally, we have the presentation of resources to end users. Presentation is an important part of device management and configuration. If a device has obtained a URL for presentation, a control point can retrieve this page and load it into a web browser. The control point can also allow a user to interact with the device using the page.

3.5.1.6   UPnP AV (Audio and Video) standards

UPnP Audio and Video (UPnP AV) is the part of the UPnP standard which focuses on home entertainment. The goal is to support a seamless multimedia home environment, in which devices find each other and are able to share media and content. The UPnP Enchanced AV Specifications define the MediaServer and MediaRenderer device classes.

3.5.1.7   NAT Traversal

UPnP implements one solution for NAT traversal called the *Internet Gateway Device (IGD)* protocol. Many routers and firewalls are Internet Gateway Devices allowing any local UPnP control point to perform a variety of actions on them. These actions include retrieving the external IP address of the device, enumerating the existing port mappings, and adding and removing port mappings. Thus a UPnP device can punch a hole in the NAT device by adding a new port mapping. This effectively guarantees NAT traversal from the private network to the global Internet.

3.5.2   Jini

Jini also addresses the challenges of device and service discovery and interoperability in decentralized environments. The aim of the Jini architecture is to make each service, as well as the network of services, adaptable to changes in the network. Service providers supply clients with portable Java-based objects that give the client access to the service. This interaction can use any type of networking technology such as RMI, CORBA, or

SOAP, because the client only sees the Java-based object [4]. Jini uses the Java serialization system in order to be able to send Java objects over the network. Serialization is a core part of the Java language, which allows objects to be saved into a format that is suitable for transmission.

When a new service, typically hosted on device, joins a Jini network, it first advertises itself by publishing a Java object that implements the service API. The implementation of this object is arbitrary. A lookup service is used to support the discovery process. The lookup service facilitates the brokering of advertisements and search requests.

A client finds a service by searching for an object that supports the given API. When the client receives the object published by the service, the Java environment will download any code needed to communicate with the service. The programmer can decide what protocols should be used in the communications. Typical communication protocols include RMI, CORBA, XML, and proprietary protocols.

Unicast communications is used when the lookup service is known to the client. When the lookup service is not known, the client uses dynamic multicast discovery. In either case, the lookup service will return a Service Registrar object to the client which is used to lookup a particular service. A lookup catalogue maintained by the lookup service can be searched based on the type, name, or description of the service. Services can be grouped together in federations. Clients can search for specific federations. Finally, the lookup service will return a Java proxy that specifies how to connect with the service.

Jini uses the concept of leases to ensure that state is eventually removed in the distributed environment. When a service registers with a lookup service, a lease is granted to the service for some duration. The service must renew the lease periodically or otherwise the entry is removed from the lookup service. This effectively ensures that with good probability services advertised in the lookup service are really active.

Jini is based on a centralized model of distributed computing, because it relies on a lookup service to support the communications between clients and services. This kind of a model has inherent scalability limitations; however, Jini has been designed for internal networks in mind so scalability to wide-area networks has not been a prime design objective.

3.5.3 Service Location Protocol

The Service Location Protocol (SLP) defined in RFC 2608 is a service discovery protocol for local area networks. The main motivation behind SLP is to allow easy discovery of devices without prior configuration. SLP is based on service announcements which are made on the local network. Each service must have a URL that is used to locate the service. Services may also have name/value pairs which are called attributes. Moreover, each service must be in one or more scopes, which are simple strings used to group services and enforce their visibility. A service template defines a special 'service': URL and the allowed attributes for the URL (RFC 2609).

A device can have one to three roles from the following set:

- *User Agents (UA)* are devices that are capable of searching for services.
- *Service Agents (SA)* are devices that can announce one or more services.

- *Directory Agents (DA)* are devices that are able to cache services. DAs are used in larger networks to improve scalability. If a DA is present, the UAs and SAs are required to use it instead of direct communications.

SLP features a number of query types that are used to locate services:

- A query can be defined for a service type or abstract service type.
- A query can be combined with a query for attributes.
- Attributes of a service can be requested given that the URL is known.
- A list of all existing scopes can be requested.
- A list of all service types can be obtained.

From the communications protocol viewpoint, SLP is a packet-oriented protocol. It typically uses UDP, but TCP can also be used for the transmission. Since UDP is not reliable, SLP repeats all multicasts several times in increasing intervals until a response has been received. All devices need to listen to UDP port 427. Multicast is a basic primitive for SLP and it is used heavily.

The presence of DA has significant impact on the protocol. Initially, clients try to query using multicast for DAs on the network. If no answers are received, clients will assume that there are no DAs present. If a DA is introduced, it will be detected using a periodic heartbeat packet originating from the DA. When a SA discovers a DA, it must register all its services there. Similarly, when a service is no longer active, the SA should notify the DA.

Security is a vital non-functional requirement in decentralized environments. SLP features a public-key cryptography-based security system. The system is based on signing the service advertisements. The system requires that the public keys of every service provider are installed on all UAs. From scalability point of view, this requirement presents significant challenges, and introduces the need for prior configuration. The system only guarantees the authenticity of an advertised URL, however it does know what device will answer to the address and how to protect the communications.

3.5.4 ZeroConf

Zero Configuration Networking (ZeroConf) is a set of solutions that automatically creates a usable IP network without special system support or a prior configuration. The goal is to allow hassle free interconnection of various devices, such as laptops, printers, desktops, and phones. DNS and DHCP are standard alternatives to ZeroConf, which typically require manual configuration. In addition, UPnP and SLP offer similar features.

The four key requirements for ZeroConf were:

- Allocate addresses without a DHCP server (IPv4 Link-Local Addressing).
- Translate between names and IP addresses without a DNS server (Multicast DNS).
- Find services, like printers, without a directory server (DNS Service Discovery).
- The solutions in the areas must coexist gracefully with larger configured networks.

ZeroConf currently provides the following services:

- Determining numeric network addresses for networked devices (link-local address auto-configuration).

- Automatically resolve and distribute computer host names.
- Automatically locate network services, such as printing services (service discovery).

The most well-known ZeroConf solution is *Bonjour* from Apple. Bonjour uses multicast DNS and DNS Service Discovery.

3.6 Mobility Solutions

3.6.1 *Mobile IP*

The current solutions being standardized by IETF for network-layer mobility support are the Mobile IPv6 and Mobile IPv4 protocols [5, 6]. MIP is a layer-3 mobility protocol for supporting clients that roam between IP networks. Upper layer protocols and applications are unaware of possible changes in network location and thus can operate uninterrupted while the host moves.

MIP6 mobility support consists of the triangle of the *Home Agent (HA)*, *Correspondent Host (CH)*, and the *Mobile Host (MH)* illustrated in Figure 3.7. MIP4 has an additional optional networking node called *Foreign Agent (FA)* that has been left out from the MIP6 specifications. In both MIP versions the HA serves as an anchor point for MNs and any CN may communicate and initially reach the MNs through the HA.

The basic MIP routing is triangular. A CN sends packets to a MN via a HA and then the HA tunnels packets to MN's current location. Finally the MN sends packets directly to the CN. In practice triangular routing is inefficient and generally also impossible due to widely used *ingress filtering*. Practical MIP deployments either route all packets via HA (reverse tunneling) or the MN and CN communicate directly (route optimization, which can be negotiated with the help of HA).

The distance between the MN and the HA may also be long both topologically and geographically. Thus routing packets between the MN and the HA may cause considerable delay. However, to improve the situation a HA may also be allocated from the visited network the MN is currently visiting. A similar way of optimizing IP mobility is to utilize some form of localized or hierarchical mobility management.

The *Hierarchical Mobile IPv6 Mobility (HMIPv6)* management solution introduces local *Mobility Anchor Points (MAP)* that are essentially Home Agents [7]. MAPs can be located at any level in a hierarchical network of routers, including the access routers. The aim of the HMIPv6 is to minimize the signaling latency and reduce the number of required signaling messages. As long as the MN stays inside one MAP domain it only needs to update its location with the MAP. The localized mobility management can also be completely handled on the network side without MN's involvement at the IP mobility protocol level. In these cases the network side needs to employ some kind of tunneling or local routing solution that is transparent to the MN.

Another network-layer mobility solution being standardized by IETF is the *Network Mobility (NEMO)*. The technical solution of NEMO is close to that of MIP6. NEMO allows complete subnetworks to change their location in a network instead of single hosts. This is realized with a mobile router that manages the mobile network. Hosts behind the mobile router do not need to be aware of mobility in any way. All packets destined to hosts behind the mobile router get routed towards the virtual home network.

Then a HA managing the virtual home network tunnels all packets to the mobile router managing the mobile network.

One practical application of IPSec based *Virtual Private Networks (VPN)* is to extend the user's home network environment to be accessible from any location. In a tunneled mode, VPNs tunnel all packets between the mobile host and a *security gateway* (that is usually located at the edge of the user's home network). Until recently IPSec VPNs have not survived the change of underlying IP addresses that are also used as the outer IP addresses of the VPN tunnel. Both *Internet Key Exchange (IKE)* and IPSec SAs (*Security Associations*) had to be rekeyed after IP addresses of either end of the tunnel changed [8]. This practically caused all existing connections to drop.

Recent developments in standardization have addressed this issue. For example, MOBIKE aims to support a way to update the IKE SA and IPsec SA endpoint addresses without rekeying the SAs. This would allow keeping the existing IKE and IPsec SAs in place even when the IP address changes.

3.6.2 Host Identity Protocol (HIP)

The *Host Identity Protocol (HIP)* addresses mobility, multi-homing, and security issues in the current Internet architecture [9, 10]. HIP is currently being standardized at IETF. HIP is located between the network and transport layers and provides a new cryptographic addressing space. Communication end-points are identified using public cryptographic keys instead of IP addresses. The fundamental idea behind HIP is to separate the identifier and locator of an Internet host. This is depicted in Figure 3.8.

The identifier uniquely names the host in a cryptographic namespace, and the locator uniquely defines the topological location of the node. Additional benefits of HIP are authentication and prevention of DoS attacks through cryptographic puzzles in the initiation phase of the protocol. HIP also defines a rendezvous mechanism for solving the bootstrap and double jump problems in mobility.

In HIP, the endpoints are identified using asymmetric cryptography. There are two main representations of a Host Identity. A *Host Identifier (HI)* is the public key component of an asymmetric key pair. The private key is owned only by the endpoint making impersonating another endpoint very difficult. As the HI is essentially a variable-size public key, it is

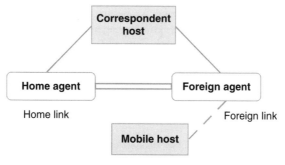

Figure 3.7   Overview of MobileIP

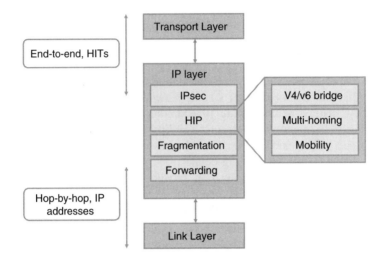

Figure 3.8 Overview of HIP

difficult to use in datagram headers and API [9]. Therefore the HIP architecture also includes fixed-size representations of the HI. A *Host Identity Tag (HIT)* is an 128-bit long hash of a HI. Public HIs are usually stored in the DNS or distributed using a mechanism, such as PKI or DHTs.

Locators, namely the IP addresses, are used only at the network layer. There is a one-to-many binding between a HI and the corresponding locators supporting mobility and multi-homing. Figure 3.9 illustrates the HIP mobility update sequence without rekeying [11]. Using the mobility protocol, a mobile node can inform its new locators to the correspondent node. The messages are authenticated via a signature or keyed hash message authentication code (HMAC). Furthermore, the peer host performs an address verification test by placing a nonce in the response message to the mobile host that the mobile host must echo back.

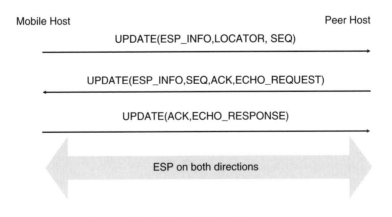

Figure 3.9 HIP mobility update sequence

3.6.2.1 Layering Model

In HIP, the application layer accesses the transport layer via the socket interface. The application layer uses the traditional TCP/IP IPv4 or IPv6 interface, or the new native HIP API interface provided by the socket layer. For legacy applications, it is possible to use HIP through IPv4 and IPv6 APIs.

The HIP layer is a so called shim or wedge layer between the transport and network layers. The datagrams delivered between the transport and network layers are intercepted in the HIP layer for processing.

3.6.2.2 Base Exchange

The new logical HIP layer, located between transport and network layers, initiates connections with a *base exchange*. The base exchange resembles the IKE key exchange, because it also uses an authenticated Diffie-Hellman key exchange. Two *Security Associations (SA)* are created, one for each direction, from the key material generated during the base exchange. The SAs are used for protecting the data traffic between hosts using IPsec ESP.

The base exchange protocol also includes a puzzle mechanism that protects the responder from some CPU-targeting DoS attacks. The initiator of a connection is required to spend CPU cycles in order to solve a puzzle sent by the responder. The puzzle mechanism allows the responder to delay the allocation of resources for the initiator until the last moment. This way, it is harder for the initiator to succeed in a resource-exhausting DoS attack against the responder. The responder may vary the difficulty level of the puzzle. The computation time grows exponentially to the number of bits.

A base exchange can be initiated without prior knowledge of the HI of the peer. This operation mode is called *opportunistic HIP*. The opportunistic mode is prone to *man-in-the-middle* attacks, because the initiating host does not know the identity of the responder. As a result, a malicious host can trick the initiator to contact an incorrect host. However, benefit of the opportunistic mode is that it does not require any infrastructure for distributing HIs.

3.6.2.3 Rendezvous Server

The *Rendezvous (RVS)* extension was developed to support HIP nodes that rapidly change their IP address [12], because the current solutions, namely DNS and DynDNS, do not support very fast updates.

HIP clients register their HIT→IP mappings with the RVS. After this registration, other HIP nodes can use the IP address of the RVS instead of the IP address of the node they want to contact. The RVS will forward the first packet of the base exchange (I1) to a destination peer based on the mapping state and the peers can subsequently perform the base exchange without involving the RVS.

Figure 3.10 illustrates the rendezvous functionality. Peers use a registration protocol to inform the rendezvous about their HIT→IP mappings. Other peers can then lookup the RVS IP address, for example from DNS, and send packets to the RVS, which then forwards the I1 packet to the peer in question. After this the three remaining base exchange packets (R1,I2,R2) are exchanged directly by the peers.

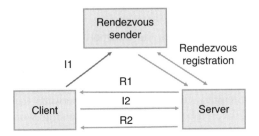

Figure 3.10 HIP rendezvous

RVS also solves the problem of the initial contact, as it always knows the location of the mobile node. When the mobile node changes its location, it informs the rendezvous server of new locators. As the rendezvous server does not change its location, the mobile and corresponding nodes always know how to contact it. This solves the problem of double jump where the peers change their location simultaneously. In a way, the rendezvous server resembles the home agent in the Mobile IP [6] architecture.

3.6.3 Wireless CORBA

The basic design principles of the *wireless CORBA* solution [13] concentrate on the client-side ORB transparency and simplicity. Transparency of the mobility mechanism to non-mobile ORBs was one of the primary design constraints. Wireless CORBA does not support solutions that would require modifications to a non-mobile ORB in order for it to interoperate with CORBA objects and clients running on a mobile terminal. This means that a regular ORB does not have to be extended for interoperability with CORBA objects residing on mobile terminals.

Wireless CORBA architecture identifies three different domains: Home Domain, Visited Domain, and Terminal Domain. Figure 3.11 presents the Wireless CORBA architecture and identifies the three different domains. The Home Domain for a given terminal is the domain that hosts the Home Location Agent of the terminal. A Visited Domain is a domain that hosts one or more Access Bridges through which it provides ORB access to some mobile terminals. The Terminal Domain consists of a terminal device that hosts an ORB and a Terminal Bridge through which the objects on the terminal can communicate with objects in other networks.

The key concepts of the specification are:

- Mobile *Interoperable Object Reference (IOR)*;
- Home Location Agent;
- Access Bridge;
- Terminal Bridge; and
- *General Inter-ORB Protocol (GIOP)* Tunneling Protocol (GTP).

The Mobile IOR is a relocatable object reference that identifies the Access Bridge and the terminal on which the target object resides. In addition, it identifies the Home Location Agent that keeps track of the Access Bridge to which the terminal is currently attached.

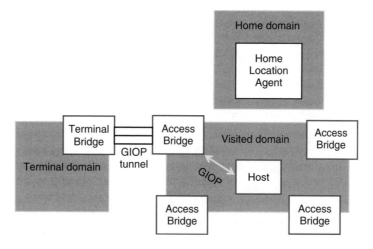

Figure 3.11   Wireless CORBA architecture

The Home Location Agent keeps track of the current location of the terminal. It provides operations to query and update terminal location. The Home Location Agent also provides operations to get a list of initial services and to resolve initial references in the home domain.

The Access Bridge is the network side end-point of the GIOP tunnel. It encapsulates the GIOP messages to the Terminal Bridge and decapsulates the GIOP messages from the Terminal Bridge. The Access Bridge also provides operations to retrieve a list of initial services and to resolve initial references in the visited domain. The Access Bridge may also provide notifications of terminal mobility events.

The signaling between the Terminal Domain and the other two domains is carried out through a GIOP tunnel. The messages are transmitted between the Terminal Bridge and the Access Bridge. The specification presents the *GIOP Tunnelling Protocol (GTP)*, which is an abstract, transport-independent protocol. The generic GTP specifies necessary control messages to establish, release, and re-establish a GIOP tunnel. The Terminal Bridge may also provide a mobility event channel that delivers notifications related to handoffs and connectivity losses.

A handoff involves a number of steps, which are illustrated in Figure 3.12.

1. The old Access Bridge starts the handoff process when the start_handoff operation is invoked on it.
2. The old Access Bridge invokes the transport_address_request operation in the new Access Bridge. This operation returns a list of transport addresses of the new Access Bridge and a Boolean value indicating whether or not the new Access Bridge accepts the terminal.
3. If the terminal is not accepted, then the old Access Bridge only reports the HAND-OFF_FAILURE status by invoking the report_handoff_status operation and the handoff is unsuccessfully completed. The old Access Bridge continues to serve the Terminal Bridge as the current Access Bridge.

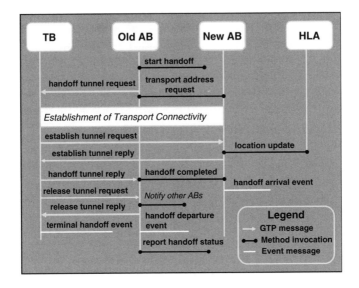

Figure 3.12   Handoff in Wireless CORBA

4. If the terminal was accepted by the new Access Bridge, then the old Access Bridge sends a HandoffTunnelRequest message to the Terminal Bridge.
5. The handoff proceeds when the old Access Bridge gets the HandoffTunnelReply message from the Terminal Bridge or when the new Access Bridge invokes the handoff_completed operation at the old Access Bridge.
6. It is assumed that the handoff status received by the old Access Bridge from the Terminal Bridge and the new Access Bridge are identical. If they are not identical, the old Access Bridge takes implementation specific actions to recover from the error.
7. The old Access Bridge notifies all other Access Bridges interested in location of the terminal.
8. If the old Access Bridge supports Mobility Event Notifications, it generates a notification of a departing terminal.
9. The old Access Bridge reports the handoff status by invoking the report_handoff_status operation.

3.6.4 Comparison

Table 3.1 gives an overview of the differences between the mobility-aware systems and protocols discussed in this chapter. We first define the key properties used in the comparison and then briefly analyse the different systems.

The *target* denotes the nature of the mobile entity. We observe that different kinds of entities can move in the computing environment. In Mobile IP and HIP the target of mobility is the host that moves. Higher level mobility protocols allow more fine grained mobility, SIP supports session mobility, Wireless CORBA facilitates the mobility of the ORB that hosts CORBA objects. Overlay networks can be used to support different kinds of mobile entities, for example i3 does not specify the target entity. Mobile pub/sub systems support the mobility of subscriptions and advertisements to various degrees.

Table 3.1 Comparison of mobility-aware systems and protocols

	MIP	HIP	SIP
Target	MH	MH	Session
Mechanism	HA	DNS/Overlay	Home registrar
Buffering	No	No	Yes (stateful proxy)
Update point	1 fixed	1	1
Location Privacy	Yes (not /w route-opt)	No	No
Authentication	Yes/no	Yes	Yes

	Wireless CORBA	i3	Pub/sub
Target	Object	Any	Subs/Advs
Mechanism	Home bridge	Rendezvous-point	Hop-by-hop
Buffering	Yes	No	Yes
Update point	1 fixed	1	≥ 1
Location Privacy	Yes	Yes	Yes
Authentication	-	-	-

In Chapter 4 we presented two key design patterns for handoff protocols, namely the rendezvous and state transfer patterns. The former pertains to the indirection points that are used to enable reachability of mobile nodes, and the latter corresponds to the protocol that is used to transfer state between physical or logical locations. In the comparison, we denote the nature of rendezvous using the term *update point*, and the nature of the handoff mechanism using the term *mechanism*.

The term update point denotes the number of indirection points and whether or not they are fixed. MIP, Wireless CORBA, and SIP have a single fixed indirection point, the HA, home bridge, or the home registrar. HIP with overlay-based address resolution and i3 have a single non-fixed indirection point. Pub/sub systems have typically multiple indirection points.

The term mechanism denotes the type of handover or handoff protocol. Examples of the mechanisms include the Home Agent (HA) approach used in MIP or the DNS and rendezvous scheme of HIP. SIP uses the home registrar for location updates and Wireless CORBA has the home bridge. The i3 overlay uses a rendezvous point that manages the triggers. Many pub/sub systems typically have to update the routing and forwarding path between the source and destination of mobility.

In many cases it is desirable to employ packet or message *buffering* during the handoff. The motivation for buffering is to prevent packet drops and thus retransmissions by the sending node. Buffering is also a useful functionality for supporting disconnected operation. MIP and HIP do not support this feature and it is up to the senders to resend any lost packets. SIP supports disconnected operation through stateful proxies that can buffer messages for short periods. SIP also copes with *network partitions* using retransmission [2]. Wireless CORBA Access Bridges maintain a list of pending invocations and pub/sub systems can buffer notifications for disconnected clients; however, this is implementation specific. The i3 overlay, on the other hand, does not buffer packets.

User privacy is a crucial concern in today's network systems. *Location privacy* hides the current IP address of the mobile entity. MIP supports location privacy with the exception of the route-optimization option. HIP and SIP do not support location privacy, because they expose the mobile node's IP address. A HIP host has access to the IP address corresponding to the public key (Host Identity) of a mobile node. SIP does not hide the IP address, it is disclosed in the session description. Wireless CORBA provides support for location privacy. Terminals are addressed using the terminal identifier and the Access Bridge hides the transport address of the terminal. On the other hand, the location of an Access Bridge is revealed. The i3 overlay supports the hiding of source addresses with private triggers. Typically mobile pub/sub systems, such as Siena and Rebeca, provide anonymous communication and only the edge brokers know the transport addresses.

Authentication of terminals and users is also an important feature for mobile systems. As such MIP does not provide specific features for authentication but rather relies on existing IP authentication methods, namely IPSec and Internet Key Exchange (IKE). On the other hand, IPSec and IKE were not initially designed for multi-homed operation and currently multi-homed operation has overhead due to additional database entries and key negotiations for each pair of source/destination address. HIP supports authentication by using the Host Identity, which is essentially a public key. The HIP base exchange performs mutual authentication of the communicating nodes. SIP supports three authentication styles: HTTP-style basic and digest authentication that uses shared secrets, and PGP using public key cryptography. Wireless CORBA may be used with the CORBA Security specification [14], and i3 and pub/sub systems may be extended to support various forms of authentication.

3.7 Advanced Topics

In this section, we present a number of advanced networking and mobile computing topics. We start with overlay networks, which are networks built on top of other networks and becoming an important instrument in scalable service provisioning. Then we examine context-awareness that currently has been mostly restricted to location-based services, but a host of other contextual attributes can also help to create future applications, which we will briefly study. Then we discuss service composition that is expected to enable value added services for next generation platforms and support both developer and customer driven compositions and mashups. Trust and privacy play an integral part in enabling an ecosystem in which users trust services to perform correctly and maintain the confidentiality of private data. Finally, we will briefly cover well-known research projects pertaining to these advanced topics.

3.7.1 Overlay Networks

Overlay networks are a promising area of research and development. Overlay networks are networks built on top of another network. Typically, they are implemented in the application layer on top of the TCP/IP networking stack. Overlays are used to add new routing and forwarding functionality to the Internet without requiring any modifications to routers. Indeed, any system that requires changes to the behaviour routers will face almost impossible deployment challenges. For example, network proposals such as IP

Multicast, IntServ, and DiffServ have not been globally deployed due to the changes they require both in routers and administration. The lesson here is that any proposed change to the network architecture must be very easy to deploy in order for it to take effect. An overlay network is a prospective technology for introducing changes to the network, because overlays can be introduced incrementally and over the current network.

Typically, overlays are based on a distributed algorithm for creating scalable routing tables. A common technique is to build an overlay using a *Distributed Hashtable (DHT)*, which are scalable lookup structures. A DHT is essentially a distributed hashtable, which partitions the hashtable over multiple machines on a network. A modern DHT algorithm can provide logarithmic lookup cost in terms of hops and the more recent algorithms can take also network proximity into account and handle so called *churn*, which features rapid joins and leaves by peers.

DHTs typically support lookup using flat labels. This technique is flexible, because any data blob can be transformed into a flat label by using a one-way hash function. The SHA-1 hash algorithm is used by most DHT systems. In addition, security properties can be added to a label through a technique called *self-certification*. A self-certified label consists of hash of data with a signature and the corresponding public key. Any entity receiving a self-certified label can check that some data actually hashes to the label, and that the public key correctly verifies the signature.

Figure 3.13 illustrates a commonly used DHT API. The API is simple and features three operations, namely *put*, *get*, and *delete*. The put operation places some data into the structure and associates this data with a key. The key can be derived from the data. We call systems that support keys that are hashes of data, data centric. Get retrieves the data based on the key. Finally, delete removes the data associated with the given key. Of course, the API operations need to be secured to be usable in open distributed environments.

Overlay networks are useful for mobile middleware, especially in the development of fixed network support infrastructure for mobile devices. As we discussed already, mobility and multi-homing are about communicating any changes to the current location to all communicating parties, and then ensuring that the location is updated so that any new communication attempts are sent to the right part of the network. It is well-known that DNS is not the best system to support mobility updates and hence a number of alternatives have been proposed. Many of the alternative systems utilize DHTs, because of their good scalability properties and support for flat label based lookups. In the following, we concentrate on some of these proposals and consider the implications for mobile hosts.

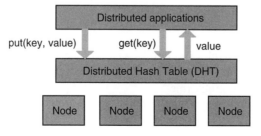

Figure 3.13 Basic DHT API

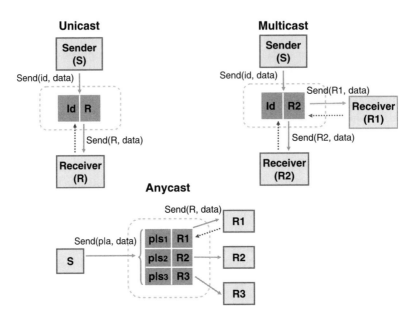

Figure 3.14   Interactions in i3

The *Internet Indirection Infrastructure (i3)* [15] is a DHT-based overlay network that aims to provide a more flexible communication model than the current IP addressing [15]. In i3 each packet is sent to an identifier. Packets are routed using the identifier to a single node in the distributed system. The server, an i3 node, maintains triggers which are installed by receivers that are associated with identifiers. When a matching trigger is found the packet is forwarded to the associated receiver. An i3 identifier may be bound to a host, object, or a session unlike the IP address, which is always bound to a specific host.

The key i3 interactions involving triggers are the following (illustrated in Figure 3.14):

- Unicast, in which a host R inserts a trigger (id, R) in the i3 infrastructure to receive all packets with identifier id.
- Mobility support, in which a host updates its trigger. The host changes its address from R1 to R2, and consequently updates its trigger from (id, R1) to (id, R2). Now, all packets with identifiers id are forwarded to the new address.
- Multicast, in which an application can build a multicast tree using a hierarchy of triggers.
- Anycast, in which i3 provides support for anycast by allowing applications to specify a prefix for each trigger identifier. Packets are then matched to the identifiers according to the longest matching prefix rule.
- Service composition, in which applications can replace an identifier with a stack of identifiers. A sender can request that all its packets be forwarded through an intermediate service point. Moreover, a receiver can also control the packet processing ny inserting a receive trigger.

The ROAM system builds on top of i3 and allows end-hosts to control the placement of rendezvous-points (indirection points) for efficient routing and handovers [16]. ROAM supports legacy applications using a user-level proxy that encapsulates IP packets within i3 packets and manages trigger related operations.

The *Delegation Oriented Architecture (DOA)* was proposed to circumvent the harmful side-effects of middleboxes [17]. DOA introduces two changes to the Internet architecture: persistent host identifiers and a mechanism for resolving these host identifiers. The DOA architecture allows hosts to explicitly delegate permissions for intermediaries. Each host has a unique EID, which is a flat 160-bit end-point identifier. DOA works in a similar fashion to HIP. The DOA header is located between IP and TCP headers and it carries source and destination EIDs. A DHT-based mapping service is provided for Internet hosts that maps EIDs to IP addresses or other EIDs. This permits the introduction of intermediaries to the routing of packets. One of the advantages of DOA is the outsourcing of intermediaries – ideally clients may select the most useful network elements to use.

DOA adheres to the architectural principles of the Internet with one exception, an intermediary may overwrite a private address with a public address. An EID is a hash of a public key and self-certification is used to check that an EID is created by its owner. Entities who perform get() operations on the DHT should check that the returned record is signed by the private key that corresponds to the public that was hashed for the EID.

Network-Extensions Boxes (NEB) are presented as example intermediaries and they leverage the DOA header in packets for demultiplexing instead of using port numbers that today's NATs use. DOA and i3 differ, because i3 also provides a new forwarding infrastructure and does not provide a primitive for receivers that allows receivers to specify intermediaries.

The DOA builds on the locator/identity split used in HIP, so in this respect they are similar. In addition, the Hi3 architecture [18] combines HIP and i3. Currently, it is envisaged that a DHT-based overlay, such as i3, may be used to distribute HI and provide rendezvous-points for mobility. HIP or Hi3 do not use the overlay for forwarding data traffic. The HIP protocol uses overlays only for the initial rendezvous and solving the double jump problem. The HIP protocol has intrinsic support for mobility and multi-homing between two participants. This feature is missing from i3.

The TRIAD architecture considered how to use NATs in the network architecture [19]. The main idea is that NATs are too valuable not to be included in the future Internet architecture. The beneficial features of NATs include end-point concealment, addressing autonomy, multi-homing, and supporting more addresses. TRIAD solutions include using end-point names instead of addresses and path-based addressing with loose source routing instead of IP addresses. This solves a number of issues, including connectivity problems and NAT state loss. They introduce a router-based name service.

Recent proposals, such as *Data-Oriented Network Architecture (DONA)*[29] aim to introduce data-centric operations to the networking architecture. DONA inserts a data-handling shim layer right above the network layer and resolves names by directly routing to data. This does not involve DNS and a lookup, but just routing on names. The architecture introduces two new network entities, namely the data handlers that operate at the data-handling layer and do name-based routing and caching, and the authoritative resolvers, which can point to the authoritative copy of a principal's data. The two new network primitives are *fetch(name)* and *register(name)*.

3.7.2   Context-awareness

Context-awareness means the ability of a system to be aware of their environment and then being able to react to changes in this environment. Thus context-awareness is a vital feature for mobile software systems. Context can be defined as *'any information that can be used to characterize the situation of entities'* [20].

An adaptive context-aware system is expected to anticipate the wishes of a user by taking the current, past, and predicted future context into account. The aim of context-aware services is to support the user in everyday tasks, such as arranging meetings, synchronizing data, supporting seamless services across variety of devices, and helping the user to obtain relevant information.

The level of context-awareness and adaptation can be divided into three categories as follows [21]:

- Laissez-faire, meaning that there is no system support for adaptation. There is no centralized arbiter that controls how resources are used by the applications.
- Application-aware, in which some changes are made to applications, and that applications and the system collaborate in realizing adaptive applications.
- Application-transparent, in which applications are not changed. This strategy makes the adaptation support solely the responsibility of the middleware.

Many research systems have addressed context-awareness. For example, Aura [22] aims to provide the users with a disruption free computing environment and to minimize the system's intrusion into people's lives. One.world [23] defines a system architecture for pervasive computing that includes discovery and migration support. The aim on this system is to cope with constant change. Gaia [24] is a metaoperating system that extends the reach of traditional operating systems to the ubiquitous environment. The goal is to make living spaces as integrated programmable environments. MobiLife [25] emphasizes self-awareness in order to support for automatic configuration arrangement of devices, services, and local connectivity in the user's local environment.

The Context Toolkit [26] aims to support context-aware operation by using widgets. The system consists of context widgets and a distributed infrastructure that facilitates their execution. Context widgets are software components that provide applications access to context information. The widgets hide the details of context sensing and acquisition. The Context Toolkit provides a number of services to applications, namely:

- encapsulation of sensors and hiding context acquisition;
- access to context data through a network API;
- abstraction of context data through interpreters;
- sharing of context data through a distributed infrastructure;
- storage of context data, including history;
- basic access control for privacy protection.

Many of the proposed middleware architectures for context-awareness do not address the specific challenges of the mobile device environment, namely the resource limited computing environment. ContextPhone [19] is an example of a mobile-phone-based tool for context-awareness developed for the Symbian platform as Open Source. The aim of this system is to make the development of mobile context-aware programs easier.

The system supports the acquisition of context data from different sources, such as cell identifier and *Global Positioning System (GPS)*, and then to process this information and communicate it using different communications techniques.

3.7.3 Service Composition

While the basic protocols of service composition are well covered in the literature, dynamic service composition is in practice still a challenge. This is, in a sense, a logical consequence of the loose coupling of web services and their compositions. In order to invoke component services seamlessly, component metadata and interface semantics must be developed and published in a heterogeneous environment. One of the starting points is to leverage best-practice SOA to support IP-based multimedia services on networking architecture such as IMS.

A key challenge is how to develop and maintain an interoperable and federated mobile service ecosystem that meets user expectations while retaining manageability, and easy charging and billing. Moreover, new tools are needed to support end users in creating new content and services. Tools for this kind of end-user and community driven effort should be easy to use, but still flexible enough to support creativity.

Web Services Business Process Execution Language (WS-BPEL) from OASIS is a business process language serialized in XML. BPEL focuses on abstract processes and messaging between these processes. Key issues are, for example, when to send a message, when to perform a state transition, and how to cope with failures or timeouts.

BPEL concentrates on one peer in the system and is thus an orchestration language. Orchestration is contrasted by choreography, which focuses on a global view of the system and by considering all entities. In order to support orchestration, the BPEL language supports programming constructs such as conditional statements, while sequences, flow control, and variable scoping. BPEL relies heavily web services and is based on XML and WSDL.

3.7.4 Security and Trust

From the mobile platform viewpoint a crucial challenge is trust. Current mobile platforms can be trusted by the end users and service-level agreements determine different parts of service access and usage. This notion of trust is very different for current Internet services, which typically have different business models than mobile services. A large part of the revenues of Internet companies stem from advertisement fees rather than actual service usage. The eluding nature of trust on the Internet makes it difficult to create pay-by-usage mashups and provide guarantees on service usage. It is also not easy to identify customers across services while maintaining privacy. Several service-level technologies have been developed to alleviate this issue, namely the Liberty Alliance and OpenID for enabling single sign-on on the Web.

Therefore, it is crucial to find ways to maintain the mobile service platform as a trusted entity by the users and, at the same time, to be able to offer the flexibility of Internet services. Technologies such as 3GPP's *Generic Bootstrapping Architecture (GBA)* [27] pave the way for identity management in this kind of converged environment.

Trust can be defined as a set of assumptions regarding the integrity, ability, trustworthiness, and other properties of an entity with respect to some set of actions. In general, it is assumed that the entity will perform some action or it will not perform some action. In a typical data communications setting, the assumption related to the identity, reliability, and availability of an entity to perform some requested action. For example, a user assumption pertaining to the capability of a server to satisfy the request, and the server assumption pertains to the user's capability to compensate for the requested action.

Privacy on the other hand relates to a personal and subjective condition of keeping sensitive data private. Privacy is very much dependent on user preferences and the context in which privacy decisions are made. This requires that any privacy aware service also takes user preferences into account. One way to define privacy is to say that it empowers the user to be able to control the existing personal and private information about the user managed by a service. This control must be done in the limits of existing regulations.

3.7.5 Charging and Billing

Accounting can be defined as the process of collecting details of a resource usage for billing purposes. The resource usage information typically pertains to the amount of time a user has spent in the network, the services a user has accessed, and the amount of data that has been transferred during a session. After information has been collected the accounting data can be further processed and then used to charge business entities.

OMA defines two charging interfaces: *Offline* and *Online*. Offline charging does not use the accumulated charging information in real time but rather stores it for later use. This model of charging is useful for post-paid users. The Online interface allows real-time charging of users. This is typically used for prepaid customers. Both charging interfaces can be categorized into two charging submodels, namely the event-based charging model and the session-based charging model.

In the event-based charging model, service usage is reported with a single *charging record (CDR)* or in a single credit control and resource usage authorization procedure. In the session-based charging model, the service usage occurs within an end-user session. The service usage is reported with several charging events and the creation of one or more CDRs in offline charging or performance of credit control session in online charging.

Service GPRS Support Node (SGSN) provides mobile clients access to packet sessions for GSM, GPRS, EDGE, and UMTS networks. The SGSN billing function is responsible for monitoring the flow of user data across the network. CDRs are generated at the SGSN before being transferred to the *Charging Gateway Function (CGF)*.

When a user has prepaid access credentials for network or service access, a prepaid server can be used to enforce access. Figure 3.15 illustrates mobile prepaid billing and charging. When a session is started, the client is authenticated and authorized. This results in an approval and some details regarding the session are communicated. The negotiation is done with a RADIUS server (Diameter is also used), which consults a prepaid billing service and a database. The client can request more quota from the RADIUS server and this will be granted if there are funds available. When funds deplete the session ends.

The charging mechanism may also need information about the resource or service being used. This is possible, for example, with SIP's Session Description Protocol that can provide additional information about media flows.

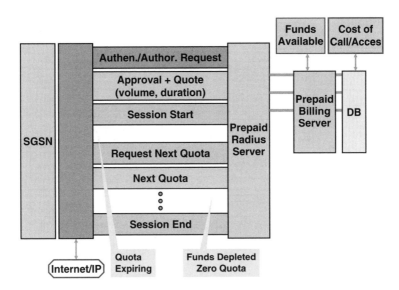

Figure 3.15 Example of a charging and billing system.

3.8 Fuego: Example Middleware Platform

In this section we present the Fuego mobile middleware platform as an example of a system that combines different services for data communications and synchronization [28]. The starting point for the middleware service set was standards-based protocols and application frameworks. Standards-compatible solutions are needed in order to support interoperability between different vendors. The base technologies for the middleware service set were the W3C specifications on messaging, namely SOAP, and web service-related specifications.

From the communication point of view, the implementor of a middleware component needs to decide many things. First, a suitable data communication protocol is selected. In our case, the only choice was HTTP 1.1. The first Java-based mobile phones supported only HTTP as the basic communication protocol. This was a severe limitation, because pushing information to the phone is difficult to implement and may be costly. The current generation of phones support TCP connections, which support limited push.

When the basic data transport is selected, a suitable message transport protocol needs to be implemented. The key questions are: how is data represented in an efficient and interoperable fashion, and what kind of interactions are possible?

The Fuego system uses SOAP as the basic communication protocol in the fixed-network. The main reasons were versatility and interoperability. However, SOAP and XML in general do not work well in the wireless environment and small devices. A custom protocol called XEBU is used for efficient XML-processing and transmission.

An alternative protocol to SOAP would have been SIP. Currently, SIP is used in the telecommunications sector for call control and related functionality. The SIP framework includes also the SIP event package, which specifies how information is subscribed and produced. Since SIP is not being used for RPC-operations, SOAP is a more versatile

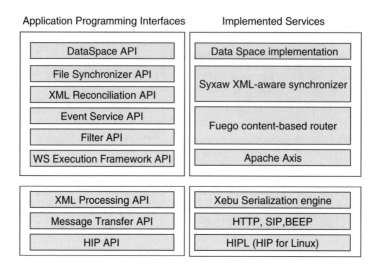

Application Programming Interfaces   Implemented Services

| DataSpace API | Data Space implementation |

| File Synchronizer API | Syxaw XML-aware synchronizer |
| XML Reconciliation API | |

Event Service API	Fuego content-based router
Filter API	
WS Execution Framework API	Apache Axis

XML Processing API	Xebu Serialization engine
Message Transfer API	HTTP, SIP,BEEP
HIP API	HIPL (HIP for Linux)

Figure 3.16   Fuego implementation architecture

choice for a generic middleware platform. A gateway component was developed that transforms messages between the SOAP and SIP technology domains.

The architecture of the Fuego middleware platform is shown in Figure 3.16. This figure is divided into APIs on the left, showing what functionality we have considered necessary, and implementations on the right, showing what component of the middleware implements which API. This selection of components is similar to that of other mobile middleware systems [22, 23].

At the bottom is the Host Identity Protocol (HIP), which provides a cryptographic namespace for identities separate from IP addresses, making it possible to identify hosts and connect to them even if their IP addresses change.

XML is used as the principal data format. XML processing is handled through an API that is better suited for messaging than common APIs. As an option a binary serialization format for XML is provided, which is much more compact than regular XML, but retains data model compatibility with it.

The Fuego event service is intended as the primary communication service of the middleware, building its broker network on top of the simpler messaging service. The main contributions of the event service are a scalable routing system and explicit mobility support.

Even now, users often have several devices that they use for network access, and in the future the number of personal devices will only increase. Thus it is vital for users to be able to synchronize their data. The system provides a file synchronizer based on optimistic reconciliation and explicit support for merging changes to XML files.

A notable characteristic of the platform is that mobility of client devices is considered at every layer, with solutions tailored to the specifics of each layer. In our view this is needed, as the issues related to mobility are different depending on the layer, so they cannot be fully solved at any individual layer.

The HIP implementation for Linux is an optional component of the architecture, which is used for secure mobility and multi-homing support. The HIP architecture is currently being standardized by IETF and it defines a new cryptographic namespace between the network and the transport layers. The Fuego middleware has a Java API for HIP, which allows applications to use HIP features.

The messaging model is based on connection-oriented one-way messaging where either end of a connection may send at any time. We further multiplex a single physical connection into several logical connections to permit multiple independent connections without the need to actually open new ones.

To provide some protocol independence, the protocol implementation is split into two layers. The bottom layer, which is specific to the underlying protocol, provides a very simple communication model with no reliability features or the like. On top of this, protocol-independent modules are provided that can be used to enhance the capabilities of the underlying protocol.

The most pressing concerns with XML messaging are the large size of messages and the time that it takes to process them. Generic compression is not always feasible, since that increases the time spent processing and does not work very well for small messages.

The data sharing model is based on establishing *synchronization links* between local files and directory trees on different devices. This link is persistently stored along with other object metadata, and may be compared to a HTML hyperlink, in the sense that it points from one object to another. When synchronizing an object *a* linked to an object *b*, we propagate changes made to these objects since the last point of synchronization. In case the objects are directory trees, the full contents, including contained objects, are synchronized. The concurrency model is optimistic, which is well suited for the mobile environment. Optimistic concurrency does, however, introduce the need for data reconciliation. The chosen design provides reconciliation services for XML data by utilizing the XML three-way merging algorithm.

Bibliography

[1] Rosenberg J, Schulzrinne H, Camarillo G, Johnston A, Peterson J, Sparks R, Handley M and Schooler E 2002 SIP: Session initiation protocol. RFC 3261, IETF.

[2] Schulzrinne H and Wedlund E 2000 Application-layer mobility using SIP. *SIGMOBILE Mob. Comput. Commun. Rev.* **4**(3), 47–57.

[3] UPnP Forum 2000 UPnP Device Architecture http://www.upnp.org/download/UPnPDA10_20000613.htm.

[4] Waldo J 2000 *The Jini Specifications*. Addison-Wesley Longman Publishing Co., Inc., Boston, MA, USA.

[5] Johnson D, Perkins C and Arkko J 2004 *RFC 3775: Mobility Support in IPv6* IETF.

[6] Perkins C 2002 *RFC 3344: IP Mobility Support for IPv4* IETF.

[7] Soliman H, Castelluccia C, Malki KE and Bellier L 2005 *Hierarchical Mobile IPv6 Mobility Management (HMIPv6)* IETF. [Experimental RFC 4140].

[8] Kaufman C 2005 *RFC 4306: Internet Key Exchange (IKEv2) Protocol* IETF.

[9] Moskowitz R and Nikander P 2006 *RFC 4423: Host Identity Protocol (HIP) Architecture* IETF.

[10] Moskowitz R, Nikander P, Jokela P and Henderson T 2008 *RFC 5201: Host Identity Protocol* IETF.

[11] Nikander P, Henderson T, Vogt C and Arkko J 2008 *RFC 5206: End-Host Mobility and Multi-Homing with Host Identity Protocol* IETF.

[12] Laganier J and Eggert L 2008 *RFC 5204: Host Identity Protocol (HIP) Rendezvous Extension* IETF.

[13] Obj 2004 *Wireless Access and Terminal Mobility in CORBA v.1.1*.

[14] Obj 2002 *CORBA Security Service v.1.8*.

[15] Stoica I, Adkins D, Zhuang S, Shenker S and Surana S 2002 Internet indirection infrastructure *Proceedings of the 2002 conference on Applications, technologies, architectures, and protocols for computer communications*, pp. 73–86. ACM Press.

[16] Zhuang S, Lai K, Stoica I, Katz R and Shenker S 2005 Host Mobility Using an Internet Indirection Infrastructure. *Wirel. Netw.* **11**(6), 741–756.

[17] Balakrishnan H, Lakshminarayanan K, Ratnasamy S, Shenker S, Stoica I and Walfish M 2004 A layered naming architecture for the Internet Proc. *ACM SIGCOMM*, Portland, OR.

[18] Gurtov A, Korzun D, Lukyanenko A and Nikander P 2008 Hi3: An efficient and secure networking architecture for mobile hosts. *Comput. Commun.* **31**(10), 2457–2467.

[19] Cheriton DR and Gritter M 2000 TRIAD: A New Next-Generation Internet Architecture `http://www-dsg.stanford.edu/triad/`.

[20] Dey AK 2001 Understanding and using context. *Personal Ubiquitous Comput.* **5**(1), 4–7.

[21] Satyanarayanan M 1996 Fundamental challenges in mobile computing *PODC '96: Proceedings of the fifteenth annual ACM symposium on Principles of distributed computing*, pp. 1–7. ACM, New York, NY, USA.

[22] Garlan D, Siewiorek D, Smailagic A and Steenkiste P 2002 Project Aura: Toward Distraction-Free Pervasive Computing. *IEEE Pervasive Computing* **01**(2), 22–31.

[23] Grimm R 2004 One.world: Experiences with a pervasive computing architecture. *IEEE Pervasive Computing* **3**(3), 22–30.

[24] Román M, Hess CK, Cerqueira R, Ranganathan A, Campbell RH and Nahrstedt K 2002 Gaia: A Middleware Infrastructure to Enable Active Spaces. *IEEE Pervasive Computing* pp. 74–83.

[25] Klemettinen M 2007 *Enabling Technologies for Mobile Services: The MobiLife Book*. Wiley.

[26] Salber D, Dey AK and Abowd GD 1999 The context toolkit: aiding the development of context-enabled applications *CHI '99: Proceedings of the SIGCHI conference on Human factors in computing systems*, pp. 434–441. ACM, New York, NY, USA.

[27] 3GP 2007 *3GPP TS 33.220: Generic Authentication Architecture (GAA); Generic Bootstrapping Architecture*.

[28] Tarkoma S, Kangasharju J, Lindholm T and Raatikainen K 2006 Fuego: Experiences with mobile data communication and synchronization *17th Annual IEEE International Symposium on Personal, Indoor and Mobile Radio Communications (PIMRC)*.

[29] Koponen T, Chawla M, Chun BG, Ermolinskiy A, Kim KH, Shenker S and Stoica I 2007 A data-oriented (and beyond) network architecture. In *SIGCOMM* (ed. Murai J and Cho K), pp. 181–192. ACM.

4

Principles and Patterns

In this chapter, we present a small set of well-known design principles and patterns for distributed systems with focus on mobile middleware. First, we consider principles, which are important in the development of coherent and reusable architectures. We then consider useful architectural and design patterns for mobile middleware and illustrate them using examples. In order to present the patterns in a useful and clear way, we first define a simple pattern format before presenting the patterns. The presented patterns define a basic pattern language for mobile middleware.

4.1 Definitions

Before going deeper into the principles and patterns, we first define the key concepts used in this section and also in this book. In the following we define the key notions:

- A *principle* signifies strong belief in a certain state or property of a subject. Principles support the formation of a rule or a norm by observing the subject. Principles have a form of minimality character, because they cannot be further divided. A rule or a norm can be reduced to a principle, but principles are not reducible.
- *Design patterns* are software engineering designs that have been observed to work well. Patterns are found in different contexts, they provide a solution for a well-defined problem area, and digress the various dimensions of the problem [1–4]. Patterns are classified into different groups based on their level of abstraction. *Architectural patterns* summarize good architectural designs; for instance the broker pattern that is used in the CORBA architecture [5]. Design patterns capture the essence of medium level, language independent, design strategies in object-oriented design. Moreover, *idioms* represent programming-language-level aspects of good solutions [4].
- An *architecture* is guided by principles and grounded on architectural patterns. An architecture consists of components, and rules and constraints that govern the relationships of the components.

To understand the relations between these notions, we can say that principles guide the development of architectures. Architectural patterns complement principles in the development of the architectural specifications. Similarly, design patterns complement principles in protocol design. A *platform* is a concrete realization of a middleware architecture. Similarly, a protocol stack is a concrete realization of a set of protocols and an architectural framework on how to use them in combination, typically using a stack pattern, although other kinds of organizations are also possible.

Patterns are typically defined in terms of their motivation, underlying problem, structure, consequences, implementations, and known users. Patterns that are applicable in a particular domain can be collected together. This kind of pattern collection is often called *pattern language*.

The following table presents important information used to define patterns:

- *Pattern name.* An informative name that uniquely identifies the pattern.
- *Intent.* Goals of the pattern and the reason for utilizing it.
- *Motivation (Forces).* A short problem statement that is presented using a scenario. The scenario illustrates the context in which the pattern can be applied.
- *Applicability.* Describes the environments and contexts in which the pattern can be applied.
- *Structure.* Describes the structure of the pattern using different graphical representations. Typically, class diagrams, sequence and other interaction diagrams are used.
- *Collaboration.* Describes how the various elements, namely classes and objects, interact in the pattern.
- *Consequences.* Describes the results that can be expected from using the pattern. Also the side effects and a description of the results, side effects, and liabilities of the pattern.
- *Implementation.* Describes an implementation of the pattern.
- *Known Uses and Related Patterns.* Examples of how the pattern has been applied in real systems. Related patterns that have been presented before.

4.2   Principles

In this section, we consider five crucial categories of distributed computing principles, namely Internet, web, SOA, security, and finally mobile computing principles. We highlight the motivation for the principles and then discuss the differences for mobile computing and mobile middleware.

4.2.1   Internet Principles

TCP/IP is generally described as having four abstraction layers (RFC 1122). This layer view is often compared with the seven-layer OSI Reference Model. An early architectural document, RFC 1122, emphasizes architectural principles over layering. Principles have had a major influence in the development of the current Internet. The original architectural principles for the Internet were [6]:

- *End-to-End Principle.* In its original expression placed the maintenance of state and overall intelligence at the edges, and assumed the Internet that connected the edges retained no state and concentrated on efficiency and simplicity. Today's real-world needs for firewalls, NATs, web content caches have essentially modified this principle.
- *Robustness Principle. be conservative in what you do, be liberal in what you accept from others.* This principle has been attributed to Jon Postel, editor of the RFC 793

(Transmission Control Protocol). The principle suggests that Internet software developers carefully write software that adheres closely to extant RFCs, but accept and parse input from clients that might not be consistent with those RFCs. As stated in RFC 1122, adaptability to change must be designed into all levels of Internet host software.

Some of these principles, namely end-to-end principle, have caused a lot of discussion recently among Internet scholars. The end-to-end principle implies that application logic is executed by endpoints of communication and follows secondary principles such as minimality, generality, simplicity, and openness. In today's Internet, logic has been distributed between end hosts, middleboxes such as firewalls and NATs, and trusted third parties, such as websites. It follows that for the end user, it is crucial that any application functionality related to the user's activities is executed in a trustworthy manner.

This observation has led to a reformulation of the original End-to-End principle called *Trust-to-Trust (T2T)*. T2T requires that application logic is executed by trusted points in the network: *'The function in question can completely and correctly be implemented only with the knowledge and help of the application standing at points where it can be trusted to perform its job properly'*. The proposal for T2T has created a lot of discussion in the networking community and it remains to be seen how trust is reflected in the future Internet architecture.

4.2.2   Web Principles

The principles of the web follow those of the underlying TCP/IP stack. Principles such as simplicity and modularity are at the very base of software engineering; decentralization and robustness are the foundational characteristics of the Internet and web. Web principles are about supporting flexible publishing of resources on the Internet and then linking these resources together. In the context of data publishing, data representation and transformations are crucial.

We are faced with the question of how to deliver information in such a way that it can be universally accessed. Here, the web follows what is called the principle of applying the *least powerful language* to do a particular job. According to this principle, one should strive towards late transformations and leaving the actual choice of the transformation language as flexibility point. The web solves this by allowing any data format to be used. The data format is identified in HTTP requests in *Uniform Resource Locators (URLs)* and in HTTP responses.

HTTP was developed by W3C and IETF, resulting in a series of RFCs, namely RFC 2616 that defines HTTP/1.1. HTTP is a request/response protocol between a client and a server. The client creates an HTTP request using a web browser, or some other tool. This client side functionality is referred to as *user agent*. The server side that serves resources receives the request and sends a response back to the client. Between the user agent and the server, there can be several intermediaries, including proxies, gateways, and tunnels. HTTP is typically implemented on top of TCP; however, some other protocol can also be used that offers reliable transport.

HTTP is stateless by nature, which has the advantage of not requiring hosts to retain information regarding users between requests. On the other hand, the stateless nature requires that web developers utilize alternative mechanisms in order to track the user's state.

Before HTTP/1.1, connections were closed after a single request/reply was completed. HTTP/1.1 introduced a keep-alive mechanism that allows a connection to be used for more than one request. This kind of keep-alive functionality is essential for the mobile environment to be able to reduce latency. HTTP/1.1 also introduced some bandwidth optimizations, namely chunked transfer encoding, pipelining, and byte serving. In the first, content can be streamed. In the second, clients can send multiple outstanding requests. In the third, server can transmit only the requested part of a resource when explicitly requested by a client.

In order to achieve modularity and decentralization, web utilizes URLs, which effectively offer separation of concerns between resources, resource locations, and the protocol that is used to access a particular resource. A *Uniform Resource Identifier (URI)* is used frequently as a synonym to URL. Technically, an URL is a URI that identifies a resource and provides means of acting on the resource (discussed in RFC 3305). URIs can be classified as a locator (URL) or as a name. In the latter case, the URI is a *Uniform Resource Name (URN)*. URIs have been phenomenally successful in applications also outside the traditional web, in for example SIP.

Representational State Transfer (REST) is a model for distributed computing. The model contrasts the normal web model in that it applies the web principles to a transaction oriented services rather than publishing oriented services. REST builds on three central web technologies, namely URIs, HTTP, and XML. URIs provide a way to universally identify and address resources, HTTP provides the universal access protocol, and XML provides the universal data description format. REST is motivated by the simple nature when compared with rather complex architectures such as the web services architecture standardized by W3C.

Typical security solutions for REST include URI-based *Access Control Lists (ACLs)*. ACLs effectively partition the URI namespace into security domains. Baseline security is typically provided by HTTPS (TLS).

One of the proposed benefits of REST is that it is compatible with the established filtering and intrusion detection practices. Web services, namely SOAP, require custom tools that understand SOAP headers and are able to parse XML messages. From the network element viewpoint, XML parsing increases the cost and reduces the scalability of the system.

One challenge is the REST's assumption of side-effect free GET is not always true. For example, Flickr's delete method is invoked with a GET. Therefore, filtering based on the HTTP method is not enough, and the whole query string needs to be filtered. This means that a number of validation operations are needed to prevent injections attacks and malformed content.

In the current development tools, both REST and SOAP are supported. For example, WSDL 2.0 and SOAP 1.2 allow both kinds of interfaces. Public web services such as Amazon, Google, and Flickr provide both REST and SOAP style interfaces; however, currently REST has more support from developers than vendors.

The principles behind REST are the following:

- Application state and functionality are divided into resources.
- Every resource is uniquely addressable using a universal syntax for use in hypermedia links.

Table 4.1   REST and WS-* principles [7]

Architectural principle and Aspects	REST	WS-*
Protocol Layering	yes	yes
HTTP as application-level protocol	yes	
HTTP as transport-level protocol		yes
Dealing with Heterogeneity	yes	yes
Enteprise Computing Middleware	yes	yes
Loose Coupling	yes	yes
Time/Availability		yes
Location (Dynamic Late Binding)	(yes)	yes
Service Evolution		
Uniform Interface	yes	
XML Extensibility	yes	yes

- All resources share a uniform interface for the transfer of state between client and resource, consisting of a constrained set of well-defined operations, a constrained set of content types, optionally supporting code on demand.
- The defining features of REST are: client-server, stateless, cacheable, and layered.

Table 4.1 presents a comparison of REST and the Web services principles [7]. The web sevices architecture is generally denoted by using WS-* indicating that it encompasses a number of specifications. Both architectural styles are layered. In REST, HTTP is an application level protocol whereas in WS-* HTTP is a transport level protocol. Both approaches deal with heterogeneity through URIs and the HTTP protocol, and both can be used for enterprise computing middleware. Moreover, both styles support loose coupling.

4.2.3 SOA Principles

Service Oriented Architecture (SOA) is a software architecture where functionality is structured around business processes and realized as interoperable services. Thus SOA encompasses a wide variety of services that can be used to support, manage, coordinate, and enhance business processes. The aim is a loose coupling of services that abstracts interoperability concerns with operating systems, programming languages, and other technologies. SOA separates functions into services that are accessible over a network. The loosely coupled nature in conjunction with interoperable service definitions allows the combination and reuse of the services in the production of business applications.

SOA is heavily based on message-oriented communications. Services communicate by sending messages to each other. This message passing is used to build complex functionality, such as distributed auctions and the coordination of multiple services.

The following guiding principles define the ground rules for development, maintenance, and usage of the SOA:

- Reuse, granularity, modularity, composability, componentization, and interoperability.
- Compliance to standards (both common and industry-specific).
- Services identification and categorization, provisioning and delivery, and monitoring and tracking.

The following specific architectural features influence the design and implementation of SOA systems:

- Encapsulation, which is used to consolidate other services to conform with SOA interfaces, such as web services.
- Loose coupling, which is the defining characteristic of SOA systems and minimizes dependencies between services.
- Contract, which defines a communications agreement between services.
- Abstraction, which allows services to hide their internal logic from the outside world.
- Reusability, which allows the division of functions into reusable services.
- Composability, which is the ability to take collections of services and assemble them into a composite service through coordination.
- Autonomy, which gives services control over the logic that they encapsulate.
- Discoverability, which makes service descriptions available so that they can be found and then used to access the service. Typically, a directory is used to store and search for services.

Web services can be used to implement a service-oriented architecture. A major focus of web services is to make functional building blocks accessible over standard Internet protocols that are independent from platforms and programming languages. These services can be new applications or simply wrapped around existing legacy systems to make them network-enabled.

4.2.4 Security Principles

Wireless networks face the same security requirements and concerns than the traditional fixed networks, for example confidentiality, integrity, authentication, and privacy. The wireless world faces additional challenges due to the resource limited devices, dynamic environment, and the broadcast nature of wireless communications. For example, resource limitations may prevent the full utilization of expensive but secure cryptographic operations. The dynamic environment makes the mobile device and user vulnerable to all kinds of impersonation, privacy [8], denial-of-service, spoofing, and identity theft attacks. Moreover, the broadcast nature requires sound encryption mechanisms to prevent malicious entities from tapping into data communications. In all, the heterogeneous and dynamic nature of the environment make mobile security a challenging topic.

The commonly agreed security aspects are the following:

- privacy;
- integrity;
- authentication;
- authorization;
- accountability;
- availability.

Inversion of control (IoC) is an abstract principle describing an aspect of some software architecture designs in which the flow of control of a system is inverted in comparison to the traditional architecture of software libraries. Control flow is expressed in imperative programming in the form of a series of instructions or procedure calls. Instead of specifying a sequence of decisions and procedures to occur during the lifetime of a process, the user of a IoC framework writes the desired responses linked to particular events or data requests. External entities then take control over the precise calling order and additional maintenance that are to be carried out to execute the process.

The principle of *least privilege*, also known as the principle of minimal privilege or just least privilege, requires that in a particular abstraction layer of a computing environment every module (such as a process, a user or a program on the basis of the layer we are considering) must be able to access only such information and resources that are necessary to its legitimate purpose.

The W3C *Platform for Privacy Protections (P3P)* working group has established the following privacy guiding principles [9]:

- *Notice and Communication.* Service providers should provide timely and effective notices of their information policies and practices. User agents should provide effective tools for users to access these notices and make decisions based on them.
- *Choice and Control.* Users should be given the ability to make meaningful choices about the gathering, utilization, and disclosure of personal information.
- *Fairness and Integrity.* Users should retain control over their personal information. Moreover, users should be able to decide the conditions under which they will share it. Service providers should treat users and their personal information with fairness and integrity. This is fundamental for protecting privacy and sustaining trust.
- *Confidentiality.* Users' personal information should always be protected with reasonable security measures taking into account the sensitivity of the information and required privacy level.

4.2.5 Mobile Principles: Device View

The requirements of the mobile environment require that we revisit principles designed for the fixed networking environment, such as the Internet, web, and SOA principles, and consider them from this viewpoint. The mobile environment is subject to various disruptions in communications, which require special emphasis on disconnected operation and security. Indeed, decoupled communications can be seen as one of the key principles for mobile computing.

We examine the principles behind the NoTa system:

- System level loose coupling. This means that loose coupling of components is built into the system. This allows replacing not only application components but also system components in a flexible manner. Decoupling of components is based on message passing, which requires a message bus (the interconnect).
- Interconnect centric. NoTA is based on an interconnect that is responsible for connecting different system components and services together via message passing. Interconnect can be seen as a common communications bus.
- Service based, which means that functionality is provided through services that have interface definitions. This allows the run-time discovery of available functionality and then subsequent execution of this functionality. Loose coupling combined with service based access can be seen as the necessary ingredients for adaptability.

- Message and data driven. Message passing is the preferred mechanism for realizing mobile applications. In addition, the communications is typically data driven meaning that a request can be forwarded based on the current system parameters and information contained in the request. This data centric operation is necessary for context-aware operation and also for allowing improved decoupling between components.
- Implementation-wise heterogeneous. The NoTA system has been designed to facilitate the integration of subsystems from various manufacturers and vendors. This requires that given interface specifications and request formats, components are interchangeable.

4.2.6 Mobile Principles: SIP Architecture

This section discusses the key SIP architectural principles. These include the usage of proxies for routing, the relegation of call state to endpoints, the usage of dialog models and not call models, endpoint fate sharing, component-based design, logical roles, Internet-based design, generality over efficiency, and separation of signaling and media.

- *Proxies are for Routing.* SIP is a distributed system, composed of user agents and proxies. Proxies usually run in the network and their role is to facilitate rendezvous between users. Each proxy on the path of a SIP request is responsible for executing the routing logic based on the desires of the entities on whose behalf it is acting. The final call routing decision is a composition of the decisions made by each of the proxies along the request path.
  There were many factors driving the decision for proxies to take on the role of rendezvous, as opposed to endpoint control and management. These included availability, scalability, and flexibility.
- *Endpoint Call State and Features.* Since the role of the proxy is to facilitate rendezvous, the rest of the capabilities needed in communications systems fall to the endpoints. In particular, SIP endpoints are responsible for maintaining call state, and for implementing features that require awareness and management of that call state.
  This particular facet of SIP's design has allowed it to be applied to problems as diverse as *Push-to-Talk* and home automation. Though sometimes these usages are not really a good fit for SIP, they are demonstrable validations of this design goal.
- *Endpoint Fate Sharing.* A benefit of the smart endpoint model is that call state and application state is co-located with the endpoints of the call itself. This means that the only way in which the call or application can fail is if the endpoints themselves fail.
- *Component Based Design.* SIP design is based on the use of protocol components and primitives. SIP does not specify a vertical communications system, instead SIP itself is just a component in any complete communications system. SIP itself provides capabilities through component functions that can be composed together to provide more complex functions.
- *Logical Components, not Physical.* SIP defines its functionality by defining the exchange of messages between logical components in a distributed system. These components – user agents, proxies, redirect servers and registrars, are logical.
- *Designed for the Internet.* SIP was designed for operation on the public Internet. The public Internet introduces a number of constraints and also a number of benefits that have been taken into account in the design of the SIP architecture.

- *Generality over Efficiency.* When designing network protocols, there is often a tradeoff between efficiency and generality. SIP generally prefers to build for generality at the expense of efficiency.
- *Separation of Signaling and Media.* One of the SIP fundamentals is that the path that any media packets follow is independent of the path followed by the signaling packets. This separation allows for the IP network to deliver the media packets using the most direct and appropriate route it can, while the signaling packets can follow a series of proxy elements needed for the processing of the request. By separating the two, more complex proxy topologies can be utilized without concern for the impact on voice quality.

4.3 Cross-layer design

As presented previously, the protocol stack is typically organized into layers. The OSI model forbids direct communication between nonadjacent layers, and communication between adjacent layers is done using procedure calls and responses.

A protocol designer has two approaches when designing a new protocol. The first approach is to follow the architecture and its layering, typically relying only on the services of layers below. The second approach is to utilize direct communication between protocols that are not on adjacent layers thus violating the reference architecture. This kind of a protocol follows the cross-layer design.

Layering offers separation of concerns, for example, reliable end-to-end (transport), unreliable packet routing (network), and hop-by-hop (link). On the other hand, layers abstract details away and may result in unnecessary overhead and redundant functionality. For example, mobility and security solutions are available on multiple layers. Moreover, upper layers may not necessarily have access to the information used by lower layers, for example, it is not possible for an application to query IPSec configuration or TCP parameters.

Various cross-layer interactions have been proposed to make the protocol stack more optimized by allowing information to be shared by layers. Cross-layer interactions can potentially improve performance; however, they also introduce new complexity into the protocol stack and communications architecture. When new cross-layer interactions are introduced, care must be taken to prevent any unwanted effects with existing interactions. A critique of cross-layer design is presented in [10]. They warn that cross-layer optimization presents both advantages and liabilities. The liabilities include spaghetti code, proliferation problems, and dependency issues.

Cross-layer interactions can be modelled using a combined dependency graph, which shows the relations between various protocols and their parameters. In this graph, each parameter of a protocol is a node and each edge denotes a dependency relation between two parameters. This kind of a graph can be used to inspect certain stability principles of protocols, for example conflicts.

Figure 4.1 presents information flows in a protocol architecture. The information can be categorized as follows [11]:

- Upward information flow, in which information is propagated from lower layers towards upper layers. Typically, an explicit notification interface is used to pass information.
- Downward information flow, in which information is propagated from higher layers towards lower layers. An interface is used to set a lower layer parameter.

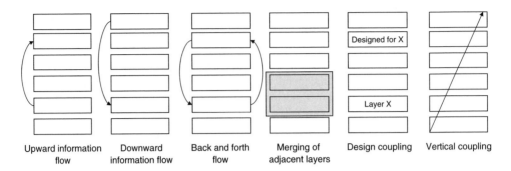

| Upward information flow | Downward information flow | Back and forth flow | Merging of adjacent layers | Design coupling | Vertical coupling |

Figure 4.1   Cross-layer interactions

- Back-and-forth information flow, in which information is propagated in both directions.
- Merging of adjacent layers, allows the combination of several adjacent layers into a super layer.
- Design coupling without adding new interfaces. In this strategy two or more layers are coupled during design time without specifying a new interface between them for information sharing at runtime. This means that the coupling is rigid and change of a coupled layer may require changes to other layers in the coupling.
- Vertical calibration across layers. This involves adjusting parameters across layers. The motivation is that joint tuning of parameters in the protocol stack can help to achieve better performance. This can be done in a static manner or at runtime.

A number of techniques have been proposed for implementing cross-layer designs. These can be divided into three categories:

- direct communication between layers;
- a shared database across the layers;
- new abstractions.

In the first technique, layers can communicate with each other. This means that variables at one layer are visible and accessible from some other layer. Interactions are realized using explicit interfaces or, for example, adding layer-specific data into packets.

In the second technique, a shared database is used to allow information exchange between layers. This is similar to the blackboard design pattern. New interfaces between layers can be created and discovered at runtime using the shared database.

The third category of techniques pertains to new abstractions that are not necessarily based on layers at all. To name some examples, a protocol stack can be organized as a heap or a single layer. These approaches, however, typically require new implementations of protocols.

4.4   Model Driven Architecture

Unified Modeling Language (UML) is the de facto modeling language for object-oriented analysis and design. UML uses a visual syntax to represent the schematics of software. These schematics are akin to architectural drawings of buildings and they use a standard set of icons to represent accepted concepts in software engineering.

UML defines a set of diagrams that are used to show relationships between objects in a system. UML diagrams include:

- Class Diagram: The class diagram shows the relationships between a set of classes. This is the most common diagram in UML and is a structural view of how classes are related.

  Object-oriented software consists of distinct modules called classes, which basically define the templates for the creation of objects. An object-oriented class has a set of attributes, which define the structural properties, and operations, which define the behavioral properties of a class. The UML notation of a class is a rectangle divided into three compartments: name, list of attributes, and list of operations.

  A UML class diagram depicts the classes and their relationships to each other. Some basic types of class relationships are association (basic), aggregation (whole/part), and inheritance (hierarchy!UML). Relationships can be further described by multiplicity (cardinality) designations, which show the number of objects that take part in a relationship.

- Component Diagram: The component diagram shows the relationship between components. A component is a physical manifestation of software, and may be a library, executable, DLL, etc. Components are built and combined to form applications.

- Deployment Diagram: The deployment diagram shows the relationship between physical entities like a CPU, printer, or workstation. It can also show where components are deployed.

- Use Case Diagram: The use case diagram captures functional requirements. It shows the relationships between actors and use cases. Actors are external entities to the system such as users and other systems. A use case is an end-to-end sequence of actions (including variants) that result in an observable and useful result.

  A use case diagram is a high-level view of the software system, as seen from a user of the system. A use case diagram depicts the main functions or usage of the software. Internal details are not depicted in the use case diagram. A use case diagram consists of an actor, which is someone who interacts with the system, and use case scenarios, which depict the various ways in which the software can be used.

- Sequence and Collaboration Diagrams: The sequence diagram is an interaction diagram that shows the communication between objects in a time-ordered fashion.

  A sequence diagram shows the chronological time sequence of events that take place between classes and objects in a particular use case. Messages that are exchanged between classes and objects are shown as arrows in a sequence diagram.

- Collaboration Diagram: The collaboration shows the structural organization of objects that communicate with each other.

- State Diagram: The state diagram contains a finite state machine that describes the behavior of a class. State machines are an excellent way to describe event-driven behavior.

- Activity Diagram: The activity diagram shows the flow of control through activities in a system.

- Other UML diagrams include Structural Diagrams and Behavioral Diagrams.

There are many automated software design tools for the creation of UML diagrams, many of which are part of integrated development environments (IDE) such as Visual Studio or Eclipse.

4.5 Architectural Patterns

Architectural patterns pertain to well-established solutions to architectural problems in software engineering. An architectural pattern describes the elements and their constraints, and how elements are used together in order to solve the problem. The pattern is a structural organization schema for a system solving the problem; however, the pattern is not an architecture in itself. Architectural patterns are more complicated and larger in scale than design patterns.

4.5.1 Overview

Well-known architectural patterns include:

- *Layers.* A multilayered software architecture is using different layers for allocating the responsibilities of an application.
- *Client-Server.* The client-server pattern is the most frequent pattern in distributed computing, in which clients utilize resources and services provided by servers.
- *Peer-to-peer.* The peer-to-peer pattern is an emerging communications model, in which each peer in the network has both client and server roles. Typically, each peer has symmetric functionality; however, this functionality can also be unevenly distributed. This is the case in several peer-to-peer systems that utilize so called super peers, which provide additional services to other peers.
- *Pipeline (or pipes and filters).* A pipeline consists of a chain of processing elements (processes, threads, coroutines, etc.), arranged so that the output of each element is the input of the next. Usually some amount of buffering is provided between consecutive elements. The information that flows in these pipelines is often a stream of records, bytes or bits.
- *Multitier.* A multitier architecture is a client-server architecture in which an application is executed by more than one distinct software agent. For example, an application that uses middleware to service data requests between a user and a database employs multi-tier architecture. The most widespread use of 'multi-tier architecture' refers to three-tier architecture.
- *Blackboard system.* In this pattern, a common knowledge base, the 'blackboard', is iteratively updated by a diverse group of specialist knowledge sources, starting with a problem specification and ending with a solution. Each knowledge source updates the blackboard with a partial solution when its internal constraints match the blackboard state. In this way, the specialists work together to solve the problem. The blackboard model was originally designed as a way to handle complex, ill-defined problems.
- *Publish/Subscribe.* Event-channel and Notifier [23].
  The event-channel and notifier patterns decouple subscribers and publishers by introducing a broker that mediates events on their behalf. The event channel and notifier also support various non-functional requirements, such as QoS and disconnected operation. The event-channel and notifier patterns are similar, but the notifier also abstracts the location and distribution of event brokers, whereas with channels the client must first obtain the reference of the channel.

The notifier pattern may be realized by using the observer pattern and mediators or proxies [12]. The event channel pattern is used in the CORBA Event Service [5] and Notification Service [13]. A separate specification defines how CORBA event channels are connected to form communication topologies [14].

- *Model-View-Control (MVC)* is both an architectural pattern and a design pattern, depending where it is used. MVC is heavily used in the Symbian OS and Symbian applications. The pattern isolates business logic from user interface issues. The aim is to allow easy changes to the user interface without changing the underlying business model. The model represents the data of the application and the business rules that are used to manipulate the data. The view consists of elements of the user interface. The controller is responsible for ensuring that any control signals from the user are properly handled using information in the model and reflected in the view.
- *Broker*, which introduces a broker component to achieve decoupling of clients and servers.
- *Microkernel.* This pattern provides the minimal functional core of a system, the micro-kernel, which is separated from extended functionality. The external functionality can be plugged into the microkernel through specific interfaces.
- *Active Object.* The Active Object pattern provides a support for asynchronous processing by encapsulating the service request and service completion response.

In the following subsection, we focus on a core set of architecture principles which are important for the mobile environment. These are the client-server pattern, three-tier architecture, Model-View-Control, Microkernel, and Active Object.

4.5.2 Client-Server

The client-server patterns divided the distributed system into two parts, namely the client and the server. The main goal is to achieve scalability and flexibility. A significant part of system functionality is performed by the server part. The benefit of this pattern is that this server-side functionality can be modified without modifying the clients. Moreover, the server can provide the functionality for a number of clients. Servers can also be distributed in some configuration to better serve the clients.

Depending on the purpose of the server, the services it provides may perform computationally expensive computations, execute proprietary algorithms or access confidential information not available to the client, encapsulate access to shared resources.

The clients issue service requests to the server. The server processes the requests either synchronously or asynchronously. In the case of synchronous processing, the control is returned to the client together with the results of processing. On the other hand, in the case of asynchronous processing, the control is returned to the client immediately, while the results of processing are delivered later, upon the completion of the processing.

Within a single host, a server typically runs in a separate thread or process than the client. Local client and server communication can be realized by using message passing or by using inter-thread data transfer.

In Symbian OS, servers are extensively used to manage resources on behalf of multiple clients. These include the file server, the window server, the font and bitmap server, the database server, the serial communications server, and the sockets server. Symbian OS features three different types of servers, namely

- constantly running system servers;
- transient servers that run until the last session is completed;
- auxiliary application servers that are started and terminated with the application.

4.5.3 Three-tier Architecture

Three-tier is a client-server architecture in which the user interface, functional process logic (the business rules), data storage and data access are developed and maintained as independent modules. The modules can be distributed on multiple platforms. The three-tier model is generally considered to be a software architecture and a software design pattern.

In addition to the traditional advantages of modular software with well defined interfaces, the three-tier architecture is intended to allow any of the three tiers to be upgraded or replaced. This is useful especially when requirements or technology changes. For example, a change of operating system from Microsoft Windows to Unix in the presentation tier would only affect the user interface code.

In the general case, a user interface runs in some device, for example a desktop or mobile phone, and is based on a standard GUI. The functional process logic needed by the application and the GUI consists of one or more separate components that can run on the same host as the GUI or a different host. In the latter case, the host that provides the separate components is a server. In addition, there is a database server that is responsible for the storage logic. The middle tier may be multi-tiered itself.

The three-tier architecture has the following three tiers:

- *Presentation Tier.* This is the topmost level of the application. The presentation tier displays information related to such services. It communicates with other tiers by outputting results to the other tiers.
- *Application Tier (Business Logic).* This tier controls the functionality of an application and performs processing. This tier typically implements the business logic of the service.
- *Data Tier.* This tier includes database servers and other storage services. This tier is responsible for supporting information storage and access.

The three tiers may seem similar to the Model-View-Control (MVC) concept; however, topologically they are different.

The three-tier pattern differs from the Model-View-Control (MVC) pattern in that clients do not communicate directly with the data tier. All communications go through all the tiers in a linear fashion. The MVC pattern, on the other hand, allows the View to send updates to the Controller, which then updates the Model. The View is subsequently directly updated from the Model.

4.5.4 Model-View-Control

The pattern is used in interactive applications in order to make the user interface easier to change. The pattern divides the application into three parts, namely the controllers handling user input, the model providing the core functionality, and the views displaying the information to the user. The pattern ensures that the user interface is formed by the view and the controller is consistent with the model [15]. Figure 4.2 presents an overview of the pattern.

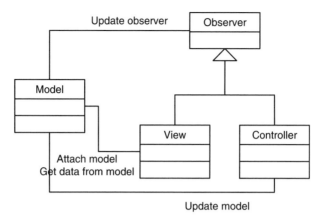

Figure 4.2 The Model-View-Control pattern

As an architectural pattern, MVC is frequently used in web applications, where the view is the HTML content, and the controller is the code that creates the HTML content. The model is represented by different data sources, namely databases, XML resources, and the business rules that govern the generation of the content. The motivation for the pattern can be seen from this example, namely decoupling models and views, which helps to reduce complexity and improve flexibility.

The MVC pattern is suitable for applications that are executed in environments in which the *User Interface (UI)* needs to be flexible or the UI is subject to frequent change. Firstly, multiple interfaces may need to be implemented for a system. Secondly, the UI may need to be changed according to changes being made to the implemented functionality. If the user interface is tightly coupled with the core functionality of an application, changing UI may be challenging. This process of changing the UI may also result in programming errors. In many cases, modifications to the UI may have implications for other system components.

The Model component encapsulates the key functions of the application, for example data structures. The Model provides interfaces for invoking application-specific services and for accessing the data encapsulated by the component.

One or more View components retrieve the data from the Model and then display it to the user. A separate Controller component is associated with each View. This component processes user input and then issues service requests to the Model. A view may provide the Controller with the functionality that allows direct manipulation of the display.

The pattern also specifies the *change-propagation* mechanism. Views and Controllers register with the Model to receive notifications about changes in the structure. When the state of the Model changes, the registered Views and Controllers are notified.

Additional Views and Controllers can be implemented without modifying the Model. These Views and Controllers can be implemented as pluggable components. This means that they can be added and removed at run-time. Due to the change-propagation mechanism, the Views and Controllers remain synchronized with the Model. The ability to change Views and Controllers without modifying the Model improves the portability of the architecture.

The liabilities of the pattern include:

- Increased complexity of the application due to new components and their interactions.
- Potentially excessive number of unnecessary change notifications. For example, not all views are interested in all changes of the Model. This can be alleviated using filtering techniques.
- Tight coupling of the View and Controller with the Model. This means that any changes in the model may require changes in Views and Controllers.
- The access to data from the View may be inefficient in terms of performance. For example, multiple calls to the Model may be needed in order to display data.
- There may be a need to change the View and Controller when the application is ported.

The pattern is widely used when developing applications for the mobile environment. Many software frameworks promote the use of the MVC pattern and provide easy creation of the relevant components. For example, the Symbian OS application framework is based on this pattern.

4.5.5 Broker

This pattern introduces a broker component to achieve decoupling of clients and servers. Servers register with the broker, and make their services available to clients through method interfaces provided by the broker. Clients access the functionality of servers by sending requests via the broker. The tasks of the broker include locating the appropriate server, forwarding the request to the server and transmitting results and exceptions back to the client [4].

By using the broker pattern, an application can access distributed services simply by sending messages to the appropriate object. Therefore applications do not have to focus on low-level inter-process communication. In addition, the broker architecture is flexible, in that it allows dynamic change, addition, deletion, and relocation of objects.

The broker pattern reduces the complexity involved in developing distributes applications, because it makes distribution transparent to the developer. It achieves this goal by introducing an object model in which distributed services are encapsulated within objects.

The CORBA architecture is based on the broker pattern. Figure 4.3 illustrates the *Object Request Broker (ORB)* that facilitates requests from clients to servers that are responsible for CORBA objects.

4.5.6 Microkernel

In order to make the system adaptable to changing requirements, the minimal functional core of the system, the microkernel, is separated from the extended functionality, which can be plugged in the microkernel through specific interfaces [15].

This pattern may be applied in the context of complex software systems serving as a platform for other software applications. The desired characteristics for such systems include extensibility, adaptability, and interoperability. The systems should be evolvable to cope with emerging technologies and standards. From the non-functional requirements viewpoint, in many cases high performance, low memory consumption, and scalability are needed. The microkernel design pattern aims to support the development of complex systems by defining a small core that is extensible with pluggable components.

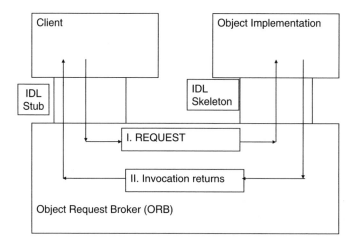

Figure 4.3   Broker pattern in CORBA

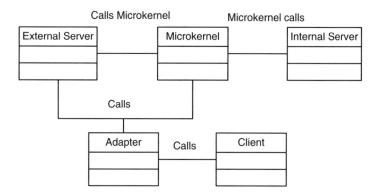

Figure 4.4   The microkernel pattern

Figure 4.4 presents an overview of the microkernel pattern. The most important core services of the system should be encapsulated in a microkernel component. The microkernel component is responsible for the system resources and it allows other components to interact with each other, and to access the functionality of the microkernel. The microkernel encapsulates a significant part of system specific dependencies that include hardware-dependent aspects.

The size of the microkernel should be kept as small as possible. This means that only part of the core functionality can be encapsulated in the component. The rest of the core functionality is placed in separate internal servers.

Symbian OS is built using the microkernel architectural pattern. The use of the pattern improves the portability and changeability of the OS. The Symbian servers follow the client-server pattern and represent the external servers in the underlying microkernel architecture. The layers and the broker pattern are alternatives to the microkernel pattern.

4.5.7 Active Object

The Active Object pattern provides support for asynchronous processing. The pattern works by encapsulating and handling asynchronous service requests and service completion responses. The pattern is useful when a server's process time is long enough for the client to be able to do something useful during the waiting time. The pattern allows the client to be notified about the task's completion and perform other tasks asynchronously with the server [15].

In asynchronous processing, a client requests the services of a server by issuing asynchronous requests. The requesting function does not need to block after issuing a request, but can continue processing and then be notified when the service is complete.

Writing asynchronous request handlers involves many low-level details. This may result in a difficult to understand code, prevent code reuse, and result in latent programming errors. In order to address the limitations of the mobile computing environment, the implementation preferably also have a small memory footprint.

The Active Object pattern is useful in this environment, because it takes the responsibility of handling and encapsulating asynchronous requests. The client makes requests by calling a specific function exposed by an Active Object. The Active Object then executes this function, and when the function is completed, the Active Object's respective callback function is called. Figure 4.5 illustrates this pattern.

The SymbianOS makes extensive use of the Active Object pattern. In Symbian, each component that implements this pattern is associated with an AsynchronousServiceProvider which offers a request function taking as a parameter a status variable. When ActiveObject's request function is called, the service request is passed to AsynchronousServiceProvider. At the same time, the status variable is updated to reflect pending status.

The service provider identified in the request eventually changes the status variable according to the results of the processing (for example that there is no error with code KErrNone) and signals the requesting thread's CActiveScheduler about the completion. Now, it is the responsibility of the scheduler to identify the ActiveObject to be resumed. The active objects and the active scheduler are situated within a same thread (different from the thread of AsynchronousServiceProvider) and provide *co-operative multitasking* within the thread. This is contrasted by *preemptive multitasking*, in which the active objects are in different threads and can be executed irrespective of each others' status.

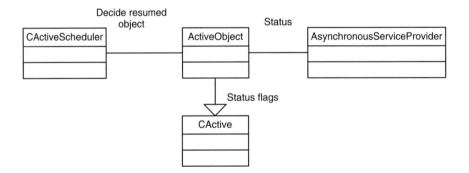

Figure 4.5 The Active Object pattern

Active Object runs in a same thread as the application. This results in reduced memory requirements, and it helps to eliminate overheads in context-switching between threads. If separate threads or processes are used to implement the same functionality, a scheduler has to control the execution of the threads and thus perform context-switching between them which incurs overheads. Single-thread implementation is simpler than multi-thread implementation, and also thread synchronization issues are avoided.

The liability of Active Object is the fact that it is non-preemptive, which means that only one event can be processed at a time. (RunL in Symbian OS). Therefore, the Active Object thread must process the callback quickly, which may be difficult to accomplish.

4.6 General Patterns

In this section, we briefly outline a set of well-known patterns. The patterns are presented in three categories, namely structural patterns, behavioral patterns, and concurrency patterns. The first pertains to the structure of system and includes patterns such as the adapter, decorator, facade, and proxy. These patterns provide mechanisms to help modify and extend system structure. The second category pertains to augmenting the behavior of components in a system. This category includes patterns such as chain of responsibility, mediator, observer, and strategy. The third category includes patterns to manage concurrency in systems. Examples of these patterns include singleton and guarded.

4.6.1 Structural Patterns

- *Adapter.* This pattern converts an interface of a class into another interface that the clients expect. Adapter lets classes work together that could not otherwise because of incompatible interfaces. The problem can be solved by implementing an adapter class responsible for converting the requests between interfaces. The pattern can be implemented as a class adapter or as an object adapter.
- *Decorator.* This pattern attaches additional responsibilities to an object in a dynamic fashion. Decorators provide a flexible alternative to subclassing for extending functionality.
- *Facade.* Provides a unified interface to a set of interfaces in a subsystem. Facade defines a higher-level interface that makes the subsystem easier to use. In order to minimize the communications and dependencies between subsystems, the pattern suggests establishing a Facade object providing a simplified higher-level interface to the subsystem's functionality. The clients can send their requests to the Facade, which translates them into one or several requests to appropriate objects inside the subsystem. Thus, the Facade makes it easier for the programmer to use the functionality provided by the subsystem, without hiding the lower-level functionality completely.
- *Proxy.* The proxy pattern provides a surrogate or placeholder for another object to control access to it.

4.6.2 Behavioral Patterns

- *Chain of responsibility.* This pattern avoids the coupling of the sender of a request to its receiver by giving more than one object a chance to handle the request. Chain the receiving objects and pass the request along the chain until an object handles it.

- *Mediator.* Defines an object that encapsulates how a set of objects interact. Mediator promotes loose coupling by keeping objects from referring to each other explicitly.
- *Observer.* Defines one-to-many dependency between objects so that when one object changes state, all its dependents are notified. The pattern explains how to define a one-to-many dependency between objects, so that all the dependent objects would be notified/updated automatically whenever the state of the object being observed changes. The Observer pattern is employed within the Model-View-Control pattern to inform the Views and Controllers about changes in the Model.
- *Strategy.* This pattern defines a family of algorithms, allows encapsulation of each algorithm, and makes them interchangeable. Strategy lets the algorithm vary independently from clients that use it.

4.6.3 Concurrency Patterns

- *Singleton.* The pattern is used to ensure that only one instance of a class can be instantiated, and specifies an access point to this instance.
- *Guarded.* In concurrent programming, guarded suspension is a software design pattern for managing operations that require both a lock to be acquired and a precondition to be satisfied before the operation can be executed.

4.7 Patterns for Mobile Computing

In this section, we focus on a small set of core patterns for mobile computing. The patterns are divided into three categories, namely distribution patterns and resource management, synchronization, and communications. The patterns have been in part inspired by the MODPA project at University of Jyväskylä, which collected a set of mobile patterns [15].

Distribution patterns pertain to how resources are distributed and accessed in the environment. This category includes patterns such as remote facade, data transfer object, remote proxy, and observer.

Resource management and synchronization category involves various patterns related to mobile data management and distributed data synchronization. These patterns include session token, caching, eager acquisition, lazy acquisition, synchronization, rendezvous, and state transfer.

The communications category includes the connection factory and client-initiated connections patterns. The former is used by applications to manage various connections. The latter is used to support communications with mobile devices over arbitrary access networks.

4.7.1 Distribution Patterns

4.7.1.1 Remote Facade

Remote facade is intended for minimizing the number of remote calls in an application. The pattern provides a coarse-grained interface to one or several fine-grained objects. The interface is provided through a remote gateway that accepts incoming requests conforming to the facade interface and then results in subsequent fine-grained interactions between the remote facade (gateway) and third party interfaces.

The remote facade pattern offers separation of concerns to a mobile application. The application and developer use a coarse-grained API and do not have to care about the fine-grained implementation details. For example, an application using the pattern does not have to know which particular servers or remote functions are used to implement a requested operation. The remote facade does not contain domain logic. It only translates coarse-grained methods onto the underlying fine-grained objects. The pattern can be combined with the *Data Transfer Object (DTO)* pattern. The remote facade may provide security features and transaction control for applications.

The following liabilities can be identified for this pattern:

- The use of pattern may result in little or no performance improvement if too much irrelevant information is sent through the remote facade.
- The remote facade may result in unnecessary indirection if the communicating objects are run within a same process.
- The remote facade may present unnecessary level of indirection, if the communicating objects are coarse-grained.

4.7.1.2 Data Transfer Object

The *Data Transfer Object (DTO)* provides a serializable container for transferring multiple data elements between distributed processes. The aim of the pattern is to reduce the number of remote method calls. A DTO can be used to hold all the data that need to be transferred. A DTO is usually a simple serializable object containing a set of fields along with corresponding getter and setter methods. The fields can be primitives, simple classes, or other DTOs. In a DTO, all the fields need to be serializable.

4.7.1.3 Remote Proxy

In this pattern, a proxy is between a terminal and the network. All or selected messages or packets from the client go through the proxy, which can inspect them and perform actions. The proxy performs computationally demanding tasks on behalf of the client terminal. The proxy serves as an adapter allowing other computers to communicate with the terminal without the need to implement terminal-specific protocols.

In the case of a terminal offering services to other hosts, the proxy is responsible for accepting incoming service requests and transforming them into a format suitable for the terminal. The proxy then delivers the messages to the terminal, and subsequently processes the results and forwards them to the requesting host.

The remote proxy controls the access from the terminal to the network. In the case of a terminal with insufficient computational power, the proxy transmits the service request coming from the terminal to a remote service provider; upon the completion of the service, the results are processed by the proxy and returned back to the terminal.

The benefits from the use of this pattern include:

- The pattern allows the applications on terminals with limited capabilities to use services that have high computational requirements.
- The amount of data that needs to be transmitted to the client terminal may decrease significantly.

- Network hosts can utilize the proxy to access any public interfaces available on the mobile device.

The following issues in the use of the pattern have been identified:

- The proxy pattern is prone to the failure of the proxy component. Failure of the proxy or the communications result in service disruption unless additional techniques, such as caching, are employed.
- The client needs to be able to discover the proxy. This requires that a service discovery mechanism is implemented or that the proxy has a fixed address.

4.7.1.4 Observer

The observer pattern explains how to define a one-to-many dependency between objects, so that all the dependent objects would be notified whenever the state of the object being observed changes.

When a system consists of a number of communicating components, consistency often needs to be ensured between the states of these components. Consistency of states can be ensured by implementing additional links between the components, so that they can inform each other about state changes. This results in tight coupling between system components, which is undesirable as it reduces the maintainability and reusability of the system. The observer pattern supports the decoupling of the components.

Figure 4.6 presents an overview of the pattern. In order to decouple components, the pattern suggests creating two objects, namely the subject and observer objects. The subject provides an interface, that allows an unlimited number of observers to subscribe and unsubscribe from the subject. The same interface also offers a method to notify all subscribed objects about a state change.

The components (ConcreteObserver) that want to be notified about the subject's state changes should implement the Observer interface. This interface offers the update method, that is used by the subject to notify the observers.

When a state of the ConcreteSubject changes, it uses the subject's interface (update method) in order to notify all the ConcreteObserver components about the change. When a ConcreteObserver component receives a notification from ConcreteSubject, it will request its current state.

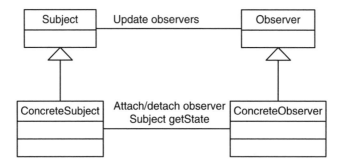

Figure 4.6 The observer pattern

The benefits of this pattern include:

- The subject and its observers can be reused.
- The coupling between the subject and observers is abstract. The components use abstract interfaces for interaction. The subject does not need to know the concrete classes of the observers.
- The pattern supports group communication. The subject does not know the receivers beforehand.

The following liabilities have been identified:

- Unexpected sequence of updates can be triggered as a result of a single state change. This stems from the fact that observers do not know about other observers and their dependencies.
- In some cases, the information that has been changed in the subject is not provided in the notification. This means that the observers need to send requests to the subject in order to obtain the changed information.

The observer pattern allows subscribers to directly register with a producer. This pattern couples the entities together and does not define how the producers are located. The pattern does not scale to large numbers of subscribers per object; however, it allows the use of a mediator that improves flexibility of the system. The observer pattern is used, for example, in the Java and Jini event models [16]. The *publish-register-notify*, a pattern similar to the observer pattern, is used in the Cambridge Event Architecture [17, 18].

4.7.2   Resource Management and Synchronization

4.7.2.1   Session Token

This pattern alleviates state management requirements of servers. A token is issued by a server to a client that contains data pertaining to the active session the client has with the server. The token contains a session identifier and possibly some security related data as well. When the client presents the token again to the server, the server can then associate the client with the proper session. This means that the server does not necessarily have to store active data pertaining to the client. The token can be cryptographically secured by the server using a symmetric key, which effectively means that only the server can decrypt the token. The token is also called cookie.

The liabilities of this pattern include the following:

- The token should contain security features and time-to-live in order to prevent replay attacks and other kinds of attacks against the server.

This pattern is heavily used in today's distributed systems. For example, many websites utilize cookies to store data on client-side. The pattern can also be used by edge gateways or proxies that manage inbound connections to alleviate state issues. A token can be stored to a lookup server to allow the association of a client to a specific connection.

Moreover, a variant of the pattern can be applied during session setup. Cookies stored in a TCP header are used by some web servers to prevent SYN attacks against them. The cookie can be combined with a cryptographic puzzle that requires a client to use

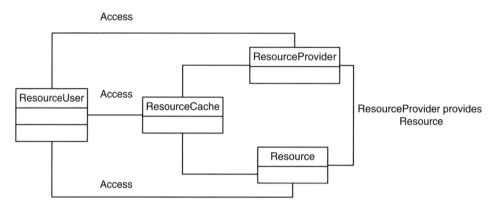

Figure 4.7 The caching pattern

resources in order to find a solution to it. This strategy is used in several protocols, such as HIP and *Datagram TLS (dTLS)*.

4.7.2.2 Caching

In order to avoid expensive re-acquisition of resources, the caching pattern suggests temporarily storing these resources in a local storage after their use, rather than immediately discarding them. This cache of elements is first checked when a resource is requested. If the element is found it is immediately delivered to the requesting application. If an element is not found in the cache, the request is performed and an entry is created in the cache for the requested object.

The solution proposed by the pattern is illustrated in Figure 4.7. The resources that have been acquired and accessed once are stored in a fast-access buffer (Resource Cache). When these resources are accessed next time, they are fetched from the Cache without the need to re-acquire them from the Resource Provider. The resources in the Cache are looked for with the help of unique identifiers, which are assigned to each resource.

The benefits of this pattern include:

- Improved availability, since resources may still remain available even if the resource provider is temporally unavailable.
- Faster access to frequently accessed resources results in improved overall performance and scalability.
- Memory fragmentation may decrease due to reduced number of resource re-acquisitions.

The liabilities of the pattern include:

- Potential losses of the changes to the cached resources if the system crashes before the cache is synchronized.
- Potential complexity of ensuring the consistency of the cache content.
- Increased memory consumption due to the cache and its content.

Data caching is used extensively in the Coda file system [19] and in the project Aura system for pervasive environments [20].

4.7.2.3 Eager Acquisition

If the resources that are needed by an application are known beforehand, a system can utilize this information and prefetch these resources. As a result, the resources are already locally available when they are needed and a remote request is not needed. The eager acquisition pattern follows this design and tries to acquire resources that may be needed later. Examples of resources include memory, network connections, file handles, threads, and sessions.

In order to achieve low and predictable resource allocation delays, resources should be acquired eagerly before they are actually requested. This can happen at start-up time or at run-time.

Figure 4.8 illustrates this pattern. Resources are eagerly acquired from a Resource Provider by a dedicated Provider Proxy which then keeps them in a container. Resource acquisition requests from the user are intercepted by the Provider Proxy which fetches the resources from the container.

The benefits of this pattern include:

- Predictability of resources availability, as resource acquisition requests are intercepted and served as they happen, and variations in delay can be avoided.
- Enhanced performance, as resources can be delivered quickly upon a request.
- Flexible customization of the resource acquisition strategy, since this strategy can be encapsulate in the Provider Proxy.
- Transparency of the solution to the user.

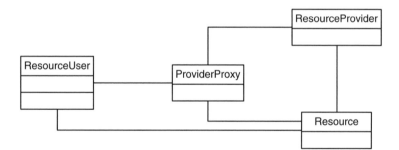

Figure 4.8   The eager acquisition pattern

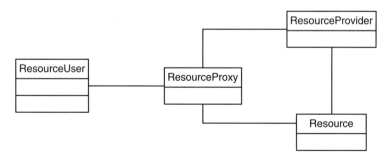

Figure 4.9   The lazy acquisition pattern

The liabilities of the pattern include:

- Management of eagerly acquired resources is important and a management policy is needed. The system may not be able to associate all acquired resources immediately with a resource user. Resource pooling and resource replacement strategies can be used to enforce a policy.
- The number of resources has to be estimated in advance.
- Possible unnecessary over-acquisition of resources. This may lead to unnecessary data communications and increased local storage requirements.
- Slower system start-up if the start-up phase involves resource acquisition.

4.7.2.4 Lazy Acquisition

In order to optimize the use of system resources the pattern suggests deferring the resource acquisition until the latest possible time.

The solution consists in acquiring the resources only when it becomes unavoidable, i.e., at the latest possible time. The pattern is illustrated by Figure 4.9. Upon initial request for a resource, a Resource Proxy is created and returned to the user. The Resource Proxy is responsible for intercepting all the resource requests issued by the User. The Resource Proxy does not acquire resources unless they are explicitly accessed by the User.

The benefits of this pattern include:

- The pattern avoids unnecessary acquisition of resources. This reduces memory requirements and alleviates possible scalability issues at the resource.
- The pattern helps in optimizing the system start-up by avoiding unnecessary resource acquisition operations.
- The pattern is transparent to the resource user.

The pattern's liabilities include:

- The pattern introduces some memory overhead for the proxy.
- A significant time delay is introduced when a resource is acquired due to deferred acquisition and due to additional level of indirection. The delay may be unpredictable.

4.7.2.5 Synchronization

In order to be able to manage multiple data items across multiple devices, this pattern advises implementing a device specific synchronization (sync) engine. The engine is for tracking modifications to data items, exchanging this information, and then updating the data accordingly when the connection is available. The engine is also responsible for detecting and resolving possible conflicts that may occur during the synchronization process. Figure 4.10 illustrates the synchronization engine.

The pattern suggests implementing the so-called Sync Engines on the devices where the data need to be synchronized. Each Sync Engine is responsible for:

- Registering the changes applied locally to the data.
- Exchanging the information about registered changes with the other Sync Engines when there is connectivity.

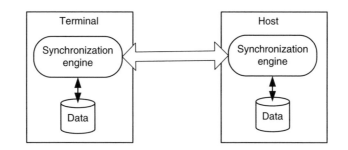

Figure 4.10   The synchronization pattern

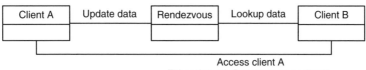

Figure 4.11   The rendezvous pattern

- Implementing (and merging) necessary updates to the data according to the exchanged information.
- If conflicts are encountered during the update process the engine is responsible for resolving them according to a predefined *conflict resolution policy*.

This pattern enables the synchronization of data stored on devices that have intermittent and infrequent connections with each other. The synchronization interaction can be based on client-server or peer-to-peer interactions. Synchronization can also be combined with the eager and lazy acquisition patterns.

A number of issues need to be taken into account when the pattern is applied:

- Users should be aware that data may not be in a synchronized state and some data may be obsolete or replaced. The users should also be aware that conflicts may arise when synchronizing the data. Application specific conflict resolvers are typically needed which, with the help of the users, resolve any conflicts.
- The synchronized data should be provided on-demand as atomic records, typically structured data. The pattern is not applicable to streaming data, for example streaming audio and video.

The synchronization pattern is employed in the SyncML mobile data synchronization framework.

4.7.2.6   Rendezvous

We propose rendezvous as a central pattern in assisting a network to cope with mobile devices. Rendezvous is a process that allows two or more entities to coordinate their activities. In a distributed system, rendezvous is typically implemented using a *rendezvous point*, which is a logically centralized entity, an indirection point, on the network that accepts messages and packets and maintains state so that it can say where a particular mobile device is located. The key elements of this pattern are presented in Figure 4.11.

A classical example of rendezvous is the DNS. The home agent of the Mobile IP specifications is an example of a fixed rendezvous point that is a point of indirection on the network. Many NAT traversal solutions involve rendezvous, for example STUN and ICE from IETF. Rendezvous is also a central function in the Host Identity Protocol architecture that aims to support mobile and multi-homing devices. Many IP multicast protocols are based on rendezvous points, such as the *Protocol Independent Multicast Sparse Mode (PIM-SM)*.

The pattern offers the following benefits:

- allows multiple mobile devices to synchronize through the rendezvous point;
- supports reachability in the cases when the mobile endpoints are not publicly reachable.

The pattern's liabilities include:

- requires a logically centralized indirection point on the network that is publicly addressable;
- rendezvous typically introduces overhead due to packets being routed via the rendezvous point.

4.7.2.7   State Transfer (handoff)

Different kinds of handoffs or handovers have been specified and implemented in the mobile computing context. Handoffs involve state transfer between access points. Handoffs are central in enabling seamless connectivity in any wireless communications system. We briefly outline an abstract state transfer protocol. Similar designs are used in mobile 2G and 3G standards to facilitate roaming between access points and access networks. In addition, the pattern is also applied in higher level mobility protocols such as Mobile IP, SIP, Wireless CORBA, and many other systems.

The state transfer pattern consists of four central entities, namely a client, two *access points (AP)*, and a rendezvous (indirection) point. The client relocates, or roams, from one access server to another, and state needs to be transferred from the old access server to new one. In order for any state transfer to happen, signaling is needed between the access servers.

Now, the state transfer is possible without any indirection point, but this does not ensure reachability of the client. Therefore, an indirection point is typically used which keeps track of the mobile nod'es current location. The indirection point can be fixed or non-fixed. In certain cases there may be many indirection points.

Figure 4.12 presents an overview of the pattern. Before the state transfer is started, the client has established connectivity with the old AP. Then the client attaches to a new AP. This new AP can then perform a location update on behalf of the client. At this point any nodes wishing to communicate with the mobile node can look up the address of the new AP. The new AP can also contact the old AP and transfer any state pertaining to the mobile node's connection. After this the mobile node can tear down the connectivity with the old AP. To support any ongoing conversations, the state transfer pattern can involve update messages from the mobile client to correspondent nodes that contain information on the active and valid addresses for the client. We note that the state transfer can be implemented in several different ways. Typically, it is either started by a mobile client

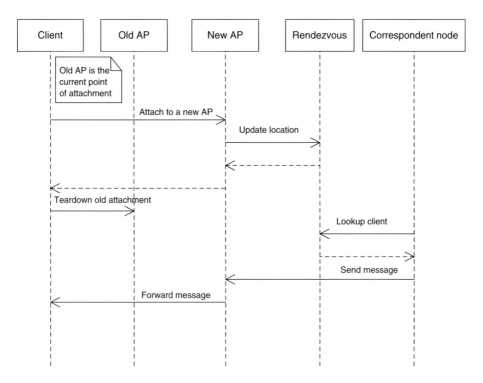

Figure 4.12   The state transfer pattern

when detecting a new AP, or the network when the mobile client has roamed to a location assigned to the new AP.

The pattern offers the following benefits:

- allows mobility between different regions that are assigned to different access servers;
- allows balance clients to be loaded from one access server to another.

The pattern's liabilities include:

- requires that state is transferred from one access server to another, and that the rendezvous point is updated.

The state transfer protocol is used in many systems, namely Mobile IP in the networking layer, and Wireless CORBA in the application layer. There are also many other examples including the Host Identity Protocol architecture and SIP home registrar.

4.7.3   Communications

4.7.3.1   Connection Factory

This pattern suggests the decoupling of the application and the underlying data communications system by introducing a component that is used to create, access, and terminate connections. The factory design pattern is utilized by the connection factory pattern in order to allow the management and reuse of connections in an efficient manner.

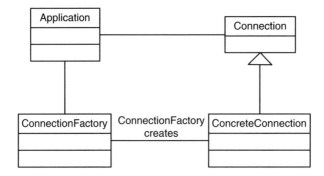

Figure 4.13   The connection factory pattern

In a typical implementation outlined in Figure 4.13, the pattern is based on a Connec-tionFactory object. An application wishing to create a connection obtains a handle to this object, and using its interfaces creates concrete connections, ConcreteConnection objects. The concrete connections follow an interface, Connection, which allows the application to use them irrespective of the nature and implementation of connection.

The connection factory pattern is used heavily in the Java architecture. The communi-cations API of the Java ME features this pattern.

4.7.3.2   Client-initiated Connection

In many cases it is impossible to reach a mobile client due to firewalls and NAT devices present on the communication path. For example, in a typical mobile access network there is a NAT device and firewall at the edge of the access network and the Internet. This means that nodes outside the access network cannot send packets to the mobile device unless the device initiates the communications.

These problems in connectivity motivate the use of a client-initiated connection to a publicly addressable server that can then push messages to the client using the connection.

Figure 4.14 presents the client-initiated connection pattern. In the figure, a compo-nent called edge proxy handles incoming client-initiated connections, performs updates on behalf of the client to lookup services, and associates incoming messages to the client-initiated connection. The edge proxy can be integrated in a single server; however, separating edge proxies from services offers better scalability.

The pattern offers the following benefits:

- allows push communications to the client irrespective of the access network and net-working conditions.

The pattern's liabilities include:

- The mobile network may have very short timeouts and they are network provider specific. This means that the client needs to probe the network to understand the timeout values.
- Requires that the client initiates a long-standing connection with a server. This involves state management at the client.

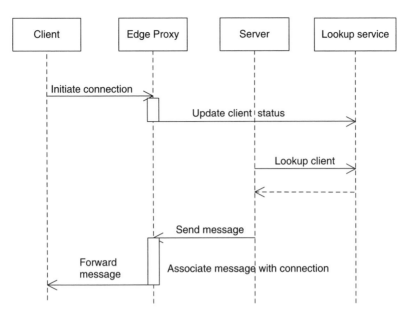

Figure 4.14 The client-initiated connection pattern

- Requires that a server can handle many stateful connections from clients. This can be a possible scalability bottleneck. This can be alleviated with flow tokens.

This pattern is widely used to push data to mobile devices. Example systems include push email, such as Microsoft's DirectPush, and the SIP Outbound specification being developed by IETF [21]. This pattern is also commonly used in AJAX applications that require push functionality from a server [22]. The AJAX push feature that uses long-standing connection is often called *Comet*.

4.7.3.3 Multiplexed Connection

It is not efficient to create many connections that may compete for system and network resources. The Multiplexed Connection pattern utilizes a single logical connection and multiplexes several higher-level connections onto it. This allows the choice of using arbitrary prioritization for messages multiplexed over the connection. Figure 4.15 gives an overview of this pattern.

The pattern offers the following benefits:

- Allows any number of higher level connections to be multiplexed over a single connection and to apply arbitrary prioritization to the flows.
- Involves only a single connection setup and teardown, and does not require separate congestion control for the higher-level connections.

The pattern's liabilities include:

- Requires additional multiplexer component that must be supported by both sender and receiver.

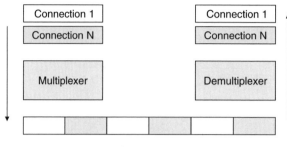

Figure 4.15   The multiplexed connection pattern

This pattern is widely used by different protocols. For example *Stream Control Transfer Protocol (SCTP)* supports multiple connections with priorities, *Blocks Extensible Exchange Protocol (BEEP)* from IETF supports multiple higher level connections multiplexed over a single TCP socket, *Real-time Messaging Protocol (RTMP)* from Adobe uses a single TCP connection for multiple media streams and their control channel.

4.8   Summary

This chapter has presented a number of principles and patterns for distributed computing with special emphasis on mobile computing. Principles guide the development of architectures. Architectural patterns complement principles in the development of the architectural specifications. Similarly, design patterns complement principles in protocol design.

We investigated mobile principles using two examples, namely NoTA for the device view and SIP for the general telecommunications view. As a common theme, decoupling of components, system interconnect with separation of control and content, and mediation can be seen to be useful principles for mobile software.

A platform is a concrete realization of a middleware architecture. Similarly, a protocol stack is a concrete realization of a set of protocols and an architectural framework of how to use them in combination, typically using a layered stack pattern, although other kinds of organizations are also possible. We considered different protocol stack organizations and the possibilities of cross-layer design.

Patterns that are applicable in a particular domain can be collected together. This kind of pattern collection is often called pattern language. We presented a number of well-known architecture patterns, some of which are especially applicable to the mobile environment, and then presented in more detail a set of mobile specific patterns.

The patterns were divided into three categories, namely distribution patterns and resource management, synchronization, and communications. Distribution patterns pertain to how resources are distributed and accessed in the environment. This category includes patterns such as remote facade, data transfer object, remote proxy, and observer.

Resource management and synchronization category involves various patterns related to mobile data management and distributed data synchronization. These patterns include session token, caching, eager acquisition, lazy acquisition, synchronization, rendezvous, and state transfer. Rendezvous is a general signaling pattern for ensuring reachability of peers when the network topology is changing due to mobility or multi-homing.

The communications category includes the connection factory and client-initiated communications patterns. The former is used by applications to manage various connections. The latter is used to create longstanding connections that can be used to multiplex messages to and from a mobile device.

Bibliography

[1] Fowler M, Rice D and Foemmel M 2002 *Patterns of Enterprise Application Architecture*. Addison-Wesley Professional.

[2] Gamma E, Helm R, Johnson RE and Vlissides JM 1993 Design patterns: Abstraction and reuse of object-oriented design *ECOOP '93: Proceedings of the 7th European Conference on Object-Oriented Programming*, pp. 406–431. Springer-Verlag, London, UK.

[3] Gamma E, Vlissides J, Johnson R and Helm R 1998 *Design Patterns CD: Elements of Reusable Object-Oriented Software, (CD-ROM)*. Addison-Wesley Longman Publishing Co., Inc., Boston, MA, USA.

[4] Schmidt D, Stal M, Rohnert H and Buschmann F 2000 *Pattern-Oriented Software Architecture* vol. 2: Patterns for Concurrent and Networked Objects. John-Wiley & Sons.

[5] Obj 2001a *CORBA Event Service Specification v.1.1*.

[6] Clark DD 1988 The design philosophy of the DARPA internet protocols *SIGCOMM*, pp. 106–114 ACM, Stanford, CA.

[7] Pautasso C, Zimmermann O and Leymann F 2008 Restful web services vs. "big" web services: making the right architectural decision *WWW*, pp. 805–814.

[8] Anthony D, Henderson T and Kotz D 2007 Privacy in location-aware computing environments. *IEEE Pervasive Computing* **6**(4), 64–72.

[9] W3C 2002 *Platform for Privacy Protections (P3P)*. W3C Recommendation.

[10] Srivastava V and Motani M 2005 Cross-layer design: a survey and the road ahead. *Communications Magazine, IEEE* **43**(12), 112–119.

[11] Kawadia V and Kumar PR 2005 A cautionary perspective on cross layer design. *IEEE Wireless Communications* **12**, 3–11.

[12] Yu H, Estrin D and Govindan R 1999 A hierarchical proxy architecture for internet-scale event services *Proceedings of 8th International Workshop on Enabling Technologies: Infrastructure for Collaborative Enterprises (WETICE '99)*, pp. 78–83, Palo Alto, CA, USA.

[13] Obj 2001b CORBA Notification Service Specification v.1.0.

[14] Obj 2001c *Management of Event Domains Specification*. http://www.omg.org/cgi-bin/doc?formal/2001-06-03.

[15] Uni 2008 *Patterns for Mobile Application Development*. http://www.titu.jyu.fi/modpa/Patterns/PatternList.html.

[16] Waldo J 2000 *The Jini Specifications*. Addison-Wesley Longman Publishing Co., Inc., Boston, MA, USA.

[17] Bacon J *et al.* 2000 Generic support for distributed applications. *IEEE Computer* **33**(3), 68–76.

[18] Hayton R, Bacon J, Bates J and Moody K 1996 Using events to build large scale distributed applications *Proceedings of the 7th ACM SIGOPS European Workshop on Systems support for worldwide applications*.

[19] Satyanarayanan M, Kistler JJ, Kumar P, Okasaki ME, Siegel EH and Steere DC 1990 Coda: a highly available file system for a distributed workstation environment. *Transactions on Computers* **39**(4), 447–459.

[20] Garlan D, Siewiorek D, Smailagic A and Steenkiste P 2002 Project Aura: Toward Distraction-Free Pervasive Computing. *IEEE Pervasive Computing* **01**(2), 22–31.

[21] Jennings C and Mahy R 2006 Managing Client Initiated Connections in the Session Initiation Protocol (SIP). Internet-Draft draft-ietf-sip-outbound-06, IETF. Work in progress.

[22] Mahemoff M 2006 *Ajax Design Patterns*. O'Reilly Media, Inc.

[23] Gupta S, Hartkopf J and Ramaswamy S 1998 Event notifier, a pattern for event notification. *Java Report* **3**(7), 19–36.

5

Interoperability and Standards

The cornerstone of any communication is *interoperability*, an agreement between the communicating parties of how information is encoded into a common format and how it is transmitted from one party to another. Examples of interoperability appear constantly even in our daily lives. For instance, when shopping, there is the expectation that the shop employees understand the dominant language of the region (how information is encoded) spoken aloud (how it is transmitted).

Especially in networking, the *layered architecture* is commonly used to separate concerns between areas that have little to do with each other. It would not make much sense to require every application programmer to understand the physics behind the actual data transmission. Thus, a designer focusing on one layer may assume that the lower layers are interoperable. The layer immediately below is expected to provide an interface with some well-specified semantics that the designer uses to form interoperable communication on their layer. Interoperability therefore reaches also into the programming interfaces of communicating components and does not remain solely in the realm of communication protocols.

Interoperability is achievable in a number of ways. The simplest of these is to have a single vendor design all of the communicating components of the system, ensuring that they all understand each other. The problem with this is vendor lock-in: all components must be purchased from a single vendor and switching vendors is practically impossible. That is why most purchasers require vendors to follow *open standards*, more or less formal documents specifying communication in sufficient detail and implementable by anyone. Due to the characteristics of large-scale distributed systems, mostly because communication is often required between independent actors, there are a large number of standards organizations and standards produced by them in this field.

5.1 Interoperability

As the example above regarding shopping showed, interoperability is a very broad concept, applicable to any situation with more than one actor. This generality is visible even in the definitions that various organizations have given for interoperability in the context of computer systems. The *Institute of Electrical and Electronics Engineers (IEEE)* defines it

Mobile Middleware   Sasu Tarkoma
© 2009 John Wiley & Sons, Ltd

as 'the ability of two or more systems or components to exchange information and to use the information that has been exchanged'. [1] On the other hand, ISO is somewhat more specific in its definition: 'The capability to communicate, execute programs, or transfer data among various functional units in a manner that requires the user to have little or no knowledge of the unique characteristics of those units.' [2]

Even in computer systems, interoperability can refer to different types of concepts. The routing information in a network-layer protocol is different from an agreement to use a specific form for the data inside a communicated message, and both differ from an application saving its data in a file formatted in a specific manner. In principle, all of this reduces to the IEEE definition, but in practice, these get treated differently. After all, a program opening a data file it saved earlier can be seen as the program communicating with itself at a later instant, but this is not considered a natural way to view the situation.

Interoperability is usually desirable from the system user's point of view, as it allows the user greater flexibility in selecting applications and increases the potential size of a distributed system, which can lead to beneficial network effects. However, it is hard to achieve this fully in practice between independently-implemented components, but often the problems appear only in some corner cases. Written specifications are rarely fully comprehensive and if the field has one sufficiently strong player, that player's extensions can serve as undocumented *de facto* requirements. Prominent examples of these are TCP, which in practice needs to be implemented based on the Berkeley UNIX, and HTML, which in practice is largely defined by the behavior of old browsers from Netscape and Microsoft.

From a component vendor's point of view, interoperability is not always desirable, as it permits customers to acquire parts of the system from other vendors. If the vendor is only supplying parts of a complete system or if it is a small vendor trying to break into an established market, making the products interoperable with the existing systems is clearly a benefit. But for an established player with a large market share, interoperability only allows other vendors to enter the field and diminish the share of the existing vendors. Interoperability is therefore often enforced by procurement standards that require acquired systems to conform to established standards. Through such policies, large organizations, especially governments, can endeavor to level the playing field and hinder monopoly formation.

The source and methods of interoperability can vary depending on the field's importance and the ecosystem of vendors. Various situations, in rough order of required collaboration, are:

Single product When a field is very specialized or of low importance, there is less pressure to break a single vendor's monopoly position. In such cases, interoperability is achieved through the vendor's internal processes that ensure the product understands its own protocol or file format and usually is also backwards compatible with previous versions of the product. Often, backwards compatibility is achieved through file or protocol versioning, so that an incompatible change can be applied by simply increasing the version number of the format.

Reference implementation Sometimes, a technical specification is accompanied by a *reference implementation* that can settle ambiguities in the written text. A similar case is when there exists an implementation of a technology with a license that permits its inclusion into products. In such cases, embedding the available implementation into different products leads to interoperability in that technology similar to the above single-product case. An additional benefit may be that the vendor has no need to acquire developer competency in that specific technology area. An example of this is the zlib compression library. Many communication protocols benefit from applying a generic compression algorithm and the zlib license permits its inclusion into any product, and zlib is indeed widely used.

Industry consortium In a field with multiple competing players, it is not always feasible for everyone to approach matters with a walled-garden tactic, ignoring others completely. An agreement for interoperability among the relevant players can be beneficial, in part because such an agreement can raise the barrier of entry of new players if an interoperable product is sufficiently complex. Nowadays, closed consortia are rarer, and it is more common for the players to do things in the open, making this case resemble more the following formal standardization.

Formal standard In a high-profile field, there is usually external pressure to formalize interoperability and also to avoid letting special interests have total control. Such cases are the purview of various standards organizations. These organizations do not have a stake of their own in the field, and typically do not themselves develop systems at all. The purpose of a standards organization is to issue precise specifications that allow someone to implement a product that is interoperable with existing deployed systems without the need to study these existing systems. If a standard is freely implementable by anyone, it is called *open*. Closed standards that may require licensing patents or something of the kind exist as well, for instance, state-of-the-art video standards require patent licensing for their implementation.

Usually, this order is also the situations' order of desirability. A system available only from a single vendor creates a lock-in issue where users are dependent on the vendor. While the reference implementation scenario is slightly better, the main problem with that is that, apart from a written specification of behavior, any independent implementations are in practice required to be bug-for-bug compatible instead of just following the feature set. A closed industry consortium is better for the participants, and if there is true competition in the marketplace, it can be good for users as well, but the ideal situation is still an open standard implementable by anyone. Of course, in many cases, the formal standard can be so complex that significant resources need to be expended in its implementation, creating a barrier to entry to completely new players in the market.

The cases of industry consortium and formal standard rely on written specifications to achieve interoperability. In general, such specifications are written in a human language like English, and while precision is the goal, ambiguities are often unavoidable. Any ambiguities in the specification may then lead to incompatibilities between independent implementations, leaving the goal of interoperability unattained.

Implementation conformance to specifications is typically verified through either of two methods. One is simply to run extensive interoperability tests between independent implementations and make sure that both implementations work correctly. This method is often somewhat *ad hoc*, and is not likely to probe every corner case. That is why many standards come with a conformance test suite, a set of cases intended to exercise every required feature. While such suites are usually more comprehensive than simpler testing approaches, and can cover all of the required features, it is not feasible to cover every combination of features. This may cause problems in very obscure cases.

5.2   Standardization

Written standards are mostly developed under the umbrella of some standards organization, a body that is chartered only to produce various technical specifications. The precise procedures that these organizations use vary, but on the general level the processes are very similar. Membership in such an organization is typically conferred on institutions such as companies and universities, but some accept individuals as members too. The more formal international organizations, such as ISO and ITU, are composed of national standards organizations or national governments.

5.2.1   Standardization Process in General

While there are differences in how standards organizations conduct their work, some general principles can be extracted from the overall process of standardization. Namely, the intent of establishing a standard is to foster agreement among the actors in a specific field. Therefore, the usual method is to gather the interested parties together and have them negotiate the specifications so that they solve the necessary problems while not leaving any participant dissatisfied with the final standard.

The process of hammering a standard together can easily take years. During this time, the people involved in the process, either as individuals or representatives of their organizations, will discuss the subject matter of the standard, determine the issues to be solved, and try to come to a consensus as to the best ways to solve those issues. Depending on the organization, these discussions can be conducted in person, over the phone, by email, or by other methods. It could be said that more 'traditional' organizations emphasize in-person meetings whereas 'modern' organizations, established due to the perceived slowness of the traditional ones, prefer email or other more immediate forms of communication.

A standards process is not intended as a place for innovation. The underlying assumption most often is that there exist multiple incompatible solutions to a specific problem, and that there is value in interoperability between these solutions. Trying to solve these issues in standardization typically leads to an overlong process with a final product 'designed by committee', that is, a complex specification that has something for everyone but is difficult to put into practice, especially for anyone who did not participate in the process.

Standardization is always both a political and a technical process. While the participants argue based on technical merit, each participant comes with a different perspective to the field, emphasizing different factors in the requirements. This variety of perspectives is both a strength and a weakness in that it permits the eventual standard to cover a wide area but may also make it very complex to take into account the full field instead of just each participant's individual area.

It is rarely the case that sole technical merit determines the eventual form of the standard. Nothing will get written into the standard unless one of the participants in the work writes it, so an idea, no matter how excellent technically, may be completely overlooked if it lacks an active proponent in the group. Examples of such 'better' ideas are not necessarily easy to find, but examples of cases where ideas have not influenced the standard due to lack of proponents abound, for example, the ad-hoc network routing protocol at the IETF where a multitude of solutions existed, but the final standard ended up being a conglomeration of the two best represented ones in the working group.[1]

Participation in a standards process is typically both time-consuming and expensive, especially if one is to have any influence on the process. The time is spent at first just establishing oneself as someone the others should listen to and afterwards in working on the standard itself. As noted above, considerations are rarely taken into account unless some participant is willing to argue for them. Depending on the organization, money needs to be spent on travelling to meetings, phone conferences, and often on participation fees. Many industry-based organizations finance their activities by membership fees, whereas others, such as ISO, acquire funding through selling the standards documents. In computing, standards are often seen to be so widely applicable that the ISO model with its generally high prices is undesirable.

5.2.2 Standards Organizations

The International Organization for Standardization (ISO, the abbreviation is by design not an acronym) is one of the more famous bodies setting formal standards. ISO standards exist for all kinds of areas of human activity, including computing. While ISO standards for areas like programming languages (e.g., C and C++) have been successful, there is a common perception that the ISO process is too slow for modern network technology that needs to move in 'Internet time'. Accordingly, there cannot be said to be any ISO standards that are relevant in the field of mobile middleware. In this field, ISO is best known for the OSI model, a 7-layer stack for modeling communication protocols. Nowadays, the OSI model is widely considered to have been made obsolete by the Internet's TCP/IP stack.

Somewhat more visible is ITU-T, the Telecommunication Standardization Sector of ITU, the International Telecommunication Union. ITU is an agency of the United Nations and has member only countries. ITU-T is the producer of standards for ASN.1 [3], an abstraction for defining structured data and a number of encoding rules for how to represent that structured data as bits. ASN.1 is widely used for data representation in the telecommunications sector. Another important standard is X.509 for public-key infrastructures (PKIs) that uses ASN.1 to define the structure of keys and certificates.

The IEEE is a professional organization for engineers, based in the United States but with a worldwide membership. The IEEE also includes standards development activities, and is most famous for its series of 802 standards dealing with local area networks. Of particular interest in this context are 802.11 (Wireless LAN), 802.15 (personal area networks like Bluetooth), and 802.16 (WiMAX).

Modern mobile phone networks are largely under the purview of two organizations, 3rd Generation Partnership Project (3GPP) and 3GPP2. Despite the similarity in names, these two organizations are completely separate, the former basing its standards on the UMTS technology with roots in GSM, and the latter on CDMA technology. As CDMA technology was never used widely in Europe on the 2nd generation cellular phone networks, 3GPP2

---

[1] This is not to disparage the final standard, which is a good one. It is intended just to illustrate how ideas find their way into the standards.

does not have European participation, but otherwise there is little difference in the two organizations' memberships.

The two primary standardization organizations for the Internet and related technologies are the Internet Engineering Task Force (IETF) and the World Wide Web Consortium (W3C). As the names indicate, the W3C is focused on the WWW and related technologies such as XML while the IETF handles practically all the other Internet protocols from the network layer upwards.

Still, as with the older more established organizations, there are also some grumbles that even the IETF and W3C are moving too slowly. Undoubtedly a part of this dissatisfaction comes from misunderstanding the purpose of a standards organization as a codifier of existing best practices and not as a trend-setter. However, it is true that many of the protocols and formats currently in use do not correspond exactly to officially-blessed IETF and W3C specifications, but mostly to pre-final versions. The primary reason is that the specifications are seen as sufficiently stable already even without the final steps of the process having been carried out.

Other important forums in the mobile and wireless communication space are Wireless World Research Forum (WWRF) and Open Mobile Alliance (OMA). WWRF is not a standardization organization as such, but more of an industry consortium focused on providing a vision of the future wireless world [4] and contributing to standardization at various organizations to help along this vision. OMA was formed as a reaction to a growing number of forums that each focused on some specific protocol. To unify the standardization process and to avoid work duplication, the OMA was formed as an umbrella organization above all these previous organizations.

In distributed computing in the late 1980s, the idea of distributed objects, a development of RPC, began to take hold. CORBA [5], a distributed object platform, was standardized by the Object Management Group (OMG). CORBA enjoyed a surge of popularity in the 1990s, but its increasing complexity relegated it out of the public spotlight and into specialized application areas. Due to the peculiarities of the mobile world, CORBA was not seen as suitable for that field, and the importance of the OMG is minor in the mobile computing industry.

Nowadays, the Organization for the Advancement of Structured Information Standards (OASIS) is perhaps more relevant than the OMG. The original focus of OASIS was on SGML-related activities, promoting the use of SGML for various document-producing needs. With the rise of XML, the focus of OASIS shifted to XML. While OASIS deals mostly with structured documents and not networking as such, its ebXML messaging platform, the UDDI language for web services, and a host of different web service standards give it a prominent role in the distributed computing field as well.

5.3 Wireless Communications Standards

Currently, there can be seen to be three different forms of wireless communication. First, there are the Personal Area Networks (PANs) that provide connectivity among a group of devices mostly in the case where one person wishes to connect all of their devices together. Second, there is Wireless LAN (WiFi) that provides network connectivity in a single building or corresponding area, with some support for terminal mobility. And third, there are the cellular phone networks that provide wide-area connectivity and allow a high rate of mobility. A fourth emerging type of wireless network is the Metropolitan Area Network (MAN) that is intended somewhere between WiFi and cellular networks, with more allowed mobility than WiFi but still higher bandwidth and lower latency than the cellular networks.

5.3.1   Cellular Networks

Cellular network technologies are usually divided into generations. The first generation consisted of analog technologies and has now been torn down in many areas across the world. The second generation were the first digital technologies, and as with the first generation, different parts of the world adopted different technologies. In Europe, where the first generation had been fragmented across countries, the second generation brought out a pan-European standard, GSM. Meanwhile, in the United States, a single analog technology fragmented into multiple digital technologies.

Current cellular phone networks are called third generation. Unlike the second generation circuit-switched networks, third generation networks have moved towards packet switching, a more scalable approach that is typically used for data communication but traditionally not for telephone calls. On the 3GPP side, the basic third generation technology is UMTS (Universal Mobile Telecommunications System) [6] while on the 3GPP2 side, it is CDMA2000.

The medium access of first and second generation technologies, apart from the latest ones, was typically based on either FDMA (frequency division multiple access) or TDMA (time division multiple access). In FDMA, each device in the wireless network is assigned its own frequencies on which it communicates, so the devices are distinguished by the frequencies on which they transmit. In contrast, in TDMA, the division is based on time slots, where time is divided into interval and on each interval each device has a specific time slot during which it can transmit.

Around the second generation of cellular networks, CDMA (code division multiple access) technology began to become practicable. In CDMA, each device transmits its signal using its specific coding method, which permits devices to transmit at the same time and on the same frequencies while still permitting the signals of different devices from each other to be distinguished. In general, CDMA utilizes the available bandwidth better than either FDMA or TDMA in common situations, but implementing it is more complex, which is the primary reason why it was not generally adopted for the second generation. Of second generation networks, IS-95 uses CDMA, and both UMTS and CDMA2000 of the third generation networks are based on CDMA.

Although referred to as generations, the technologies inside each generation are not as static as the terminology would imply. Already in the second generation GSM, there have been technologies called 2.5th generation and 2.75th generation. Such designations are intended to convey that they are a significant step forward in characteristics like bandwidth or latency but still compatible with the baseline technology.

The 2.5th generation technology on top of GSM is called GPRS [7], for General Packet Radio Service. GSM itself is fully circuit-switched, and the pricing model of such a technology is not well suited for accessing the packet-switched Internet, so GPRS was designed to offer packet-switched connectivity without requiring massive changes in the infrastructure. A further development, EDGE, for Enhanced Data rates for GSM Evolution, as its name says, enhances the data rate of GPRS, and is sometimes referred to as 2.75th generation because of its position in the GSM family, even though it has been formally placed in the third generation.

Similarly, UMTS has spawned a 3.5th generation, a technology called HSPA, for High Speed Packet Access. Changes in packet scheduling and coding increase the theoretical data rate of the network to 14.4 Mbps downlink and 5.76 Mbps uplink, which makes HSPA the fastest cellular network access technology currently available in the market. It must be noted, though, that these are theoretical maximums only attainable in perfect conditions if then. Measurements over real cellular networks typically achieve at most half of the theoretical maximum, and the less established the technology, the larger the gap between theory and practice usually is.

The next step in cellular networks is, unsurprisingly, referred to as the fourth generation or *Beyond 3G(B3G)*. As this is upcoming technology, there is not yet a precise technology that can be pointed at as fourth generation, but one expected characteristic is a move toward IP-based networking. While current third generation systems already support packet switching and can run IP, the integration between IP and the cellular network is expected to become much tighter than currently.

The clearest example of fourth generation technology development is 3GPP's Long Term Evolution (LTE). While not intended to be a fourth generation technology itself, LTE is intended as a bridging technology between current third generation systems and future fourth generation systems. LTE will be released as a part of the 3GPP release 8 at the end of 2008, while its evolution to a true fourth generation technology is expected by 3GPP release 10, estimated to occur in 2010. Downlink and uplink data rates are aimed at 100 Mbps and 50 Mbps, respectively, both of which have been demonstrated to be possible. Throughput and latency are also expected to improve several-fold.

5.3.2 Local Area Networks

A local area network (LAN) is formed by a group of computers all sharing the same physical network. A LAN is typically at the edges of the Internet, with hosts connected to a LAN that also has a single router with an uplink to the Internet. With the modern office requiring network access in meetings, even in ad-hoc meetings in the corridor, a Wireless LAN (WLAN) is required in addition to the normal fixed LAN to allow people to connect anywhere in the building without requiring cables or network outlets.

In standardization, most LAN standards come from the IEEE 802 family of standards, the most famous of which are the groups 802.3 (Ethernet) and 802.11 (Wireless LAN) [8]. In total, there are 22 working groups under the 802 family, many of which have already been disbanded or simply become inactive, such as 802.4 for Token bus LANs.

Nowadays, Wireless LAN is typically called WiFi. The primary standards for WiFi are the IEEE 802.11 set, with the a, b, and g versions being commonly available, allowing data rates of 11 Mbps or 54 Mbps, and the n version still being prepared. Products complying with 802.11n drafts are already available, allowing data rates of up to 300 Mbps, but as the standard has not yet been finalized, there is no guarantee of interoperability between devices from different manufacturers.

For Internet access, WiFi is not a competing technology with cellular networks, but rather a complementary technology. Namely, its effective range in normal use is short enough that widespread mobility with WiFi, especially at the speeds supported by cellular networks, is not really feasible. Instead, the preferred approach, at least in the future, is to use the cellular network for outdoors mobility, switching to WiFi for faster access when reasonably stationary, say, in a building.

In addition to the infrastructure mode typically used for Internet access, WiFi also supports an ad-hoc mode where there is no central base station or server to provide configurations but the devices configure themselves statically. As long as all the devices have been configured to use the same WiFi channel and the same network prefix in their addresses, they can communicate among each other. Such ad-hoc interactions are expected to become a large part of the social applications of the future: when two people meet, their devices could discover each other somehow and allow them to share data in addition to traditional communication.

In between cellular networks and WiFi lie technologies such as WiMAX, for Worldwide Interoperability for Microwave Access. WiMAX is sometimes lumped in with technologies in preparation of fourth generation cellular networks, though it cannot quite manage the cell size that cellular networks can, making it less suitable for rural areas. Indeed, an alternative name for WiMAX is Wireless MAN, for Metropolitan Area Network.

5.3.3 Personal Area Networks

A common problem in home computing is the proliferation of cables to connect all the various devices to the computer. Furthermore, sometimes the available slots just get filled, like the few USB slots of somewhat older, but still usable, computers could easily run out in the modern day, with all the USB-using devices currently on the market. To alleviate this, one could try for some wireless technology to run the communication between the main computer and its peripherals. Such an arrangement is called a *personal area network*, or PAN.

Bluetooth [9] is a technology originally designed for exactly the above kind of application. It is a wireless communication technology based on *piconets*: small, at most eight-node,[2] networks consisting of one master node and several slave nodes. This organization also belies Bluetooth's original goal, as it was expected that the computer itself would be the master node and its peripherals would be slave nodes in the piconet.

Bluetooth has definitely broken through in this application, though not completely. Peripheral devices such as mice, keyboards, and headphones are available in cordless versions using Bluetooth, but larger peripherals such as printers or scanners are still mostly connected by cords in common use.

As Bluetooth is, however, a general-purpose networking technology, it is not limited to the case of a computer and its peripherals, but it is naturally possible to send arbitrary data over it. Bluetooth's low power requirements make it an ideal choice for short-range communication between devices. Furthermore, as Bluetooth is more widely available than WiFi in smaller devices, its applicability range is correspondingly greater.

An additional feature of Bluetooth that makes it attractive is its Service Discovery Protocol (SDP). A Bluetooth device does not need to know the address of the device it wishes to communicate with, but it can first perform a device discovery to locate nearby devices and then run a service discovery on these discovered devices to locate its desired service. In typical use, these services are general services like file transfer, but a program can use custom services to denote a specific application, easily permitting

---

[2] In fact, Bluetooth allows for more devices on the piconet, but at most eight can be active.

short-range network applications to discover instances of themselves on other devices and communicate with them.

While the features of Bluetooth make it attractive for short-range communication applications, the practical reality is still lagging behind. Experience with real devices suggests that general network applications written over Bluetooth communication need to include several workarounds for problems in different devices that cause disruptions or breaks in the communication. Furthermore, these problems are not consistent even across the same manufacturer's devices, so each different device requires a slightly different set of workarounds.

In theory, it is also possible to connect Bluetooth piconets together to form a *scatternet*, where a device is a part of two piconets.[3] These scatternets would allow Bluetooth to be used in larger networks than the maximum allowed for piconets and at longer ranges than Bluetooth is capable of. But while this is an interesting concept, practicalities of the Bluetooth technology make it infeasible in practice, relegating it to the domain of research alone.

Bluetooth is not the only short-range network technology, merely the most popular and widely-implemented. A promising recent technology development is ultra-wideband, a technology long being used in military applications, but nowadays being considered for commercial use as well. Unfortunately, standardization efforts at the IEEE for an ultra-wideband technology were cancelled in 2006, so it is in question whether widely interoperable systems using this technology will appear.

As the name suggests, ultra-wideband operates on an extremely wide frequency spectrum. This allows it to use a very low power mode, lower than the usual regulated minimum-power transmissions, so ultra-wideband can operate on normally-restricted frequencies, since its transmissions are low-powered enough not to trigger the restrictions.

As the IEEE is the primary body for standardizing physical networking technologies, the IEEE 802.15 group for Wireless PANs is of much interest. Bluetooth falls under the purview of 802.15.1, ultra-wideband was being considered by 802.15.3, and low-rate PANs, suitable for devices like sensor nodes that need a long life, are standardized by 802.15.4.

5.3.4 Comparison

The various wireless communication technologies are not directly comparable as each is designed for different conditions. However, a comparison of the characteristics of each will reveal what kind of performance an application can expect in each of these conditions. Rough figures for bandwidth and latency, measured in real conditions, are given in Table 5.1. The latency measurements are given for a TCP-level round trip and data rates are based on the transfer of a large file.

The perception of cellular phone networks as slow ones with high latency is getting more inaccurate as third generation networks become more common, as the UMTS data rate and latency are approaching tolerable levels. When WiFi access is available, taking advantage of that is still the sensible policy, though. Comparing Bluetooth to anything except the ad-hoc-mode WiFi is somewhat unfair, as Bluetooth is not intended for network access but for short-range non-routed communication.

---

[3] Bluetooth technology dictates that the device be a master in at most one of the piconets it is participating in.

Table 5.1 Characteristic figures for
wireless networks

Network	Data rate (kbps)	Latency (ms)
Bluetooth	35	450
WiFi (infra)	1160	10
WiFi (ad hoc)	1050	340
GPRS	32	700
EDGE	210	300
UMTS	330	200

Energy consumption is a different matter altogether. Measurements indicate that UMTS and WiFi consume approximately the same amount of power, around 1 W, but since WiFi data rate is higher, sometimes much higher, it still consumes clearly less energy in total for data transmission. Bluetooth at its shortest range consumer only 2.5 mW of power, making it the clear winner in energy consumption even when taking into account the longer transmission time. Other PAN technologies, such as the upcoming ultra-wideband, are promising even smaller energy consumption.

5.4 W3C Standards

The web is a prominent application and service platform in the modern Internet. Since mobile devices should be full-fledged members of the Internet, web-related technologies and standards will need to feature in a major way when considering their prominent role in the current field of distributed applications and services.

As a standards organization, the W3C is divided into Activities, each focused on a particular area of web standardization. Activities are further divided into groups, of which working groups produce the actual standards. Currently, the W3C has 23 Activities, but many of these are composed of one group alone. The most significant larger Activities for our purposes are XML and web services, with the smaller Mobile Web, XForms, Synchronized Multimedia, Security, and Graphics being important as well. We shall cover web services and Security in Chapter 6.

In W3C parlance, a standard is called a Recommendation, which is the end result of the Recommendation track process. This process begins with a succession of public Working Drafts, each of which will possibly generate some comments. When the working group decides that the specification is complete, it issues a Last Call, which is the final stage at which comments can affect the future specification, at least in theory.[4]

If the Last Call is successful, the specification is allowed to proceed to the Candidate Recommendation phase. This is where a call for implementations is issued, as the specification is considered feature-complete. The W3C rules require two independent interoperable implementations for each feature in the specification (but not two independent implementations of the complete specification, which may, in theory, lead

---

[4] For instance, the SVG specification has gone through five Last Call periods.

to specifications unimplementable in full). After interoperability has been verified, the specification becomes a Proposed Recommendation. The Proposed Recommendation is reviewed by the W3C's Advisory Committee, a body composed of representatives from all member organizations which oversee the technical work at the W3C.

5.4.1   The Mobile Web Initiative

Users of mobile devices have also become visible to the W3C. The Mobile Web Initiative (MWI) is an Activity established to improve the experience of users browsing the web on their mobile devices. Their purpose is to identify interoperability and usability problems, caused by too desktop-centered a view of the web, and address these issues through developing technologies or recommendations for best practices so that web access from mobile devices would be practical in the future.

There are three working groups under the MWI umbrella: Best Practices, Device Description, and Test Suites. In addition, an interest group for Social Development is also a part of MWI. Interest groups are W3C groups that do not aim to produce Recommendations, but only Notes, which are W3C official documents that are not intended to function as specifications.

The purpose of the Best Practices working group is to establish, well, best practices for developing websites that work for users on mobile devices. They have a Recommendation out on the best practices for web content to be accessed from mobile devices. In addition, the group has produced a brand, mobileOK, that is intended to be a set of executable tests to determine whether a website is conformant with the best practices.

The Device Description working group is based on the idea that web content needs to be adapted according to the device requesting it, and thus, a method for accessing a repository of device descriptions needs to be specified. This view has also drawn criticism, since it implies that special versions of sites would be needed for mobile devices, which is unlikely to happen due to the effort required, but which also serves to fragment the web, as different devices suddenly do not have access to the same content.

The Test Suites working group develops tests for browsers on mobile devices. The primary purpose is to establish which web technologies are sufficiently well supported on mobile devices so that they can be used on the mobile web. This group collaborates with OMA, who have plenty of experience in the wide variety of mobile devices on the market currently and who also have a number of test suites for those devices.

5.4.2   Basic XML Specifications

Perhaps the most important W3C standard is that of Extensible Markup Language (XML) [10]. While HTML (Hypertext Markup Language) also falls within the W3C's purview and it could be argued that HTML is more visible on the web, the practicalities of HTML make it unattractive for most other applications, in particular its rather ad-hoc in-practice semantics caused by years of browser extensions and bugs becoming codified into required behavior. In fact, this undesirable quality was the reason for requiring 'draconian' behavior from XML processors: quitting processing with an error immediately when encountering malformed content.

XML is, like its predecessor SGML, a generic markup language for describing structured documents. The original goal was to strip down SGML sufficiently so that writing

```
<?xml version="1.0" encoding="UTF-8"?>
<person nationality="DE">
  <name>
    <first>Richard</first>
    <last>Wagner</last>
  </name>
  <occupation>Composer</occupation>
  <born>1813-05-22</born>
  <died>1883-02-13</died>
</person>
```

Figure 5.1   An example XML document

processors would be much easier than in the SGML days, permitting XML to be used over the web easily. In this goal, XML has succeeded, as witnessed by the sheer number of applications and systems built on top of XML.

The basics of XML are best understood by example, shown in Figure 5.1, which presents a complete XML document describing a person. The first line is the *XML declaration* that in this case shows the version of XML used by the document and the character encoding used to encode Unicode characters into bytes. The values shown are the defaults, so the XML declaration is actually needless in this document.

Primarily, an XML document is composed of *elements*. Each document contains a single root element, here called person. An element begins with a *start tag*, surrounded by angle brackets <>, and ends with an *end tag*, indicated by a slash / as the first thing inside the angle brackets. An element may also have *attributes* placed inside the start tag, like nationality here. The contents of an element, placed between its start and end tags, consist of text and other elements. XML does not have any data types, so all text is just text, even though an application may place some semantics on the text.

While plain XML is sufficient as a method of representing structured data, standardized interoperable communication requires more than the plain XML specification can provide. For instance, in a distributed application, it is common that some parts of a message are standardized boilerplate, applicable to many different kinds of applications, so such parts would be candidates for independent standardization. Plain XML does not easily permit this, since the element and attribute names are global, so global agreement would be required to make sure the same name is not used for two different concepts.

This problem is solved by XML namespaces. Namespaces partition the space of all XML names into independent subspaces, each identified by a URI. Since URIs, and especially URLs, are typically hierarchical, based on administrative boundaries, each independent entity can create its own URIs and be certain that they don't clash with URIs created by other entities. Thus, URIs are a suitable naming technology for a global scope.

XML namespaces do not break XML in any way: each document with namespaces is also a plain XML document. Figure 5.2 shows how namespaces can be added to XML. Here, the person element of Figure 5.1 has been placed into the namespace http://example.org/people by using the xmlns:p attribute. The colon and p signify that this namespace is to be identified with the prefix p throughout the attribute's element. This is illustrated by all of the elements inside person being prefixed by p and a colon.

```
<?xml version="1.0" encoding="UTF-8"?>
<favorite-composers xmlns:p="http://example.org/people">
  <p:person nationality="DE">
    <p:name>
      ...
    </p:name>
    ...
  </p:person>
  <p:person>
    ...
  </p:person>
</favorite-composers>
```

Figure 5.2 An example XML document with namespaces

One subtle point, visible in Figure 5.2, concerns attributes. Attributes can be placed into namespaces in the same way as elements. But, since an attribute is typically associated with a single containing element, it is rarer to place attributes in namespaces because the containing element already uniquely identifies their purpose. On the other hand, attributes that are intended to be used in elements not defined in a specific namespace, placing the attribute in a namespace makes sense. For instance, there is an attribute `lang` to indicate the language that the text in an element is written in. This attribute is located in the XML namespace, since it can in principle appear anywhere.

The XML specification itself defines XML only at the level of individual characters, but this is hardly sufficient for the needs of all applications. At the very least, some sort of information on higher-level structure is needed to properly map between data structures and their XML representations. The grammar in the specification naturally provides one view through its products, but as is commonly known in compiler implementation, such concrete syntax trees, or parse trees, are not very useful in actual processing. Rather, an abstract syntax tree is needed.

Any API for XML processing provides one way of looking at the information in an XML document. There exist streaming APIs (SAX, StAX) that treat an XML document as a sequence of events, with events existing for things like start tags, text content, comments, and the like. There are also tree-based APIs (DOM) that build an actual tree, usually in computer memory, based on the element relationships in the XML document.

APIs are a rather poor basis for thinking about XML at the abstract level, though. An API is usually designed for a single programming language and therefore reflects that language's quirks.[5] The design of a good API is also influenced by efficiency in using it, without consideration for the abstract properties of the processed language. These points make talking about XML documents in API terms somewhat difficult.

The W3C has defined two data models for XML. The first, older, one is the XML Information Set [11], commonly called Infoset. The Infoset specification is written to

---

[5] The DOM specification is based on Interface Definition Language (IDL), which is language-independent, so this concern does not fully apply to it. Still, IDL is largely influenced by C++, so the properties of C++ are probably reflected in the DOM API.

capture all 'useful' information in an XML document, primarily for the purpose of other specifications based on XML, so that they have a consistent set of terms to which to refer. The Infoset does not capture everything about an XML document, for instance, the amount of whitespace used within markup or whether quotes or apostrophes are used for attribute values are not available in an Infoset.

The other data model for XML from the W3C is the XPath Data Model, produced on the side of the XQuery specification process, and was primarily designed for the needs of XML querying and transformations. It is based on the Infoset and is very close to it, but it has two significant additions. First, there is support for data types, required by the database community that was the driving force behind XQuery. Second, there is support for sequences and collections, whereas the Infoset concerns itself only with complete XML documents.

5.4.3 XML Schema Languages

In applications, the XML documents that are produced and consumed usually follow some set patterns in their element and attribute structure. For instance, a purchase order will contain information on the purchaser and the list of items to purchase, and these data always have the same structure even if the specific content changes from one order to the next.

To describe the element and attribute structure of an XML document, there exist various schema languages. The original schema language for XML, Document Type Definition (DTD), was inherited from SGML, but as with XML itself, it also got simplified along the way. Figure 5.3 shows a possible DTD for the XML document of Figure 5.1. The DOCTYPE defines the root element and ELEMENT and ATTLIST declarations define the content and attributes of an element, respectively.

DTD is a perfectly fine language as a schema language, but it suffers from a few problems:

- There is no support for namespaces.
- An element with the same name must always have the same content model, regardless of its position in the document.

```
<!DOCTYPE person [
  <!ELEMENT person (name,occupation?,born,died?)>
  <!ATTLIST person nationality CDATA #IMPLIED>
  <!ELEMENT name (first,middle?,last)>
  <!ELEMENT first (#PCDATA)>
  <!ELEMENT middle (#PCDATA)>
  <!ELEMENT last (#PCDATA)>
  <!ELEMENT occupation (#PCDATA)>
  <!ELEMENT born (#PCDATA)>
  <!ELEMENT died (#PCDATA)>
]>
```

Figure 5.3   An example DTD for the example XML document

- The syntax is not XML, necessitating a different parser.
- DTDs are tightly coupled with XML, being placed inside the document either in full or by reference.

The one strong point of DTD is that it is still the only language for XML that makes it possible to define entities, which are kind of macros in XML documents. So if entities are needed, some parts of the DTD language are required. This is mostly the case for human-edited documents where entities can be used to save repetitive typing and provide uniformity for the defined case.

These problems make DTDs undesirable in some circles, for instance, most modern XML applications need namespaces and use them as extensibility tools, but using namespaces with DTDs is difficult and requires care. The weakness of the content model specification might not be that serious; some studies have found that more than 80% of schemas written in more complex languages could still be expressed in DTDs. The requirement for implementing another parser is a real burden, especially where code size is a concern. Finally, a common viewpoint in the XML community is that validation, that is, verifying a document against a schema, should be completely separated from the actual document, for instance, in case someone would like to validate against a different schema from what the author intended.

To alleviate these concerns and some others, for instance, the lack of data typing, the W3C specified XML Schema. The XML Schema language was explicitly intended to fix the problems of DTDs, in particular the lack of support for multiple vocabularies through namespace definitions.

The full XML Schema is a very complex language, with a variety of different structure descriptions. Being XML, it is also very verbose, so the schema shown in Figure 5.4 is only a partial schema for the document in Figure 5.2. Note further that this schema is not a complete schema for the document, as XML Schema allows only declarations for a single namespace (the `targetNamespace`) in one schema.

This example shows how optional components are handled: each component has a `minOccurs` and a `maxOccurs` property, and changing these from their default values of 1 allows the construction of optional components or repeating components. The `born` element illustrates the use of a data type, in this case the `date` type specified in XML Schema.

Still, XML Schema is also perceived to be faulty in some circles. Its main failings are:

- lack of formal theory behind it, leading to poorly-understandable specific constraints;
- a somewhat clunky syntax;
- its perceived emphasis on data types, devoting the complete second part to them; and
- a rather ad-hoc collection of predefined data types.

The lack of theory behind the language is partially caused by the early inception of the language, as there was little practical experience in applying the theory of tree languages to XML schema languages. The data type problems are most probably caused by the influence of the database community on the language design; XML has become popular even in that world.

The third popular schema language for XML is called RELAX NG, an OASIS and ISO standard. It has a formal basis and allows the description of document structures that

```
<?xml version="1.0" encoding="UTF-8"?>
<xs:schema xmlns:xs="http://www.w3.org/2001/XMLSchema"
           elementFormDefault="qualified"
           targetNamespace="http://example.org/people"
           xmlns:p="http://example.org/people">
  <xs:element name="person">
    <xs:complexType>
      <xs:sequence>
        <xs:element ref="p:name"/>
        <xs:element minOccurs="0" ref="p:occupation"/>
        ...
      </xs:sequence>
      <xs:attributeGroup ref="p:nationality"/>
    </xs:complexType>
  </xs:element>
  ...
  <xs:element name="born" type="xs:date"/>
</xs:schema>
```

Figure 5.4   A partial XML Schema for the example XML document

XML Schema prohibits. It does not include native support for data types, but can use an external data type library; the XML Schema data types are a popular choice for such a library. RELAX NG also supports co-occurrence constraints: the concept that, say, an attribute value in an element start tag can determine the element content model. These are perceived to be so useful that the next version of XML Schema is also going to include them. RELAX NG does not support Infoset augmentation, as the designers felt that changing the content of the document is not in the purview of a schema validator, but belongs to a transformation.

In addition to an XML syntax,[6] RELAX NG includes a compact syntax that is more familiar to people used to typical programming languages and perhaps also easier to author by hand. Figure 5.5 shows a RELAX NG schema in compact syntax for the document in Figure 5.2. As can be seen from this example, RELAX NG does not require everything in a single file to be in the same namespace.

The grouping and repetition constructs in this example are similar to those in the DTD example of Figure 5.3, but RELAX NG also supports an interleave operator that is responsible for much of the power of RELAX NG. The `nationality` attribute declaration illustrates the possibility of constraining data types using regular expressions, as is also possible with XML Schema.

Considering tool and specification support for the schema languages, RELAX NG is probably the loser in that specifications requiring a schema language mostly just refer to XML Schema and while RELAX NG validators exist, there is less integration in systems than with XML Schema.[7] It also seems to be the default assumption when

---

[6] Even the RELAX NG XML syntax is seen by many to be cleaner than XML Schema's.

[7] DTDs, as a required part of the XML definition, are naturally in a class of their own in this respect.

```
namespace p = "http://example.org/people"
start = favorite-composers
favorite-composers = element favorite-composers {
    element p:person {
      attribute nationality { xsd:string { pattern = "\w\w" } },
      element p:name {
        element p:first { token },
        element p:middle { token }?,
        element p:last { token }
      },
      element p:occupation { token }?,
      element p:born { xsd:date },
      element p:died { xsd:date }?
    }+
}
```

Figure 5.5   A RELAX NG Compact Syntax schema for the example XML document

writing new schema-requiring specification to refer to XML Schema, especially at the W3C. In summary, RELAX NG is spreading mainly through grass-roots efforts of its proponents.

5.4.4 XML Querying and Transformations

Accessing the information in an XML document at the application level is most often performed through an API such as DOM that reads the document into memory and provides a tree-based data structure view of it. The application uses the access methods provided by the DOM API to drill into the interesting parts of the XML document. Much of the code needed to access parts of an XML document using an API is essentially boilerplate, which makes it tedious to write. A better way to access the data in an XML document would be to use a special-purpose language designed for matching XML's tree structure, just as regular expressions are used in many systems as a special-purpose language for string matching.

In the XML world, the primary language for addressing individual pieces of data in an XML document is called XPath. An XPath expression consists of a sequence of tests, each test being performed at certain levels of the XML document when viewed like a tree through the XPath data model. Each level may also contain branches that allow content in other subtrees of the document influence what will be matched and picked up from the document.

Two example XPath queries are shown in Figure 5.6. These are intended to match the document in Figure 5.2. The first query explicitly shows no branching, but drills down into the p:person elements, of which the first one is picked (the condition in square brackets), and the child element p:name of that is returned as the result of the query.

The second query is more complex. First, it starts with a descendant step, //, which will find any element at all in the document with the name p:person. There are then two branches from these elements, denoted by square brackets as well, first is to test for

```
/favorite-composers/p:person[1]/p:name
//p:person[p:died][@nationality="DE"]/p:name
```

Figure 5.6   Example XPath queries

the existence of the `p:died` child element, and the second is test that the value of the `nationality` attribute is DE. Finally, all the `p:name` child elements of the `p:person` elements meeting the two conditions are returned. In summary, this query returns the names of all the dead German people (presumably composers) in the document.

One point about XPath that can be illustrated through the queries in Figure 5.6 is its handling of namespaces. Namely, the queries use prefixes just as in XML documents, but no mapping is defined for these prefixes anywhere in the queries. This is because XPath is designed to be embedded into XML processing languages, which are expected to establish the namespace prefix mappings.

The most prominent language that can natively process XML is XSLT, for XSL Transformations, a W3C Recommendation. As the name indicates, it is a programming language for transforming XML documents based on their content.[8] XSLT is mostly a declarative language, with the transformations described by matching content with XPath expressions and then describing what kind of output to produce based on the content inside the matched part.

The combination of XPath with XSLT can be likened to the Unix world where regular expressions play the role of XPath and a language like Awk functions as the transformation language. Thus, these languages could be seen as the building blocks for transforming the line-oriented Unix thinking into a tree-oriented version, more suitable for structured data.

While this view has somewhat materialized, it is still nowhere near the ubiquity that regular-expression-based approaches enjoy nowadays. This might be caused by the clunky syntax of XSLT: a programming language with an XML syntax is not a very good idea. Another cause might be that XPath processing performance is not even close to that of regular expression matching, a problem caused by XPath's more complex requirements as well: matching trees is a more difficult problem than matching strings.

In theory, it would be possible to define an XML application to a large extent using, say, XSLT to transform the incoming XML into a form needed by the next phase. This output could be either XML, HTML, or plain text, depending on its consumer. The issues illustrated above are probably in large part the reason why this is not commonly done, but instead XML processing is handled through low-level APIs and special-purpose transformations written for each single case separately.

While XSLT is perhaps not widely usable for distributed applications, XPath has some useful properties. Namely, it provides matching functionality for XML documents. With message content moving toward XML, using XPath allows a system to make decisions based on a message's content without first converting it to another format, which is especially useful for application-layer routers. To make full use of such functionality, it is necessary to have an efficient XPath matcher that can match simultaneously against a

---

[8] The XSL stands for Extensible Stylesheet Language, which comes from XSLT's origins as a language for applying presentation styles to XML for different kinds of output. The other XSL is XSL-FO (formatting objects), which can be used to produce printable documents from XML.

large number of XPath expressions and process documents in a streaming manner, to the extent possible.[9]

5.4.5 User Interfaces and Multimedia

When considering the web and mobile devices, the most pressing concern is perhaps the user interface. A mobile device's small form factor does not allow a large screen, and much of the web is often 'best viewed with 800×600 resolution', making it difficult to use on a mobile device. To compound the problem, the mobile devices are not very uniform in their capabilities either, with screen sizes ranging from tiny to not-so-tiny. These concerns are perhaps less relevant from the middleware perspective, but the specifications that we cover here are making definite inroads into the mobile device space, so it is expected that they will play a significant role in future mobile applications, which may affect middleware design as well.

As noted in Chapter 4, the Model-View-Controller (MVC) pattern is often used for user interface design, especially on mobile devices. The XForms language is a form description language built on this paradigm. A further development of regular HTML forms, XForms aims to be a platform-independent web form description language.

The main idea behind XForms is the separation into a *template*, an arbitrary XML document structure, and the *form*, which is displayed to the user and has *controls* that are bound to parts of the template, filling out the needed data as the user interacts with the form. Here, the template, form, and controls correspond to the Model, the View, and the Controller, respectively. The use of user-specified XML as the data document allows XForms to integrate with pre-XForms applications that expect specific formats for the XML documents that they consume.

Multimedia applications are also making inroads into mobile devices. Long considered too bandwidth-heavy or processor-intensive, advances in both networking and processors have brought advances graphics to the mobile world as well.

Scalable Vector Graphics (SVG) is a language for describing two-dimensional graphics. Being a vector-graphics language, the images drawn with it can be scaled arbitrarily without concern for the actual pixel width or height of the screen. It is thus well suited for creating graphics for mobile devices, and that has indeed been one of the specified application areas of SVG, with mobile profiles being defined alongside the full profile. The 3GPP has selected SVG to be a mandatory vector graphics format for multimedia in its release 5.

Another multimedia-related XML language specification is SMIL, for Synchronized Multimedia Interaction Language (pronounced like 'smile'). The purpose of SMIL is to act as a 'glue' language for multimedia, bringing independent objects together into a coherent presentation, with features like specifying temporal events, layout, and hyperlinks. Like with SVG, SMIL has a mobile profile and the 3GPP has adopted it as a part of the 3GPP platform for multimedia.

5.4.6 The XML Stack

XML serves as the basis of a vast array of technologies, often referred to as the 'XML stack'. The intent is to evoke an image of a stack of technologies, like the networking

---

[9] The full XPath language cannot be matched in a fully streaming manner, though large subsets can be.

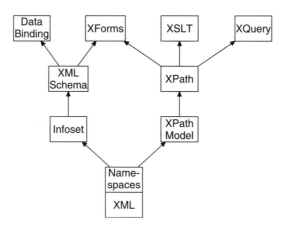

Figure 5.7   A partial XML stack

stack, built on top of the XML specification, implementing each piece of functionality in only one place, using existing solutions for common functionality instead of developing specific techniques for each different case.

A rough picture of how all these technologies (and some others) fit into the XML stack is given in Figure 5.7. The figure leaves out many XML technologies, and does not even state all the dependencies of the ones shown (for instance, XForms also depends on XML Events).

It is also simple to criticize this idea of solving each problem only once. For instance, take XPath, a technology for addressing XML document content. Version 1.0 was a reasonably simple language, though implementations efficient in all cases have taken a long time to appear. But further requirements on systems using XPath necessitated adding new features to the language, making XPath 2.0 a much larger and more complex language. Now, a specification that only needs the ability to address XML elements can either adopt the full XPath, with minimal hassle, or subset it, with more work for the specification authors but with a better fit of the addressing language to the specification's requirements. The latter view gains further credence when we note that it is extremely rare for a specification to become simpler in new versions, but becoming more complex is very common indeed.

5.5   IETF Standards

The IETF is still the most relevant standards organization regarding Internet technologies, even if the W3C has received its part of the spotlight due to the increasing focus on XML. But apart from XML technologies, the IETF is responsible for practically all other Internet standards, including the HTTP protocol and URIs, both central components of the web.

The work in the IETF happens in working groups. Communication inside a working group is primarily handled through email, not by telephone conferences or frequent meetings as is the case at the W3C. The IETF does have meetings as well, but these are meetings of the whole IETF and participation is not mandatory to be able to participate in a working group. Nevertheless, like with any standards organization, attending an IETF meeting provides much better opportunities to discuss and agree on the standards than remote communication such as email.

The documents published by the IETF are called RFCs, for Request for Comments. RFCs that are intended to become standards are said to be on the standards track. Such documents start out as Internet Drafts, several versions of which are usually produced. As a draft expires in six months, frequent work is needed to keep drafts existent. After a draft is judged sufficiently mature, it is published as an RFC. This RFC is labeled a Proposed Standard. An RFC is never modified; if revisions are required, a new RFC is published and the old one marked obsolete by the new one.

A Proposed Standard graduates to Draft Standard status after two independent inter-operable implementations have been demonstrated, and sufficient deployment experience gathered. This is part of the IETF motto 'rough consensus and running code', intended to illustrate that working implementations are more important than following strict procedures in agreeing on features. As more experience is gathered and deployment becomes wider, a Draft Standard graduates to Standard status, and is added to the list of Internet Standards.

Like in the W3C, all phases of the IETF process happen in public, so it is likely that implementations begin to appear in early stages of the process. Going through the whole process to a Standard (or Recommendation in the W3C case) can take a long time and as implementations are available, it is likely that a promising technology gets adopted early in the process. However, it is possible for the formal process to get stalled in matters that need to be resolved, so that while the technology is usable and multiple interoperable implementations exist, it is still not formally a standard.[10]

The IETF is intended only to standardize existing practice, not create completely new technologies or research new areas. These tasks fall under the responsibility of the Internet Research Task Force (IRTF), a separate body also under the Internet Society. However, there is much crossover between IETF and IRTF because after research has been completed at the IRTF, it can be ready for standardization at the IETF, and conversely, when the IETF finds out that a technology is not yet sufficiently mature to be standardized, the IRTF can take responsibility for exploring the unfinished parts.

5.6 Emerging Internet Standards

There are a number of current developments in standardization that promise improved support in Internet computing for different fields, in particular mobile computing. Some of them have been covered in Chapter 3. Here we cover two standards that are currently making their way through the IETF and the W3C, namely Peer-to-Peer SIP and Efficient XML Interchange. Since these technologies are currently making their way through the standards process, some information in this section may have become outdated by this book's publication. For current information, see the web pages http://www.p2psip.org/ and http://www.w3.org/XML/EXI/.

---

[10] For instance, formally SMTP, the Simple Mail Transport Protocol, is still standardized by RFC 821, since its replacement, RFC 2821, is only a Proposed Standard. Regardless, trying to use only RFC 821 as a specification for implementing SMTP on the current Internet would probably lead only to trouble.

5.6.1 Peer-to-Peer SIP

SIP, covered in Chapter 3, relies on registrar servers to allow user agents to contact other user agents. Such an arrangement implies some form of centralized control of the server, requiring resources to cover issues like fault tolerance. While this works reasonably well over the Internet, an infrastructure-based approach is not always suitable for a mobile environment, especially if the mobile devices are all there is.

The purpose of Peer-to-Peer SIP (P2PSIP) is to distribute the responsibility for registration and mapping in SIP throughout the network so that there is no centralized point that handles registration information. This makes the network more robust in that there are no single points of failure and also makes setting up things easier in an *ad hoc* environment.

To implement the required functionality, the hosts on the network run some peer-to-peer [12] content searching algorithm, in this case a distributed hash table (DHT). P2PSIP divides the hosts on the network into two classes, peers and client. The peers are the ones that run the DHT algorithm and thus implement the registration database whereas clients are ordinary SIP user agents that contact the peers when they need the registration functionality.

Currently, the Peer Protocol run by peers to implement the DHT is called RELOAD, for REsource LOcation And Discovery. RELOAD does not specify any particular DHT algorithm to use. Rather, it allows for pluggable DHTs, though as with most specifications of this kind, there is one DHT algorithm that is mandatory to implement, in this case, Chord [13].

5.6.2 Efficient XML Interchange

The main problem with XML is that it consumes much more space, and thus bandwidth, than is warranted by the actual information it contains. The causes for this extra verbosity are at least three-fold: first, being based on text, XML requires more bytes to represent data than a binary format; second, XML's generality and extensibility require explicit tagging of content instead of implicit as is common with special-purpose formats; and third, the need for simpler automatic processing caused the dropping of many space-saving features from SGML.

It must also be noted that XML was originally designed as an SGML replacement, that is, a language for representing structured documents. The use of XML for representing general structured data came later, and could be argued to be an unsuitable use of XML. Still, the design goals of XML do not prescribe XML only to the world of structured documents, and with the wide popularity XML has enjoyed, it was perhaps inevitable that everything would move toward XML.

One concept that separates XML from other formats used for structured data exchange, most prominently ASN.1, is its optional validation. The point is that while XML documents are often described by schemas, this is not a required part of the XML stack and can be left out if need be. This, in conjunction with XML's text nature and clean structure, permits evolution of the data formats without requiring lockstep upgrades between parties. Naturally, this view can be criticized by noting that if two communicating parties do not have complete and shared understanding of the messages sent, there cannot be an actual agreement as to their meaning. Still, in practice, such evolution seems to be acceptable and is possibly one reason for XML's popularity in structured data exchange.

The verbosity of XML is not a new problem in wireless communication. Practically immediately upon XML's publication the WAP Forum designed a 'binary XML' format to represent the WAP Markup Language in a form more compact than the XML used to define it. The term 'binary XML' gained popularity, perhaps due to the WAP format name, as a designation for binary serialization formats for various XML data models, such as the Infoset or XML represented through the DOM or SAX processing APIs.

Recently the W3C decided to work on the field of binary serialization formats itself. A W3C-approved format specification would, one hopes, bring some unity into the fragmented world of binary formats, which is one purpose of standardization. The Efficient XML Interchange (EXI) working group is chartered to produce a format specification, the requirements for which were laid out by a previous group, the XML Binary Characterization working group.

Many of the existing binary formats had been special-purpose ones and there was much doubt in the XML community that a general format would even be possible. Considering the characteristics of XML, a set of stringent requirements was established that a proposed W3C standard format would have to meet. Decreased size and improved processing speed were the ones to be paid the most attention, but the format also had to be fully compatible with XML so that existing XML-based systems and technologies would also require at most simple modifications to adopt the new format, and not have to be completely overhauled.

The above-mentioned property of XML's optional validation was also taken into account. Namely, while one requirement on the format was that it be able to use schema information to decrease document size even further, it would also have to be prepared for the case where documents did not conform to the schema. Such cases could happen through two paths: either a document was produced that did not conform or a schema was modified and documents conforming to the old schema did not conform to the new one. Both cases happen frequently enough in practice to be necessary requirements. This requirement eliminated many of the proposed candidate formats.

After running extensive measurements and performing analysis on the candidate format specifications, one was chosen as the basis for the future EXI standard. The specification is currently in Last Call, with the expectation that Recommendation status could be achieved some time during 2009. Several implementation efforts are already in existence, but it is too early to say whether the expectation is realistic.

The basics of the proposed EXI format are reasonably simple. At heart, it always requires schema information, but if that is not available for the document, a fully-permissible schema is substituted. The view of an XML document in EXI is as a stream of events, not unlike the information items of the Infoset. Each event is encoded separately from the others, making it possible to process streaming EXI documents, though there is some concern on how EXI will handle partial documents required by, e.g., XMPP, because of its bit-oriented nature.

The schema information that EXI requires provides a content model for each type of element. With DTDs, it is sufficient to consider the element name alone to determine its type, but with more complex schema languages it is necessary to take into account its context, that is, the sequence of its parent elements. The content model described by the schema is represented by a grammar that describes the content model in terms of the

EXI events. Thus, the language produced by the grammar is precisely the set of event sequences corresponding to schema-conforming content.

In addition to the precise grammar, it is possible to add productions to the grammar to allow for events not permitted by the schema. This allows EXI to handle non-schema-conforming documents without penalizing the case of conforming documents over much, as the expected events are encoded in a smaller amount of space than unexpected events, as suggested by information-theoretic considerations.

The grammars produced for EXI are all regular grammars to ease implementation, and cannot therefore describe a full XML document. That is why each grammar describes only the content of a single element. The full processing of a document relies on a stack of grammars, with one grammar being active at each point and the other grammars in the stack corresponding to the currently-processed element's context.

EXI also includes two types of content learning. First, all names occurring in an XML document are tokenized, that is, assigned a compact identifier, which allows repeated occurrences to be encoded more efficiently. Second, when a fully-permissive schema is used for some element's content, that content is learned in a rudimentary fashion as the document is processed. This allows some degree of schema utilization even when a schema for the document is not available. After all, even when an explicit schema is not written, an XML document is intended to have some semantics and this semantics dictates what elements and content can appear where, creating a form of an implicit schema.

Bibliography

[1] IEEE 1991 *IEEE Standard Computer Dictionary: A Compilation of IEEE Standard Computer Glossaries*.

[2] ISO 1993 *ISO 2382-1:1993. Information Technology – Vocabulary – Part 1: Fundamental Terms*.

[3] ITU 2002 *Abstract Syntax Notation One (ASN.1) Specification of Basic Notation*. ITU-T Rec. X.680.

[4] WWR 2004 *Technologies for the Wireless Future: Wireless World Research Forum*. John Wiley & Son Ltd.

[5] OMG 2004 *Common Object Request Broker Architecture (CORBA/IIOP), version 3.0.3*.

[6] O'Mahony D 1998 UMTS: the fusion of fixed and mobile networking. *IEEE Internet Computing* **2**(1), 49–56.

[7] Cai J and Goodman DJ 1997 General packet radio service in GSM. *IEEE Communications Magazine* **35**(10), 122–131.

[8] IEEE 1999 *IEEE Std 802.11 – Wireless LAN Medium Access Control (MAC) and Physical Layer (PHY) Specifications*.

[9] Blu 2004 *Specification of the Bluetooth System, Core Package version 2.0*.

[10] W3C 2006 *Extensible Markup Language (XML) 1.0* 4th edn. W3C Recommendation.

[11] W3C 2004 *XML Information Set* 2nd edn. W3C Recommendation.

[12] (ed. Steinmetz R and Wehrle K) 2005 *Peer-to-Peer Systems and Applications* vol. **3485** of *Lecture Notes in Computer Science*. Springer.

[13] Stoica I, Morris R, Liben-Nowell D, Karger DR, Kaashoek MF, Dabek F and Balakrishnan H 2003 Chord: a scalable peer-to-peer lookup protocol for internet applications. *IEEE/ACM Trans. Netw.* **11**(1), 17–32.

6

Mobile Messaging

This chapter covers the fundamentals of middleware communication, which we have dubbed 'messaging'. Messaging includes the protocol used by the middleware, which is considered an application-layer protocol in the TCP/IP model, and the representation of application data inside messages. We cover the general communication model, show requirements and challenges posed by mobility, and common patterns used in messaging systems. We also examine existing systems and demonstrate how they fit into the framework, as well as consider some emerging technologies in the field.

6.1 Messaging Fundamentals

In our parlance, messaging refers to any communication above the transport layer. That is, in the Internet stack messaging is strictly an application-layer function, whereas in the OSI stack messaging may encompass all of session, presentation, and application layers (Figure 6.1). This is mostly by reasons of convenience: the Internet protocol stack is so entrenched on the network and transport layers that practical deployment of distributed systems is mostly limited to using the Internet stack.

6.1.1 Messaging System Components

A messaging system consists of several components, all of which combine to form the full system. The components that we consider here are:

Architecture The overall architecture of the complete messaging system.
Message syntax The kind of container that is used for application data and how it is encoded for transmission.
Message protocol The application-layer protocol that parties use to exchange messages.
Locator The address of a messaging party that allows sending messages to it.

Often, syntax and even locator are subsumed under the concept of the protocol, but we believe it is better to separate them. In the modern world, there are a number of generic options for each of the three, all combinable with others, so the messaging system designer has more freedom with this division than they would if they were all bundled together.

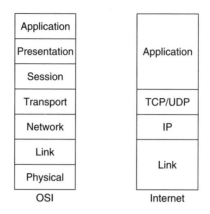

Figure 6.1   The Internet and OSI protocol stacks

Figure 6.2   A generic messaging architecture

The way that these components combine to form a complete messaging system is illustrated in Figure 6.2. Here, the locator identifies the message receiver. The sender queries a resolver to map the locator into something usable for transmitting the message. The message is constructed by the sender and then routed through the messaging system using the messaging protocol.

6.1.1.1   Messaging Architecture

The principal component is the general messaging architecture. As with any software architecture, this consists of the components of the system and the communication links between them. Unlike in software engineering in general, where components are modules of the program and communication links are function call across module boundaries, in a distributed system architecture the components are independent applications usually running on separate hosts and communication links denote actual network communication.[1]

One of the most important principles in architecture design, in both software and distributed systems, is *loose coupling*. In software architecture this means following precisely each module's public interface, not relying on any specific implementation, and designing the interfaces so that they do not constrain the implementation overly. The concept is similar in distributed systems, though the interface-implementation distinction is not

---

[1] Most messaging systems optimize the case where two components are on the same host or share the same address space, but this has no bearing on the architecture.

the prevailing concern, but rather the ability to replace components individually without needing to replace the complete system every time one component needs to change.

In distributed systems, loose coupling is promoted by relying on open, standard protocols and formats, as this permits the most freedom in choosing and evolving components. This, however, is not typically sufficient for all purposes. Namely, as the system evolves, new requirements are often discovered that need additional features from either the protocol or the message format. Thus, there also needs to be some form of extensibility to allow for experimentation of new features.

A simple way to achieve extensibility is to embed version identifiers in the messages. Then, unused versions can be used to experiment with new features, though unrecognized versions will need to be treated as errors. This means that versions will be completely incompatible with each other, so even if the change between versions is just something small, there will be no formal compatibility between them.

A better way is to embed extension points in the formats, both in the protocol and messages. Here, an extension point refers to some specific construct of the protocol design that permits adding new kinds of information into specific places. For instance, many protocols use different types of messages for different purposes, so defining a new message type lets one experiment with some features without needing to overhaul the whole protocol.

Backward compatibility, that is, allowing implementations of new versions to support old versions, is usually not too difficult to achieve, unless the original design has been optimized so far that there is little possibility in ever extending it. On the other hand, forward compatibility, that is, allowing old versions to function even in the presence of a new version, is often much more difficult, and usually achieved by having the newer versions revert to the old behavior.

One type of forward compatibility in message syntax, especially associated with loose coupling, is to specify the extension mechanism so that an implementation not understanding the extension can still make sense of the rest of the message. This allows at least partial interoperability in an open environment, but it is a controversial approach, since, as critics ask, if the extension part is not important to the message, why was it put there in the first place? And will the reader misunderstand the message in a critical manner if it did not read the extended information? In practice, however, things are not so black-and-white, so this approach can work in some cases.

The conceptual difference between 'regular' software architecture and distributed system architecture also brings out some differences in how to approach design in the two different worlds. While loose coupling is important, in typical software design there is not usually much concern for the connectivity of the components (except to the extent that it hinders understanding the architecture) or their number of interactions. On the other hand, in a distributed system where communication links go over the network, it is imperative to minimize the use of those links, and therefore prefer local processing to communication.

On the transport layer and below, data is mostly just bits (well, bytes on the higher layers). The layer protocol wraps the data into its own headers and possibly trailers and does not care at all about any potential semantics of the data. This does not hold anymore when considering middleware. Middleware provides a model for communication and a part of this model is usually some internal structure in the data.

Still, middleware is not the application, so it is not appropriate for the middleware to dictate any specifics about the data being transmitted. The usual method of middleware is to define an API for generic structured data and let the application package its own data into this predefined structure. Depending on how the application is structured and how much the middleware hides its internal representation, this may require conversions between formats in the application before passing any data to the middleware.

6.1.1.2 Message Syntax

For handling the marshaling and unmarshaling of data, that is, converting from the programming language data structures to bytes and vice versa, there are several methods. The following ones are in use in current widespread platforms:

Application-implemented This method requires the application to provide marshaling and unmarshaling functions for each kind of data it wishes to use in messaging. Usually, marshaling functionality is provided for the primitive types of the programming language and the programmer needs to worry only about structured types.

Generated code This method is typically based on using an interface description language (IDL). The IDL defines the data structures that are possible to use in communication, and code is automatically generated from it to perform marshaling and unmarshaling. The middleware then defines language mappings that determine how each kind of IDL structure is represented in the target language. It is the programmer's responsibility to make sure any data passed to the middleware is in this form, but the generated code takes care of any conversion to and from bytes.

Introspection In more dynamic languages that allow run-time inspection of data structures without knowing the precise structure, it is also possible to use introspection. Here, there are no constraints on the programmer, but rather the middleware inspects any data passed to it, determines the structure using the language's introspection capabilities, and marshals the structure on its own. This method is the easiest for the programmer, but introspection is usually a heavy operation compared to using marshaling functions that know the precise structure.

There exist a few popular general-purpose formats for representing structured data. These can differ completely or be somewhat similar to each other. The clearest dividing line is between binary and textual formats. A binary format will encode any structure as binary tokens and also encode any primitive data in a binary form, usually mirroring a computer's internal representation. A textual format, on the other hand, attempts to mimic the way a human would write the structure out when editing a text file.

There are arguments for and against both approaches. In the past, with network bandwidth and computing power being scarce resources, binary formats ruled the roost due to their compact nature and better processing speed resulting from the use of close-to-the-machine representations. Nowadays, with large-capacity networks and powerful computers, text formats are generally preferred because of their perceived greater debuggability and experimentation capabilities, namely, direct packet inspection is usually sufficient to understand the format and at a pinch it is even possible to write the data directly without going through the middleware.

Binary formats also have the additional problem of incompatible machine formats. Direct dumping of the machine's internal format will not suffice if the receiving party does not use the exact same machine. Therefore, a binary format will need to specify sizes of integer types, endianness, and especially any floating-point formats used.

There are three ways to handle any encoding issues in binary formats:

Specification The specification of the data format can define precisely all the data type sizes and how they are encoded. This is the method of binary Internet protocols that always specify *network byte order*, that is, big-endian.

Negotiation If the messaging system includes some form of session establishment, the establishment messages can include negotiation of the formats. One way to implement this is that each party announces its native formats and the specification defines rules on how to decide the common formats based on this.

Receiver makes right In this method, the sender simply uses its native formats and metadata in the message indicates which formats these are. It is then the receiver's responsibility to convert to its native format.

All of these have their advantages and disadvantages. Specification is easiest for the implementors, but usually the formats chosen are the native formats of the dominant platforms at the time, and if the platform distribution changes, this might end up causing needless conversions. Both negotiation and receiver-makes-right avoid such problems in that if the two parties use common formats, these formats will be used. However, they are more complex to implement, and negotiation does not fit into many interaction patterns.

Of the data formats specified independently of any particular application or system, the most widespread binary one is ASN.1 (Abstract Syntax Notation One) [1]. ASN.1 defines a syntax for specifying structured data and a number of different encoding rules for how the structure is represented as bytes (so typically it falls under the generated code category above). The encoding rules vary depending on desired properties, and there are rules for each of maximal compactness, some extensibility, and unique representations.

On the textual-format side, recent years have witnessed the dominance of XML [2], which we covered extensively in the previous chapter. Due to its document-world roots, XML is not a perfect fit for structured data representation, since it includes all kinds of excess baggage that is simply not needed in a message syntax. Because of these additional unneeded features, there has been some movement towards JSON (JavaScript Object Notation) lately. JSON is a very simple syntax that is able to represent a few primitive types and uses arrays and dictionaries to represent structure, with syntax directly from the way these concepts are represented in the JavaScript programming language.

6.1.1.3 Protocols

As on the lower layers, the concept of a protocol is fundamentally the same on the application layer. Namely, there exists some envelope, that is, metadata for the protocol, and the actual payload. As noted above, some syntax for the payload may also be dictated by the messaging system, unlike on the lower layers, but we consider that to be fully the purview of the message syntax, and mostly assume that the protocol treat the payload as opaque. The metadata usually has very similar purposes as lower-layer metadata, for instance, the metadata is typically used to implement any message exchange patterns supported by the middleware.

There are a variety of purposes to which application-layer metadata can be put, and usually more are invented as usage experience with the system is gathered. Therefore a common pattern in protocols is to follow some form of header-body structure, with the header consisting of a number of fields. The boundaries of a field are defined in the general specification, so even if an application does not recognize a field, it can still extract all the information it does understand. This provides comfortable extensibility to the protocol, allowing new features to be implemented while still being compatible with the old versions.[2]

In addition to the protocol metadata, the messaging protocol also defines the types of messages that can be sent, when they can be sent, and by which party. Usually on the protocol level, there are not many messages that have payloads, since the semantics of the payload are mostly just to deliver it to the application identified by the locator. In contrast, of course depending on the design of the middleware, there can be several different kinds of messages for, e.g., connection management.

In messaging protocols that reflect the transport layer roles directly, connection management is also often relegated to the transport layer. That is, a client opens a transport connection and that is a signal that application-layer connectivity is also established. Similarly, connection closure is signaled by simply closing the transport-layer connection. This often suffices for simple messaging needs that are happy with 'traditional' client and server roles.

In many cases, and especially when considering mobility, connection management is not that simple. In many middleware platforms, it is common that transport-layer connections are only used when data needs to be sent, and there is no need to maintain them just to retain the identity of a connection. Such platforms can be said to implement the fifth layer of OSI, the session layer, by providing persistent session functionality on top of an intermittent transport-layer connection.

Session functionality is especially necessary because transport-layer reliability guarantees do not always hold in the presence of mobility. When looking at TCP, if the underlying IP address changes during mobility, all existing TCP connections will break, and there is no way to know which segments have actually made it to the recipient. This is one reason why mobility middleware cannot relegate all reliability to the lower layers, as systems like Mobile IP cannot be relied on to exist everywhere.

An interesting approach to messaging protocol syntax is provided by MIME (Multipurpose Internet Mail Extensions) [3]. MIME began its life as an email protocol and syntax to allow 8-bit characters and binary attachments in email messages. Since then, it has developed into a more general purpose transmission method, with many concepts also being adopted by other protocols.

The protocol parts of MIME include a set of headers to be used in an email message (or other MIME-compliant protocol). These headers indicate the type of the content, which includes the character encoding used to encode any characters into bytes, as well as the transfer encoding, which signifies the conversion that was applied to 8-bit data to make it

---

[2] Similar structure can be found in IP, especially IP version 6, which allows extension headers to be defined with greater flexibility than version 4.

7-bit-clean, that is, have all 0's as the high bits of all bytes. Modern systems are usually 8-bit-clean, permitting the use of the full 8 bits of a byte.

6.1.1.4 Locators

On the network layer, a locator is just a network address. On the transport layer, network addresses are still used, but something else is required to properly multiplex several different transport communications between the same hosts.[3] But when moving to the application layer or even to middleware, a locator can become arbitrarily complex.

In Internet-based applications, it is common to use DNS names as a part of the locator. Email addresses have a host name component after the @ sign, HTTP URLs include a host name, and so on. This is really just a way to include a network address in the locator, and systems typically also support using IP addresses directly in place of domain names. However, the information additional to the host name is typically much more complex than just the port numbers of the transport layer. For instance, in HTTP, the path component of the URL is much more than just a port number.

Some middleware systems abstract away from using network addresses directly. This gives them some independence from the underlying communication technology, and allows them, for instance, to use different network layers. With the current trend towards all-IP networking, this capability is no longer as necessary as it was even 10–20 years ago, since a locator directly based on IP addresses will likely work in a number of future networks. Still, communication in, say, PANs is not always IP-based, so network independence can perhaps help middleware systems intended also for short-range communication.

In modern systems, there is usually some desire for transparency in locators. Just as message syntax is moving towards XML, so are locators increasingly being defined as URIs, sometimes with custom schemes to permit the middleware protocol to be used. The usual cited benefit for this is that locators are easier for people to handle, which also increases the system's debuggability, since it is not always necessary to fire up the full middleware to test the locators.

A contrast to the use of URIs is provided by the idea of opaque locators, most prominently in CORBA. The idea behind an opaque locator is that the full locator is not supposed to be in any way inspectable, but some structure is defined so that the middleware can extract the relevant information for sending messages. In principle, this promotes extensibility as new interaction methods can be added to the locator.

Opaque locators have the problem that there is no way to construct them without the middleware, and even then, constructing a locator for a remote party can be impossible with only the information that is typically available (protocol, network address, etc.). Thus, a middleware using opaque locators also requires a form of naming service that can map to either names or properties.[4] This all adds complexity to an opaque-locator system, and is a likely reason why they are not as popular as their extensibility and generality would seem to call for.

---

[3] On the Internet, both TCP and UDP use port numbers as this additional information.

[4] These are often referred to as 'white pages' and 'yellow pages', respectively, in analogue with the different kinds of telephone catalogs.

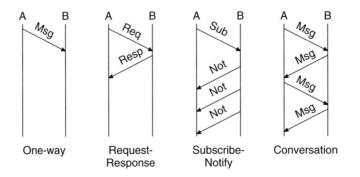

Figure 6.3 Common message exchange patterns

6.1.2 Message Exchange Patterns

The interaction between communicating parties in messaging is frequently called a *message exchange pattern*. At the most fundamental level, message exchange patterns can be divided into two kinds:

One-way In its most basic form, one-way messaging is just one party sending a single message to the other party. Usually one-way messaging is not considered reliable, but it can be extended to provide various levels of reliability. And some message exchange patterns call for several or repeated one-way messages being sent.

Two-way Two-way messaging is any messaging where both parties send messages to the other. Note that we use this pattern only for the cases where messages in both directions contain information semantically meaningful to the application; if traffic in the other direction is simply acknowledgments of messages received, that is counted as a form of reliable one-way messaging. Furthermore, the expectation is that messages being sent are somehow dependent on the previous messages; otherwise the exchange is simply a series of one-way patterns in two directions.

There is not much variability in one-way patterns, but there are several different two-way patterns, depending on who communicates and when. In general, a two-way pattern can include arbitrary messaging between the two parties, but in practice there exist only a few simple basic patterns that are used for all applications (illustrated in Figure 6.3):

Request-response This is the simplest two-way pattern. It consists of one party sending a message, called a request, to which the other party sends another message as a response.

Subscribe-notify This pattern began by one party sending a message, called a subscription. The intent is that the sender of the subscription is interested in knowing whenever a specific event happens. The other party will then send a message, called a notification, every time this event of interest happens, until the time that the subscriber sends a message to cancel the subscription.

Conversation In this pattern, the parties communicating alternate sending the messages until the last message of the conversation is sent. This pattern is rarer than the other two, and is most often considered to be a repeated request-response. The main difference between these two is that in a conversation there is more of a link between the consecutive messages than when considered as individual request-responses, and of course a conversation can terminate with an odd number of messages being sent.

Any other patterns are extremely rare. For instance, it is possible to imagine a kind of 'reverse' subscribe-notify where one party first sends a number of messages and finally gets a single response from the other, but in practice such a situation is handled by a repeated request-response.

When considering message exchange patterns, it is important to make a distinction between the roles of the parties at the messaging level and at the transport layer. With connectionless transports, all communication is just single one-way messages that may be reliable depending on the transport protocol. This does not, however, preclude the messaging layer from providing additional reliability guarantees or two-way patterns of its own using the message syntax or protocol to communicate any needed metadata for such additional functionality.

With connection-based transports, the situation can get even more confusing. Namely, a transport-layer connection has two roles: *initiator* and *listener*. The initiator is the active party, initiating the establishment of the connection. The listener is the passive party, waiting for any connection establishment attempt from initiators. In this book, the terms 'client' and 'server' always refer to application semantics, and the terms 'initiator' and 'listener' are used when transport-layer semantics are covered.

The roles at the transport layer do not need to have any connection to the roles at the messaging layer. In one-way patterns, it is not necessary for the initiator to be the one sending the messages. This can easily happen with the subscribe-notify pattern if transport-layer connections are not kept continuously open. Similarly, for instance, in the request-response pattern, the initiator does not also have to be the requesting party. An example of this is provided by the FTP active mode, where the response (file) is sent on a new transport connection, initiated by the server.

It must be noted that, apart from the subscribe-notify pattern, all of the patterns shown above are easily implemented without special functionality on the application layer. Namely, responses to messages are usually produced quite soon after the message has been received, so it is possible, with a connection-based transport, to simply keep the initial connection open and return the response over that.

With subscribe-notify this strategy is no longer feasible, though. Since the notifications can come arbitrarily far in the future, it is not wise from the resource-conservation point of view to keep the connection open indefinitely. Thus, the sender's locator needs to be included in the message itself to allow the notifier to send notifications to the subscriber.

An important consideration in message exchange patterns, especially the two-way ones, is whether they are synchronous or asynchronous. In a synchronous pattern, the sender of a message will block if the pattern dictates that a message is coming in its way, and in an asynchronous pattern, the sender will continue processing, so when the next message arrives, it needs to be notified somehow.

There are two principal ways to implement an asynchronous message exchange pattern, both of which have their advantages. The *polling* method gives the sender of a message an object called a promise, or a future, that represents a potential response message. The sender may inspect the promise to determine whether the promised message has already arrived, or it may claim the promise. If the message has not yet arrived, the claim operation will block until it does.

The other asynchronous interface is based on callbacks. The sender is given a function specification that it needs to implement. When the expected message arrives, this callback function is invoked by the messaging system. It is dependent on the messaging system's implementation and the programming language used as to how the arriving message is integrated into the main processing logic of the application. Usual methods are to use signals or to use a separate thread for invoking the callback function and using inter-thread communication to inform the main thread of the message.

The callback style is the more general one, as the promise style can be implemented by making the promise object be, or contain, the callback function. However, programming in the callback style is somewhat more difficult than in the more straightforward promise style, as it requires thinking about asynchronous events happening. The polling style, on the other hand, is very easy to even retrofit into an existing application programmed in the synchronous style.

6.2   Messaging Architectures

The text above considered messaging as application-layer functionality, communication between two applications running on (potentially) different hosts, as illustrated in Figure 6.4. Such communication consists of the applications establishing a transport connection between them (or, with connectionless transports, just sending datagrams or other messages to each other) and using that to communicate. The transport layer, in turn, relies on the routing functionality of the network layer to deliver its data between the endpoints through all the data links between them.

However, this can be limiting to application developers. The network layer does perform routing, but that is limited to the information available to the network layer. For instance, if the system architecture calls for mediators, or middleboxes, between communicating parties,[5] this view just splits the communication in two, as illustrated in Figure 6.5. Thus, the pure view does not permit considering the communication between the far ends of Figure 6.5, even though in many cases that is precisely what the application semantics means by communication.

Because of reasons like the ones detailed above, we do not limit messaging to simple point-to-point communication but will consider general distributed system architectures that consist of multiple components with application-layer connections between them, forming an application-level overlay network on top of the actual network built out of physical connections. We still limit the term 'messaging' itself to only general

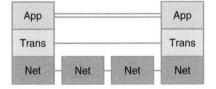

Figure 6.4   'Pure' view of the protocol stack

---

[5] A familiar example is provided by the web where proxies and gateways have become a practical requirement as traffic has increased to current levels.

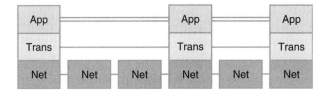

Figure 6.5   End-to-end messaging

considerations of such architectures, and leave out any specifics. For instance, the next chapter on publish/subscribe covers one method of routing; in our view the end-to-end communication there is messaging, but the specifics of route selection are out of scope for this chapter.

While the messaging overlay does not have to respect the actual network topology, it still makes sense to structure it in a similar manner. Namely, the basic concept is that there is core network dedicated to routing the messages and the actual applications connect to this core network at the edges.

Network-layer routing is usually focused on simply getting the information from the sender to the receiver. In contrast, application-layer routing can be more flexible, to take into account special needs of the full system.

One specific case of building messaging semantics that have no commonly-used counterpart on the network layer are store-and-forward systems. A store-and-forward system consists of the usual structure of a core network and messaging applications at the edges, but the applications in the core network are not reduced to simply forwarding information. Rather, they possess storage to store the sent messages, which they use when the next hop is not available (for instance, it could have been rebooted or crashed). Then, when the next hop does become available, any stored messages can be sent to it to continue on their journey towards their final destinations.

One benefit of the store-and-forward architecture is that it decouples the sender and receiver in time. There is no need for both to be available simultaneously as long as the receiver is available some time after the message is sent, since the message will wait inside the system for as long as is needed, or at least as long as storage space does not run out. This helps especially with mobile devices that typically do not maintain a continuous connection with the network, so sending messages between mobiles does not require synchronizing their network access.

Another benefit comes from optimizing reliable delivery, if such capability is built into the forwarders. Namely, there is no need to do full end-to-end retries if the message is still within the network moving towards its destination, since it is enough for the last forwarder holding the message on the path to perform any retries. Such functionality does require application knowledge at the end-hosts as well, and can be said to violate the end-to-end principle of networking, that is, that application functionality is not placed inside the communications subsystem.

6.3   Mobile and Wireless Communication

There are two main problems in communication when speaking about mobile devices. The first, mobility, is embedded directly in the name. The second, wireless networking, is an implication of the kind of mobility we are considering. It must be noted that neither strictly implies the other: there can be mobile hosts that do not communicate wirelessly and there can be wireless communication without mobility. In the cases we are interested in, though, both of these issues manifest themselves, and must be covered.

6.3.1 Mobile Hosts

When people speak about mobile hosts, there are a number of different things that it could mean. From a bird's eye view, there are two principal ways to be mobile ([4] uses the umbrella term *nomadicity*):

Portability Intermittent connectivity to the network, stationary while connected. An example is a traveller accessing the Internet from their hotel room.

Mobility Mobile while connected to the network, requires uninterrupted connectivity even when moving. An example is a traveller on a train[6] needing network access.

This bird's eye view is not the whole story, though. While portability is pretty much always the same, subdivisions in the concept of mobility can easily be made, as different environments call for (or allow) different technologies. Therefore, we subdivide mobility into three kinds:

Local mobility Mobility strictly limited in location, such as inside a single building.

Urban mobility Mobility in a town or city, slightly limited in range and with abundant infrastructure.

Wide-area mobility Any mobility, with extensive range and possibly in areas with little or no habitation.

The main concerns, from the middleware point of view, in these different kinds of mobility are related to the characteristics of the technologies that permit them. For instance, WiFi is pretty much limited to local mobility and, more recently, urban mobility, whereas cellular networks are designed from the start to support wide-area mobility. These mostly reflect themselves in the visible efficiency of the connection, as noted in Chapter 5.

In theory, a messaging system should be pretty immune to the problems of mobility. When looking at the message exchange patterns shown earlier, there is very little need to require long-term connectivity. One-way patterns should not need it at all, and the typical two-way patterns only need intermittent connectivity. In practice, however, if mobility is not considered in system design, there will typically be an implicit assumption of stationary hosts, with potential subtle problems that manifest when mobility is brought in.

It can be argued that mobility should be handled only on the network layer. Namely, since it is the purpose of the network layer to get the data where it is supposed to go, that should be the only place that cares about the location of hosts. That is why the claim that Mobile IP is sufficient to handle all mobility issues is sometimes seen.

This view is not without its problems, though, especially when considering where mobile applications are going. Context-aware applications will need to be aware of their surroundings and therefore of the device's mobility. Relegating all mobility functionality

---

[6] There are characteristics of mass transportation that make the mobility problem there slightly different from what it is when individual users are moving independently. We shall not dwell on these differences.

to the network layer will make applications unaware of it, making profiting from mobility needlessly difficult. This is the basic contrast, between seeing mobility as a nuisance to get over and as an opportunity for new kinds of applications.

Perhaps the main problem posed by mobility is treating the mobile hosts as second-class citizens. Absent systems such as Mobile IP (the deployment rate of which is less than stellar), a mobile host is going to change its network address as it moves. As the locators of many messaging systems are predicated on the assumption of stationary hosts, a changing address usually means that the locator is also going to change. This makes it hard for the mobile host to be the listener end of transport traffic.

This would not be such a problem if messaging systems were designed to handle this problem. First of all, there is no need to require the locator to be tied to the network address. Second, as we noted above, while the mobile host might not function well as the listener of transport traffic, there is no need for the messaging layer to mirror the roles on the transport layer. However, either separating the messaging locator concept from the network address or making sure that messaging is possible even when only one end can open transport connections are both extra work for the system designer, and since using network addresses and mirroring transport layer roles works 'well enough', it is rare to do anything more complicated.

6.3.2 Wireless Links

The problems caused by wireless links are not similar to the problems caused by mobility. In the case of mobility, the issues are mostly on the semantic level: they require explicit design in the messaging system architecture to take mobility into account. On the other hand, wireless communication links usually just cause poor performance, which is not a mobility-specific issue. Still, designing for the possibility of wireless links may be required in a messaging system, but it is not always necessary.

Network performance can roughly be said to consist of three components: data rate, latency, and error rate. Table 5.1 presented data rate and latency figures for some common wireless network technologies. For comparison, a typical wired local area network has a data rate of 100 Mbps or 1 Gbps and sub-millisecond latency, and these characteristics are similar for short-range Internet connections as well. Bit error rate for Ethernet can be as high as 1 in 10^8 but is more commonly around 1 in 10^{12}, whereas in wireless technologies the upper bound is more commonly something like 1 in 10^5.

One point to consider when discussing error rate is where error correction happens, and this is where the two major technology families differ. In cellular networks, the data link layer includes error detection and retransmission functionality, so packet loss and errors are rarely visible on the transport layer. In contrast, the WiFi link layer is much simpler, so errors there reflect on the transport layer more often.

The place where errors or data loss are detected and corrected is not only of academic interest. When using TCP, lost segments are interpreted as congestion, which triggers the slow start phase. Therefore frequent transport-visible data loss can lead to severe underutilization of the network capacity.

The interesting measurements to perform for a messaging system are typically just data rate and latency. A measurement that has real significance for the messaging system designer needs to be performed at the transport layer, so the data rate and latency measurements will include all the necessary error correction to handle the networking technology's error rate already. This is why we have not spoken much about the networking technology's error rate: our primary interest is in the effective data rate and latency that we can get from the transport layer.

Messaging systems are also not used to transfer data in bulk. To many, it is implicit in the term 'messaging' that the sent data consists of smallish messages, and other protocols, such as direct FTP or HTTP, are used to transfer large amounts of data. If needed, the messaging system may act as a control channel for the data transfer. Something similar is seen in SIP where the proxies are involved in managing the sessions, but the sessions themselves are typically directly between the user agents.

Because of this use of messaging systems, the most critical characteristic of the underlying communication channel is typically latency. While data rate is often cited as a reason for reducing message sizes, in many cases the real reason is actually latency. Namely, when using TCP, which is often used to get certain reliability guarantees, the initial slow start used by TCP will increase the number of round trips needed when the message sizes are large. This effect can be ameliorated by keeping TCP connections open, but that may then run into issues with mobility, as changing IP addresses during mobility lead into TCP connections breaking, possibly causing problems with the application.

There has been much work done in improving TCP performance over wireless links [5]. Most of the original work, however, was done in the context of Wireless LAN, where the problems are quite different from those of cellular networks, though more recently it has been acknowledged that cellular networks are interesting as well.

One common thread through much of the TCP over wireless work has been to use a proxy-based approach (see [6]). The primary concept behind this is that since a wireless link and a fully-wired connection have completely different characteristics, it is possible to improve performance by splitting the TCP connection so that one part is strictly wireless and the other strictly wired. Such an approach can indeed lead to performance improvements, and, when fully split, could also permit the use of alternate protocols over the wireless link that are more suitable for those conditions.

The main drawback of proxy-based approaches is that they may break TCP's end-to-end semantics. The idea is that the proxy is transparent and it looks to the end hosts as if they are communicating directly with each other using TCP, but if the proxy cannot obey TCP semantics so that congestion control in the end hosts works properly, this can lead to problems.

6.3.3   Two-way Communication

A very common current problem on the Internet, especially prominent in mobile computing is the use of NAT (network address translation). With NAT, a host does not receive a routable address, but instead an address in a private address space. A NAT gateway in the ISP's network provides translation of this address into a publicly routable address and ensures that traffic back is forwarded to the host itself.

Such a solution is of course feasible only when the hosts behind NAT act only as clients to services and do not attempt to provide services themselves. In fact, even some client applications may be affected. For instance, FTP in active mode requires connections to be opened back to the client, and IRC file transfer also requires ports to be opened for listening. Some NAT gateways, for instance, the one in the Linux kernel, have extensions that let them recognize certain common protocols and rewrite the IP addresses and port numbers inside the protocol data. This solution, of course, requires special code in the NAT gateway for each different protocol to be supported.

NAT is also incompatible with the Internet's basic principles of every host having its own IP address and being contactable. The exhaustion of IP version 4 addresses and the increase of dynamic connection methods are probably the main causes of NAT becoming so widely used. Arguments about security and firewalling are mostly red herrings, as there is no need to use NAT to implement a firewall policy. If IP version 6 is ever deployed all the way to consumers, these concerns with NAT may disappear, but that is unlikely unless consumers begin demanding all the features possible with IP version 6.

6.3.4   Designing for Mobility

We noted that it is not typically sufficient for middleware to rely on lower layers to handle mobility and wireless links. Instead, the design of the middleware needs to take these possibilities into account. This can easily come into conflict with the desire to maintain compatibility with existing systems or with using existing technologies and standards as much as possible. Considering the pieces of a messaging system as we have defined them, potentially all of them are affected by host mobility.

The messaging system architecture is obviously affected by mobility. It is not sufficient to rely on the connections between hosts to be functional forever, or even functional intermittently. Instead, the architecture needs to prepare for the case that a host leaves some location and turns up somewhere completely different.

Depending on how the locator has been designed, it may be affected by mobility as well. If the messaging system has been designed as a thin layer on top of the underlying network infrastructure, it may well be that network addresses are embedded inside the locator. This means that either the network addresses need to support mobility or the components need to understand that locators may change.

With a more opaque locator that only identifies the components and not their network locations, the locator itself may not need to be modified to take mobility into account. In this case, however, there needs to be some form of a resolver that translates locators into information usable for transmitting messages. Bringing mobility into play can therefore affect the design of the resolver, for instance, by requiring it to handle rapid changes in associations between individual locators and their network addresses.

The messaging protocol itself is perhaps the most affected by mobility and wireless communication. First, if we look at the wireless communication, the protocol needs to be lightweight to avoid causing too much overhead. Also, it needs to avoid needless round trips in implementing the required message exchange patterns.

The effect of mobility depends on the specifics of the protocol. A protocol that can transparently reopen transport-layer connections is preferable as there should be no need to modify such a protocol to take mobility into account. To distill this into a more general requirement, a messaging protocol must be able to complete a message exchange pattern even if one or both parties move during the execution of that pattern.

The only thing to consider with message syntax is that it needs to be compact to ameliorate the problems of wireless communication. Compression is in many cases feasible, so even heavier syntaxes can be acceptable if compression functionality is available. Also, compression often actually reduces total processing time, since the network is often the bottleneck in wireless communication, so reducing the amount of data transferred will have more of an effect than increasing processing time by adding compression. However, in many cases, especially on lower-end hardware, the application footprint of a compression library may be prohibitive, so bolted-on compression cannot be seen as a panacea that solves all the problems.

6.4 Security

Security is perhaps the most crucial of the so-called non-functional requirements of messaging systems. Since typically messaging is seen as a large-scale operation with independent actors, there are no centralized guarantees of who the actors are. All three aspects of security, confidentiality, integrity, and availability, find their place in messaging.

Often, security is confused with the technologies that make security possible, thinking that adding encryption or digital signatures to a system make it secure. This, of course, is not entirely correct, even though the specific technologies are needed to achieve the needed security properties, but it is also necessary to know what are the attacks to protect against and where to apply each technology to guard against the attacks.

In a distributed system with routing between communicating peers, it is important to distinguish between two notions of security. These notions apply more widely, but for simplicity we formulate them in the context of messaging as defined above:

Hop-by-hop security Any security properties that apply between peers that communicate through the transport layer, that is, peers that do not require application-layer routing.

End-to-end security Any security properties that apply between peers communicating in the application sense, that is, peers that the whole distributed application, not the messaging system itself, considers communicating.

Both of these notions may have their place in a system. It is important to note that neither can be substituted for the other, even though a system possessing one may accidentally possess the other as well. Instead, the security analysis must identify whether each is needed, independently of the other, and design them in the system if necessary.

In particular, a common security protocol nowadays is SSL (Secure Sockets Layer). This is used at the transport layer to provide a 'secure session' between the communicating parties. This concept includes optional authentication of the parties and encrypted communication throughout the session. Many messaging systems base their security on the ubiquitous SSL, but we will argue below that this is not sufficient.

6.4.1 Confidentiality

Confidentiality is the property of messages not being readable by unauthorized parties. The usual way of achieving confidentiality is by encryption, and over the public Internet there is no other way.[7] Still, proper design of confidentiality is more about where to apply encryption than in, say, selecting encryption algorithms.

Very often in networking, confidentiality is treated only as protection against network-level eavesdroppers, either people who control the network routers or people

---

[7] In private, controlled networks, there is less need for encryption of communication. And of course, avoiding communication altogether also obviates the need for encryption, even though confidentiality may still be a desired property.

who have some access to the physical networks used in transmission. This works in the 'pure' model of Figure 6.4 where the transport endpoints are also the ones communicating, but fails in the messaging system architecture that we consider here.

Instead, a messaging system needs its own confidentiality protection, applied to individual messages instead of to transport connections (see more below in section 6.4.3). Naturally it may be valuable to be able to protect the messages by securing the network traffic as well, but in a messaging system confidentiality design this should be treated as secondary, since it has much less impact on the design than achieving proper confidentiality.

6.4.2 Integrity

Integrity has two components: first, an assurance that messages have not been modified after sending, and second, that the communicating parties are who they claim to be. These properties are typically provided by digital signatures. For the former, a secure hash of the contents would be sufficient, except for the fact that someone modifying the message can easily compute a new hash value as well.

Integrity is a common requirement in systems, especially ones where auditing of events is a required part. As with confidentiality, in a messaging system it is not typically sufficient to have integrity as a part of the communication, but it must be included in the actual messages as well. Another reason for not relying on the integrity of the communication is that, in a routed system, the sender of the message may well be different from the one who eventually delivered it, and auditing requires the actual sender to be verified, not (necessarily) the hops in its transmission.

The main problem with authenticating the communicating parties is the question of how to trust the identities. If the two parties are previously unknown to each other, as is likely the case in Internet-scale communication, how can they 'bootstrap' trust between them? There are ways to do this, but unfortunately none are very good.

The most common in use is the public-key infrastructure (PKI) typically used in conjunction with SSL on the web. It is based on there existing a number of certificate authorities that are implicitly trusted. These certificate authorities sign the public keys of various parties, which is called providing certificates of the identities, and the signatures can then be checked to ensure that the correspondent really is what it claims to be.

The problem with this is that it does not work to provide security [7]. It works in an enclosed system where it can really be assumed that everyone trusts the certificate authority, for instance, an organization's internal communication. But when trying to scale it to the Internet, any trust is lost. There is no way for a user to verify each of the common certificate authorities used on the web and to vet their processes to ensure they have appropriate safeguards against issuing incorrect certificates.

An alternate approach, used in PGP [8], is the Web of Trust. Here, there are no certificate authorities. Rather, each user is able to sign other users' identities and each user can determine themselves how much to trust other users' signatures on keys. Thus, each user has two trust values for a public key: do they trust that the key belongs to the person identified, and how much do they trust that key in identifying others. PGP uses a simple algorithm to calculate trust paths to keys that have not been explicitly marked as trusted using the signature, or owner, trust values.

The PGP Web of Trust is much more complicated than the idea of a hierarchical PKI. The need for two different trust values is particularly complex, but absolutely necessary, since, for instance, there is no way to prevent a user to certify absolutely every key; such a user should never be trusted to verify anything, but there is no obstacle in trusting that the user's key does correspond to the specified identity.

6.4.3 Message-level Security

The most common way of achieving security in messaging currently is to use SSL to establish a secure session between the sender and the receiver. The main problem with this approach is that it follows the view in Figure 6.4 by assuming that the transport-layer connection is end-to-end whereas we have noted that application-layer routing is often necessary. Using only SSL means that all the intermediate routers will need to be as trusted as the receiver, since they are indistinguishable in this method.

The SSL-only method can be called *connection-level security*, since it secures only the connection. On the other hand, *message-level security* is for securing the message itself. This separation of concerns between the message and the connection allows us to consider security on two different layers, as depicted in the case of encryption in Figure 6.6.

Here a message consists of a header and content, with the content shown divided into public (P) and secret (S) parts. With no encryption, everything is in plain sight. The differences between connection- and message-level security are evident: with connection-level security the whole message is obscured in transit but fully visible at the first receiver whereas with message-level security the secret part is obscured throughout. Naturally these two approaches can be combined to fully protect the content while in transit (from eavesdroppers) but also to protect the sensitive parts from any application-layer intermediaries.

We have shown the message to be somewhat structured in that there are independent parts that may need to be protected independently. One could imagine a half-way point between this and connection-level security, namely that any security is applied to either the whole message or nothing. Such a scheme is required when the middleware does not know about the message structure or when there is no specified way on how to protect

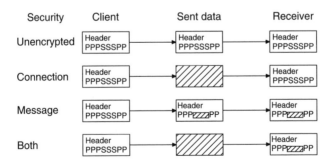

Figure 6.6 Different types of security in messaging

only parts of the structure. However, in many use cases for message-level security, this simpler scheme essentially reduces to connection-level security.

There are two major specifications (or specification families) that provide message-level security. S/MIME (RFC 3851) is for MIME messages, specifying how the multiple parts in a MIME message can be individually signed or encrypted. XML Signature and XML Encryption are for XML, allowing any element or element content inside an XML document to be individually signed or encrypted. In particular, the XML security specifications are going to have a major role in the flexible messaging systems used in the future that require message-level security.

6.5 Reliability

Depending on the application design, reliability can be a crucial property, as making sure that messages get to their intended destinations often has significant business consequences. Like with almost any property, reliability is not a simple on-off matter, but has various levels, depending exactly on what is desired.

Reliability has two components: reception and integrity. Reception means that the message is guaranteed to reach its receiver. Integrity means that the message is not corrupted during transit. Note that integrity in the context of reliability is a weaker property than in the context of security, since reliability only considers accidental errors whereas security considers a malicious adversary that can arrange errors to his best advantage. This is a common difference between security and other properties: security must tackle an active adversary whereas other properties only combat random faults when probabilistic guarantees are sufficient.

First of all, since we are considering routed communication, there are two kinds of reliability:

Hop-by-hop This refers to the message being reliably transmitted across each individual link in the routing chain. When considering the network layer, this refers to the link layer, but with a messaging system as we have defined it, it rather refers to the application-layer hops between messaging routers.
End-to-end This refers to the message being reliably transmitted to its ultimate receiver.

Even though a communication path may possess hop-by-hop reliability for each hop, this is not sufficient for end-to-end reliability. For instance, consider the Internet, where link layers are typically reliable. Still, it may happen that inside a router a packet is dropped due to congestion, so end-to-end reliability requires mechanisms on the transport layer.

Achieving integrity is typically a matter of computing checksums. Since only random faults are considered, this checksum can be very simple, like just a sum of the bytes or, if it is more complex, still efficiently computable like a CRC method. It is actually rare for middleware to even consider integrity of messages, since the transport layer usually has sufficient integrity protection, and middleware routers often preserve only the semantics of messages, not their byte representations, so computing end-to-end checksums would be futile.

Achieving reliable reception has one basic technique: acknowledgments and retransmissions. That is, when a receiver receives a message, it sends back a special message, an acknowledgment, to confirm to the sender that the message was received. If such an acknowledgment does not appear within a certain time, the sender resends the message. Thus, messages need to possess some unique identifier so that the sender can recognize which message is being acknowledged. Most often, this is a running sequence number.

There are some refinements to the basic acknowledgment idea. If there is message traffic in both directions, the acknowledgments can *piggy-back* on the message traffic, avoiding the need to create separate acknowledgment messages. If the sequence numbers are consecutively increasing, a *cumulative* acknowledgment allows acknowledging all messages up to a specified sequence number. And finally, *negative* acknowledgments may be used to indicate positively that some message was never received, which can be detected if there are gaps in the sequence numbers.

As examples of different reliability mechanisms, TCP by default uses cumulative acknowledgments and can use piggy-backing as well. TCP does not have specific negative acknowledgments, but it has a feature called *selective* acknowledgments, which is often used in conjunction with cumulative acknowledgments to acknowledge messages arriving out of order. Such an acknowledgment can also serve as an indication that some messages were lost. Other protocols exist that are explicitly based on negative acknowledgments: since networks are mostly reliable, sending positive acknowledgments just creates more traffic, so such protocols only send negative acknowledgments when a message has gone missing.

End-to-end reliability also has a number of levels. The simplest level is to ensure that the receiver host has received the message. In this case an acknowledgment is sent by the messaging protocol. But since messaging systems can experience congestion, another, more useful level would be when the application has accepted the message for processing. Finally, even requiring the receiving application to process the message before sending an acknowledgment could be considered as a reliability feature. These different levels can be seen in Internet email: the usual reliability of an SMTP connection only means that the message has been accepted by the receiver, but it is often important to know that a message has been read, for which there also exists possibility for automation. Finally, a reply message from the receiving human will provide acknowledgment of the message having been processed.

In-order delivery is a special case of reliability. A messaging system that guarantees in-order delivery will deliver a sequence of messages to the receiver in the same order that the sender sent them. Some messaging systems go even further, so that if the messages were the requests in a series of request-response interactions, the responses will also be delivered in the same order. This has both good and bad sides, as the guarantee helps in application design but the implementation requires buffering of messages on the receiver end. Many systems treat each individual message as a separate interaction, sacrificing this property, as they do not consider the upsides to be worth enduring the downsides.

Full reliability is not always needed, as many applications can function with best-effort service. Including reliability adds complexity to the messaging system, so if expected applications will not need it, that is just a burden for both the implementor and application developer.

Several application-layer protocols on the Internet are specified to run, or are commonly run, on top of UDP, which does not provide any reliability. Examples are DNS, a critical part of the Internet infrastructure, NTP that cares more about timely delivery than dropped messages, and SIP, which can use TCP, but is commonly run on top of UDP.

Reasons for selecting unreliable messaging vary, as do the methods of handling any problems caused by it. The above protocols provide a few typical reasons. Both DNS and SIP care about being implementable with limited resources, and a full TCP implementation might have too much overhead, both in code size and in the amount of network traffic. DNS also cares about server resource consumption. By being UDP-based, DNS requires no connection state to be kept on the server, which leads to much higher scalability. Finally, as noted, NTP requires timely delivery of messages, so a retransmitted message would actually be useless and could even lead to incorrect time information.

Usually the way to handle unreliability in the above protocols is to still perform retransmissions. But instead of the retransmissions being hidden inside the protocols, they are fully visible to the application. While superficially this looks the same, it is in fact a completely different situation. The application retransmission is treated as a new interaction, so any problems associated with protocol-level retransmissions do not apply. And modern networks are sufficiently reliable in themselves that best-effort service works pretty well with a very small number of retries required.

6.6   Java Message Service

Java Message Service (JMS) [9] defines a generic and standard API for the implementation of message-oriented middleware. The JMS API is an integral part of the Java Enterprise Edition (Java EE). JMS is an interface and the specification does not provide any concrete implementation of a messaging engine. The fact that JMS does not define the messaging engine or the message transport gives rise to many possible implementations and ways to configure JMS.

JMS supports a point-to-point (queues) model and a publisher/subscriber (topics) model (Figure 6.7). In the point-to-point model only one receiver is selected to receive a message, and in the publisher/subscriber model many can receive the same message. The JMS API can ensure that a message is delivered only once. At lower levels of reliability an application may miss messages or receive duplicate messages. A standalone JMS provider (implementation) has to support either point-to-point or the publish/subscribe approach, or both. Normally, JMS queues and topics are maintained and created by the administration rather than application programs. Therefore the destinations are seen as long lasting. The JMS API also allows the creation of temporary destinations that last only for the duration of the connection.

The point-to-point communication model consists of receivers, senders, and message queues. Each message queue is addressed to a particular queue, and receivers extract messages from the queues. Each message has only one consumer and the client acknowledges the successful delivery of a message to the component that manages the queue. In this model there are no timing dependencies between a sender and a receiver; it is enough that the queue exists. In addition, the JMS API allows the grouping of outgoing messages and incoming messages and their acknowledgments to transactions. If a transaction fails, it can be rolled back. In the publish/subscribe model the clients address messages to a topic. Publishers and subscribers are anonymous, and messaging is usually one-to-many. This model has a timing dependency between consumers and producers. Consumers receive messages after their subscription has been processed. Moreover, the consumer must be active in order to receive messages.

The JMS API provides an improvement on this timing dependency by allowing clients to create durable subscriptions. Durable subscriptions introduce the buffering capability of the point-to-point model to the publish/subscribe model. Durable subscriptions can accept messages sent to clients that are not active at the time. A durable subscription can have only one active subscriber at a time. Messages are delivered to clients either synchronously or asynchronously. Synchronous messages are delivered using the receive method, which blocks until a message arrives or a timeout occurs. In order to receive asynchronous messages, the client creates a message listener, which is similar to an event listener. When a message arrives the JMS provider calls the listener's onMessage method to deliver the message. JMS clients use JNDI to look up configured JMS objects. JMS administrators configure these components using facilities specific to a provider (implementation). There are two types of administered objects in JMS: ConnectionFactories, which are used by clients to connect with a provider, and Destinations, which are used by clients to specify the destination of messages. JMS messages consist of a header with a set of header fields, properties that are optional header fields (application-specific, standard properties, provider-specific properties), and a body that can be of several types.

Message selection is supported by filtering the message header against the given criteria using an SQL grammar. A JMS message selector allows clients to define the messages they are interested in. Headers and properties need to match the client specification in order to be delivered to that client. Message selectors cannot reference values embedded in the message body. An example is JMSType='stock' AND company='abc' AND stockvalue > 100. JMS supports five different messages types: Map, Object, Stream, Text, and Bytes. MapMessage is a set of name/value pairs, where names are strings and values are primitive Java types. ObjectMessage is a message containing a serializable Java object. StreamMessage is a stream of sequential Java primitive values. TextMessage represents an instance using the java.lang.String class and can be used to send and receive XML messages. BytesMessage is a stream of bytes.

Typically a JMS client creates a Connection, one or more Sessions, and a number of MessageConsumers and MessageProducers. Connections are created in the stopped mode. After a connection is started (start() method) messages start arriving to the consumers associated with that connection. A MessageProducer can send messages while a Connection is stopped. A Session is a single-threaded context for consuming and producing messages. Sessions act as factories for creating MessageProducers, MessageConsumers, and temporary destinations. JMS defines that messages sent by a session to a destination must be received in the order in which they were sent.

Messages are acknowledged automatically in the transactional mode (supported by the Java Transaction API), however, if a session is not transacted there are three possible options for acknowledgment: lazy acknowledgment that tolerates duplicate messages, automatic acknowledgment, and client-side acknowledgment. In persistent mode delivery is once-and-only-once, and in non-persistent mode the semantics are at-most-once. JMS messaging proceeds in the following fashion:

1. Client obtains a Connection from a ConnectionFactory.
2. Client uses the Connection to create a Session object.
3. The Session is used to create MessageProducer and MessageConsumer objects, which are based on Destinations.

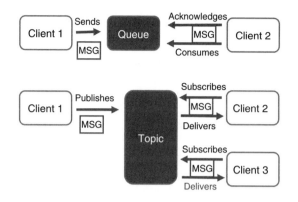

Figure 6.7   Overview of queue and topic based communications

4. MessageProducers are used to produce messages that are delivered to destinations.
5. MessageConsumers are used to either poll or asynchronously consume (using Mes-
   sageListeners) messages from producers.

The JMS API does not address load balancing, fault tolerance, error notification, administration, or security. JMS implementations are available from many vendors, such as IBM
(it is supported in MQSeries), Sun Microsystems (J2EE), The ExoLab Group (Open-
JMS), SoftWired (iBus//Mobile), and Oracle (8i and later). The latest JMS version is 1.1,
which incorporates changes approved by a Java Community Process program Maintenance
Review that closed on 18 March 2002. In JMS 1.0.2 client code must use the queue and
topic interfaces, and it is impossible to reuse queue clients with topics. JMS 1.1 supports
client code that works simultaneously with either the point-to-point or publish/subscribe
domains. Queues and topics can be accessed through the same session and thus in the
same transaction.

6.7   CORBA and CORBA Messaging

Common Object Request Broker Architecture (CORBA) [10] is a specification for distributed objects. Distributed object technology is an object-oriented extension to remote
procedure calls, where the remote invocation targets are treated as objects in a class-based
language with fields and methods. The basic model for CORBA comes from C++.

CORBA also includes a Messaging Service, which provides more facilities for the
kinds of messaging that we are interested in, so this section is going to cover both regular
CORBA and its messaging system. We covered Wireless CORBA already in the previous
chapter, but we will point out below how that fits into the messaging vocabulary we are
using.

6.7.1 Architecture

As the name indicates, CORBA is a Broker architecture. That is, there is a logically
centralized entity called an ORB (for Object Request Broker) that mediates all communication. CORBA is a distributed object platform, meaning that the components of the

system are objects as defined in object-oriented programming. The method invocations on objects thus go through the ORB, which is responsible for locating the target object, invoking the method, and returning the result.

CORBA also includes a number of different services, divided into horizontal (needed by CORBA applications in different domains) and vertical (of interest primarily to applications in a single domain, such as telecommunications or finance). The primary ones of such services are the Naming Service, mapping hierarchical names to object identifiers, Trading Service, mapping service types to object identifiers, and Event Service, providing simple publish/subscribe functionality.

The basic CORBA architecture is based on single-invocation RPC: the client sends an invocation and gets a value or values out of that. Furthermore, in accordance with local procedure call semantics, the client will block waiting for the return values. As we have noted above, this is rarely the correct thing to do in mobile computing, so CORBA definitely has some problems in this domain.

Current CORBA versions have incorporated the CORBA Messaging specification inside. The Messaging specification defines, primarily, two different parts, the Asynchronous Messaging Interface (AMI) and Time Independent Invocations (TII). AMI provides the possibility of asynchronous invocations and TII provides message routing functionality.

As is common with CORBA-related specifications, AMI includes both options for asynchronous request-response interaction, namely polling and callbacks. CORBA also takes advantage of the fact that an invocation always looks the same to the server, so an AMI implementation is purely a client-side matter. Also, in the callback style, the callback object provided in the invocation is a CORBA object like anything else, so it doesn't have to live in the same address space or even host as the client.[8]

TII is a way for the client to specify certain CORBA objects that act as routers for the message. The idea behind this is that if the client and server are only intermittently connected to the network, it is possible that they are too rarely connected together, making direct communication between them infeasible. Thus, the object reference can specify a number of routers through which the message can be passed. These routers form a store-and-forward network, allowing intermittent connectivity of the client and server.

6.7.2 Locator

CORBA objects are addressed through IORs (Interoperable Object References). An IOR contains a number of different profiles, each typically associated with a single protocol binding (such as the IIOP profile for communicating over the Internet). A profile contains all the information needed to contact an object over the specific protocol used by that profile. For instance, in the case of an Internet protocol, the profile would contain the target host name and the port that the server process is listening on. In addition, the profile contains an object key that lets the server process identify precisely which object the invocation is to be redirected.

The IOR is defined in IDL as a structured type, but naturally, when transmitted, it needs to be encoded using CDR. This is why it is defined in an extensible manner, since it cannot

---

[8] Of course, since AMI is purely on the client, the invocation response still has to travel via the client host even if the callback object is somewhere else.

be assumed that each ORB using an IOR knows all the profiles contained in that IOR.[9] Namely, the data in a profile is actually defined to be a sequence of bytes, accompanied by a tag identifying the profile type. Thus, even if an ORB does not recognize the profile type, it knows how to extract the sequence and go on decoding the rest of the IOR. On the other hand, when an ORB does recognize the profile type, it can apply CDR decoding to the sequence of bytes to get an appropriate profile structure.

6.7.3 Syntax

We already alluded a couple of times to CORBA's message syntax, CDR. CDR is a binary format that is capable of representing different kinds of typed data, the ones definable in IDL. The representation of primitive types like booleans, integers, and floating-point numbers, use typical machine representations so that marshaling and unmarshaling CDR messages is usually quite efficient.

For choosing the endianness of multi-byte data, CDR uses the receiver-makes-right model, that is, each message contains a byte-order flag that tells the receiver which byte order was used in encoding the message. Some people consider this to be a mistake that needlessly complicates CDR decoding.

One peculiarity of CDR is that all data is aligned on a suitable boundary. That is, a 4-octet value such as `long` is always aligned on a 4-octet boundary. The original intent was that this would potentially permit even faster encoding and decoding if a machine could simply copy data in bulk between its main memory and the CDR buffer. In the end, it seems that the only effect was a more complex implementation of CDR encoding and decoding, as apparently the potential gain in efficiency was seen to be too small for the necessary checks needed in implementing the optimization.

6.7.4 Protocol

CORBA is nothing if not flexible in its use of protocol. Originally, there was no protocol specified by CORBA, so ORBs from different vendors would not interoperate unless the vendors made a specific effort to do so. This changes with version 2 and the interoperability specification for CORBA. Still, CORBA allows any communication protocol to be used, and provides only some guidelines for implementing protocols.

One specific protocol was defined to be mandatory for interoperable CORBA implementations, though. This is the General Inter-ORB Protocol, or GIOP for short. GIOP is a message-based protocol that is intended to be run over some transport protocol. The requirements that GIOP places on the transport protocol are pretty much those that are provided by TCP, and accordingly, the Internet Inter-ORB Protocol, or IIOP, which is GIOP running on top of TCP, is also a mandatory part of CORBA interoperability.

The primary messages of GIOP are Request and Reply. These comprise the regular invocation behavior in CORBA. It is also possible for the client to send a LocateRequest message, the reply to which will tell the client where the object is located. This allows the client to locate the object for certain without needing to invoke with the full request,

---

[9] In fact, some ORB vendors might want to embed proprietary profiles to make communication with their ORB more efficient.

as an invocation reply can be a redirection. This is especially useful for load balancing with a so-called Implementation Repository.

Other parts of GIOP are less often used. It is possible to fragment a GIOP message, similarly to what IP does, but here the intent is to save buffer space of the ORB if the message is too large to hold in memory while being constructed. A client can also cancel a request at any time before receiving a reply. Finally, there is some connection management in the form of the ability to close the transport-layer connection from the server end. This is required at the GIOP level because, while a closed transport connection can be detected, it is possible that a client request had been initiated when the server decided to close the connection, so the client needs to be informed of this to be able to resend any outstanding requests on a newly-opened connection.

The roles in basic GIOP are rigidly defined: the party opening the transport connection is the client and the other party is the server, and the client is allowed to send only requests and the server only replies. This causes problems, for instance, in callback scenarios, which are very useful in distributed systems. For one, opening a second connection is wasteful of resources. But more importantly, it is not always even possible for the server to open a connection to the client, as firewalls may be blocking the communication. This is why GIOP now also includes bidirectional functionality: the ability to use the same transport connection for invocations in both directions. This is commendable, but the original design does show quite clearly how people become fixated on the idea of reflecting transport-layer roles on the application layer as well.

6.8 XMPP

The *Extensible Messaging and Presence Protocol (XMPP)* [11] (RFC 3920 based on the Jabber protocol) has been designed for instant messaging with support for extensions. Today XMPP can support different message-based communication styles. XMPP extensions include publish/subscribe mechanisms, presence and status updates, alerts, feature negotiation, service discovery, and other features that make it suitable as an asynchronous middleware solution. XMPP is becoming increasingly popular on the Internet with companies such as Google, Twitter, and Facebook using XMPP as a general API.

XMPP can be implemented using both peer-to-peer and client-server interaction models. In the basic case, a client connects to a server and opens an XMPP stream to the server. After negotiating connection parameters, such as authentication and encryption, the client and server can send XML stanzas over the connection. If a client sends a stanza to an entity that is not known to the server, the server will negotiate using a server-to-server protocol with the targeted foreign domain. The message will be sent to the corresponding foreign server for delivery to the ultimate destination. Figure 6.8 presents an overview of the XMPP protocol and the interaction models.

XMPP defines three core stanza types, each with different semantics:

- The message stanza is very similar to email. A message stanza is pushed by one entity to another entity.
- The presence stanza is a basic broadcast or publish/subscribe mechanism. Multiple entities can receive information to which they have previously subscribed.
- The Info/Query stanza is a request-response mechanism. This is similar to HTTP and enables an entity to make a request and receive response to it.

An XMPP server is responsible for managing XML streams and XML stanzas. Typically XML streams use long-lived TCP connections and XML stanzas are first-level child elements sent over the stream. Thus XMPP is based on incremental parsing of XML. According to RFC 3920, a server implementation must support TLS for channel encryption, SASL for authentication, and DNS SRV records for port lookups. Servers stamp XML stanzas with validated 'from' address to prevent bogus messages.

XMPP-based instant messaging and presence servers provide more services including IM session management, contact list storage, presence subscription management, and management of block lists (RFC 3921) network.

XMPP is an interesting candidate for mobile communications due to its popularity. However, XMPP was not designed for mobile networks and there are a few problems related to XML processing and energy management. Particularly problematic are reconnects to the XMPP server which are more frequent in a mobile network. One of the limitations of XMPP is that the protocol needs to re-establish a new session on every connection. This involves exchanging presence data in XML and parsing this data. This means that the amount of data exchanged and the related processing result in significant overhead on mobile devices. This has led to the adoption of binary non-XML-based protocols. For example, Android uses a binary protocol that avoids the need for creating a new session for each connection.

6.9   Web Services

While the term 'web services' is nowadays used to refer to many different kinds of services used through web technologies, in this book we will limit ourselves to the 'original' meaning, that is, a collection of technologies based around the SOAP protocol [12], as depicted on a general level in Figure 6.9. Regarding messaging, our primary interest is SOAP itself, though the architecture is somewhat more general.

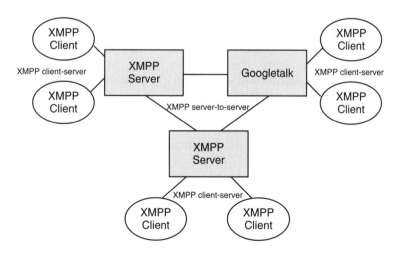

Figure 6.8   Overview of XMPP

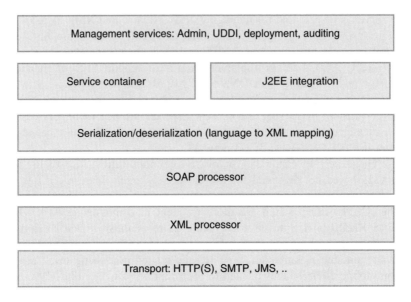

Figure 6.9   Detailed view to web services stack

6.9.1 Architecture

The Web Services Activity at the W3C originally included an Architecture working group chartered to define an architecture for web services. This work partially fizzled, as the group determined that it is not feasible to produce such a document on the W3C Recommendation track. The effort was not a total loss, however, since it did produce a Note describing web services architecture [13].

The W3C Web Services architecture is in fact a collection of different models, most probably because the group could not come into agreement on which model is the fundamental one (a common ailment in standardization groups trying to define some high-level concepts). These four models are:

Message-oriented This model focuses on the actual messages, their structure, syntax, and transport.
Service-oriented This model focuses on the services, how they are described, who provides them, and who controls them. Messages might be used to invoke services but this is relegated to a lower importance.
Resource-oriented This model mirrors the web architecture, see section 6.10.
Policy This model focuses on policies, constraints on the behavior of services and their invokers.

All of the focused components are typically present in each of these models; the differences between the models are limited to which component is the primary one.

When looking at things on the SOAP level, as we do in this section, the model to use is the message-oriented one. In the SOAP message processing model, SOAP processors are divided into *senders*, *receivers*, and *intermediaries*. These roles are illustrated in Figure 6.10. The sender sends the message, it passes through zero or more intermediaries,

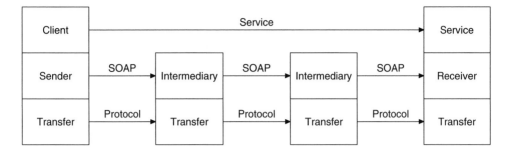

Figure 6.10   SOAP message processing

and arrives at the receiver. The underlying message transfer layer is responsible for transferring the SOAP message.

SOAP intermediaries function as application-level routers (recall Figure 6.5). The ultimate view of the client application is that it invokes on the service located at the receiver node, in accordance with the principles of layered architecture.

6.9.2   Locator

Being that SOAP is intended for something called web services, its locators are those of the web, namely URIs and most often HTTP URIs, since HTTP is often configured to get past firewalls. In SOAP, the entity identified by the locator is a service, which can perform a number of different operations on behalf of its client.

The usual method these days is to use document-style SOAP, where the whole SOAP message is something to be processed by the service. This is in contrast with the older RPC-style SOAP where the entity identified by the locator was an RPC target that could perform a number of different functions, the function being specified in the message.

6.9.3   Syntax

SOAP uses XML as its message syntax, with some structure defined but most of the structure left to the application. A SOAP message has an envelope, consisting of an optional header and a mandatory body, shown on a general level in Figure 6.11. To dig deeper into the actual XML, the parts defined by the SOAP specification are shown in Figure 6.12, with ... replacing all the parts that are application-specific.[10]

Any element inside the header is called a header block. SOAP defines certain attributes that such header blocks can have, to control the processing at intermediaries. Each header block is always targeted at some SOAP processor, which is expected to extract blocks targeted at it from the message, process them, and send the message onwards.

The `role` attribute can be used to override the default target for header blocks (in the example, it is useless, since the next processor in the chain *is* the default). The `mustUnderstand` set to true forces the processor to generate a fault if it does not understand the header block. And finally, the `relay` attribute set to true tells the processor that the

---

[10] Actually, the element names inside the `soap:Header` element are also application-specific, but we could hardly show the attributes without inventing some names for the elements too.

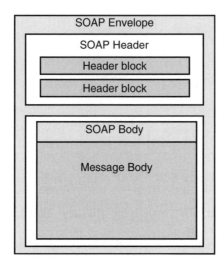

Figure 6.11   SOAP message structure

```
<soap:Envelope xmlns:soap='http://www.w3.org/2003/05/
   soap-envelope'>
 <soap:Header>
    <target soap:role='http://www.w3.org/2003/05/soap-envelope/
       role/next' soap:mustUnderstand='true'>
       ...
    </target>
    <priority soap:relay='true'>
       ...
    </priority>
    ...
 </soap:Header>
 <soap:Body>
    ...
 </soap:Body>
</soap:Envelope>
```

Figure 6.12   The SOAP message structure

header block is to be retained in the message instead of removed, which is the default behavior.

6.9.4   Protocol

As specified, SOAP includes a protocol binding framework that can be used to specify how SOAP messages are carried over any underlying protocol. The only protocol binding in the SOAP specification itself is a binding for HTTP, using either a one-way or request-response message exchange pattern.

Other protocol bindings for SOAP have been defined, including email and XMPP, but these have not caught on. We consider it unlikely that non-HTTP protocol bindings will ever acquire a significant portion of SOAP traffic, as the perception that SOAP is used with HTTP is strong, and many web service implementations can typically only be relied on to support HTTP, so interoperability dictates that HTTP be used for SOAP.

The style in which SOAP uses HTTP has drawn much criticism from the web community. The basis of this criticism is that HTTP is an application-layer protocol, with its own semantics, and SOAP ignores all of HTTP's functionality by treating it merely as a transport protocol. This is a correct view, since SOAP indeed uses only the POST method of HTTP and relies more on SOAP error messages than HTTP status codes to indicate error conditions, but on the other hand, novel ways of using existing protocols are constantly being thought up, so it is not obvious that SOAP is doing something wrong. Especially in a mobile environment, where HTTP might be the only protocol available or the only protocol that can be practically used due to NATs and firewalls, (ab)using HTTP as a transport protocol would appear perfectly legitimate.

6.10   The Web and REST

Around the time that SOAP-based web services were gaining in popularity, an approach closer to the architectural principles of the web was also gaining momentum. Critics of the SOAP style were pointing out that SOAP treated HTTP as a transport protocol whereas it is properly an application protocol with several kinds of messages instead of just the POST method that was always used with SOAP.

The architectural principles of the modern web were codified by Roy Fielding [14] and called REST, for Representational State Transfer. The web is not, strictly speaking, a messaging system, but its architecture has many interesting points that are rarely seen in other architectures, so we feel exploring it can be beneficial for messaging system designers. Still, it must be kept in mind that the web was designed for transferring documents, and some of the features were explicitly designed to permit efficient transfer of large documents.

6.10.1 Architecture

REST is defined as an architectural style based on fulfilling constraints. This design style is explicitly cited by Fielding as providing for an architecture design where excess creativity is not required, but rather the designer's hands are tied to satisfying the given constraints.

There are a number of constraints in REST:

Client-Server Client-server is the simplest distributed architecture, yet it does provide some benefits, including the separation of concerns between data storage and user interface.

Stateless Communication between the client and the server needs to be stateless in that every request contains all the necessary information for understanding it, and no state is required on the server. This promotes both scalability, since the server does not need to retain any interaction state, and reliability, since recovery from failures is simplified.

Network efficiency is decreased, since consecutive requests must repeat information that a stateful server could store.

Cache To again improve network efficiency, it must be possible to cache server responses and serve them to the client from the cache, avoiding a request being transmitted all the way to the server.

Uniform Interface Every component of the system must support the same interface. This promotes simplicity and flexibility at the cost, again, of network efficiency, since a component-specific interface can usually be made more efficient.

Layered System Making the system layered has the same benefits as making the network stack layered: components are decoupled from each other since they only see one immediate successor and only a well-defined interface for that. Again, efficiency is decreased because cross-layer design is made impossible. In a distributed system case, it is not even possible to describe the system using the layering abstraction while implementing it more efficiently as non-layered, since in a distributed system the components live in separate address spaces on separate hosts.

Code-on-Demand In REST it is possible to extend client functionality in previously-unforeseen ways by downloading executable code from the server. This promotes flexibility at a severe cost to transparency. As pointed out in [15], the least powerful language needed for expressing some information should be used, so the code-on-demand functionality of REST should be used only when absolutely necessary.

Architecturally, REST is based on resources. The definition of a resource is intentionally somewhat vague, but the intent is that a resource is a piece of nameable information. Since resources are nameable, each resource also has an identifier. This identifier is intended to be a persistent identifier, always uniquely identifying the same resource, even if the content of that resource changes.

Resources are not actually transferred anywhere. Instead, for transfer, a representation for a resource must be created. Such representations will be sequences of bytes transmittable over the network. One benefit of splitting the consideration into resources and representations is that a single resource may have multiple representations, and the system can select the most appropriate one as circumstances indicate.

Of the architectural constraints above, the Uniform Interface constraint also places some requirements on the interface itself:

Identification of resources To be a part of the web, a resource needs to have a unique identifier. This is perhaps the most trivial of REST's constraints.

Manipulation through representations Since resources are only accessed through representations, their content is also manipulated only through representations.

Self-descriptive messages Each message must contain all the information necessary to understand it. This is a stricter constraint than the previous statelessness constraint: not only must no state external to the message be assumed, but also each message must contain information on its semantics.

Hypermedia as the engine of application state Even though no interaction state exists, a distributed application is naturally going to have a state. This state in REST is stored in hypermedia documents, with the transitions between states described by hyperlinks in the documents.

The self-descriptive message constraint is one of the most important. It implies not only that the message must describe itself but also that the receiver must be capable of understanding the message. This requires well-defined languages for representations, agreed-upon identifiers for them, and a method for negotiating which languages are understood by each end.

The last constraint, often referred to by its acronym HATEOAS, is an important one when building REST-based services. Namely, it, along with the other constraints of REST, leads into a service design explicitly focused on resources and the links between them. New application states are defined by creating new resources with the appropriate links. Because of this focus on resources, REST is also sometimes referred to as Resource-Oriented Architecture, in intentional parallel with the term Service-Oriented Architecture often used of SOAP-based services.

The way that all these architectural constraints come into play is illustrated in Figure 6.13. Here a song collection service is being invoked by the client. The client first requests the list of songs, and based on that, requests the song identified as 700. The client then creates a new resource for a new song using POST and transmits its content using PUT. Finally, the client deletes the just-added song with DELETE. All of these operations succeed, as indicated by the server using a 2xx status code. The 'No Content' response simply says that the response message itself contains no body.

6.10.2 Locator

The locators in REST are URIs (Uniform Resource Identifiers). Most commonly on the current web, this means HTTP URIs. Such a URI has several components: the network

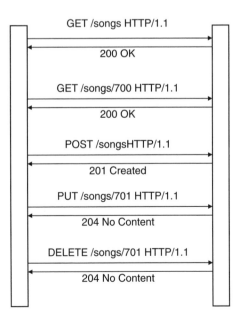

Figure 6.13 An example of REST interaction

address, and possibly port number, of the server, the path of the resource being accessed, and optional query parameters.

The benefit of using HTTP URIs is their hierarchical nature. Each domain administrator is fully responsible for assigning host names in their domain, and each host administrator is free to create URI paths. On systems where multiple users share a single server, each user typically has their own space, indicated by a path prefix specific to them.

6.10.3 Syntax

The syntax of a 'message' on the web is defined by the protocol, and, considering only the specified parts, is a simple one consisting of three parts:

Request or status line One line at the very beginning identifying the request method or the status of processing the request.

Header A sequence of name-value pairs, each on a single line.

Body A piece of binary data, opaque to the messaging system.

The message structure is very flexible, as the header can be split into lines without any knowledge of the header fields themselves, so it is possible to define arbitrary header names, and applications not familiar with these names will simply disregard them while being able to process the rest of the message.

The headers contain three kinds of metadata:

Resource metadata Information on the resource. Such information will be in any message containing that resource.

Representation metadata Information on the representation chosen for this interaction. Such information will not be in all the messages containing the resource, only in the ones using this particular representation.

Control data Information on the specific message. Such information is specific to this particular interaction and might be different in a different message even if the same resource and the same representation are used.

An example of resource metadata is the `Last-Modified` header, which indicates when the resource was last modified. An example of representation metadata, in fact the primary example of it, is the `Content-Type` header, specifying the Internet media type (MIME type) of the representation. An example of control data are the various header fields for cache control, specifying whether and for how long the provided representation is cacheable.

6.10.4 Protocol

The protocol used on the web is HTTP [16], and its current version, 1.1, has been explicitly designed according to the principles of REST. HTTP is an explicit request-response protocol, and semantically only a single interaction is supported, due to the statelessness constraint.

Each request of HTTP uses a specific request method. There are eight methods defined in the HTTP specification, but only four of them are of interest for the actual protocol:

GET Retrieve a resource identified by the provided URI.

PUT Store the given representation as the resource identified by the provided URI.
DELETE Delete the resource identified by the provided URI.
POST Provide a representation to the target of the request.

The other methods are for querying capabilities or properties (HEAD, OPTIONS), special-purpose connections (CONNECT), or debugging (TRACE).

Of the HTTP methods, GET, PUT, and DELETE provide the CRUD (create, retrieve, update, delete) functionality familiar from the data storage world, with PUT providing both creation and update facilities. POST is the Swiss army knife of HTTP methods: its semantics are defined by the server for each request URI and it can do pretty much anything. This is why the HTTP binding of SOAP uses POST, since the other methods cannot have semantics that permit invocation of remote procedures.

All the metadata related to the requests and responses is carried in the HTTP headers. REST defines three different kinds of metadata:

Resource metadata Data about the resource itself, for instance, the timestamp of the latest modification of the resource.
Representation metadata Data about the representation being transferred, for instance, its media type.
Control data Data about the specific message, for instance, information on whether caching is permitted and for how long.

The HTTP header is not divided in any way, so determining the kind of each HTTP header field depends on that field's specification.

One specific feature of HTTP enabled by the headers is content negotiation. Current HTTP content negotiation is embedded in the request-response interaction: the client, in the request, indicates all the types of content it can support and the server picks from these according to its capabilities. The negotiable parameters include the actual type of the representation itself (for hypertext documents, this could be, for instance, HTML or XHTML), the encodings applied to it (for instance, compression), its language (English, German, ...), and the encoding applied to transfer it (again, could be compression, but also includes the chunked encoding that permits a server to start sending information prior to knowing its length but also delimits messages so the client can know where a message ends).

Bibliography

[1] ITU 2002 *Abstract Syntax Notation One (ASN.1) Specification of Basic Notation*. ITU-T Rec. X.680.
[2] W3C 2006a *Extensible Markup Language (XML) 1.0* 4th edn. W3C Recommendation.
[3] Freed N and Borenstein N 1996 *RFC 2045: Multipurpose Internet Mail Extensions (MIME) Part One: Format of Internet Message Bodies* Internet Engineering Task Force.
[4] Kleinrock L 1996 Nomadicity: Anytime, anywhere in a disconnected world. *Mobile Networks and Applications* **1**(4), 351–357.
[5] Balakrishnan H, Padmanabhan VN, Seshan S and Katz RH 1997 A comparison of mechanisms for improving TCP performance over wireless links. *IEEE/ACM Transactions on Networking* **5**(6), 756–769.
[6] Border J, Kojo M, Griner J, Montenegro G and Shelby Z 2001 *RFC 3135: Performance Enhancing Proxies Intended to Mitigate Link-Related Degradations* Internet Engineering Task Force.
[7] Ferguson N and Schneier B 2003 *Practical Cryptography*. Wiley Publishing, Indianapolis, Indiana, USA.
[8] Garfinkel S 1994 *PGP: Pretty Good Privacy*. O'Reilly, Sebastopol, California, USA.

[9] Sun 2001 *Java Message Service Specification*.

[10] OMG 2004 *Common Object Request Broker Architecture (CORBA/IIOP), version 3.0.3*.

[11] Saint-Andre P 2004 *RFC 3920: Extensible Messaging and Presence Protocol (XMPP): Core* Internet Engineering Task Force.

[12] W3C 2003 *SOAP Version 1.2*. W3C Recommendation.

[13] W3C 2004 *Web Services Architecture*. W3C Note.

[14] Fielding R 2000 *Architectural Styles and the Design of Network-based Software Architectures* PhD thesis University of California, Irvine.

[15] W3C 2006b *The Rule of Least Power*. W3C TAG Finding.

[16] Fielding R, Gettys J, Mogul J, Nielsen HF, Masinter L, Leach P and Berners-Lee T 1999 *RFC 2616: Hypertext Transfer Protocol – HTTP/1.1* IETF. http://www.ietf.org/rfc/rfc2616.txt.

7

Publish/Subscribe

This chapter presents the publish/subscribe (pub/sub) paradigm [1], which is a frequently used communication paradigm that supports loosely coupled systems. Pub/sub builds on asynchronous communication and typically adds the ability for subscribers to specify what kind of information should be delivered to them. As such, pub/sub is a good candidate for supporting mobile applications and services, because it allows them to be decouple in multiple dimensions. Pub/sub is an enabler for data-centric communications in which subscribers subscribe to the data they need and publishers make this data available.

The chapter builds on the messaging techniques described in the previous chapter, and illustrates how service components can be combined in a dynamic fashion using publish/subscribe. Concrete technologies considered include: Notification Service and Data Distribution Service from OMG, Java Message Service (JMS) from Sun [43], .NET communication framework, WS-Eventing, SIP Events, and several research system proposals including SIENA and Rebeca.

7.1 Overview

Figure 7.1 presents an overview of the pub/sub paradigm. The key idea is simply to allow subscribers to specify their interests to publishers or producers and when these previously subscribed events happen, deliver them to the proper subscribers. This means that pub/sub is inherently a one-to-many or many-to-many communication paradigm. In order to be able to build distributed pub/sub systems, a system is needed that contains a notification engine and subscription manager functions illustrated in the figure. These are typically provided by a logically centralized pub/sub service.

As discussed in Chapter 7, a number of design patterns can be used to realize pub/sub. Frequently used patterns include the observer, event-channel, and notifier patterns [2]. Observer is a basic pattern that is typically combined with the broker, mediator, and proxy patterns to create more complex distributed pub/sub systems.

The event-channel and notifier patterns decouple subscribers and publishers by introducing a broker that mediates events on their behalf. The event channel and notifier also support various non-functional requirements, such as QoS and disconnected operation.

Mobile Middleware   Sasu Tarkoma
© 2009 John Wiley & Sons, Ltd

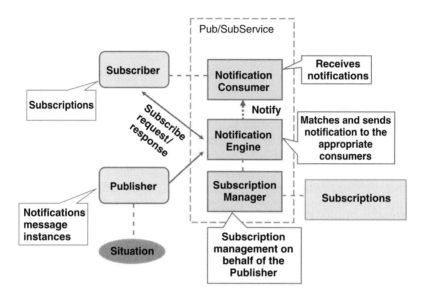

Figure 7.1 Overview of publish/subscribe

The event-channel and notifier patterns are similar, but the notifier also abstracts the location and distribution of event brokers, whereas with channels the client must first obtain the reference of the channel.

The notifier pattern may be realized by using the observer pattern and mediator or proxies [3]. The event channel pattern is used in the CORBA Event Service [4] and Notification Service [5]. A separate specification defines how CORBA event channels are connected to form communication topologies [6].

Pub/sub is realized by using message queuing and message-oriented middleware. Message queuing is a frequently used communication method because it supports disconnected operation. When a client is disconnected, messages are inserted into a queue, and when a client reconnects the messages are sent. The distinction between popular message-queue-based middleware and notification systems is that message-queue-based approaches are a form of directed communication, where the producers explicitly define the recipients. The recipients may be defined by the queue name or a channel name, and the messages are inserted into a named queue, from which the recipient extracts messages.

Notification-based systems extend this model by adding an entity, the event service or event dispatcher, that brokers notifications between producers of information and subscribers of information. This undirected communication supported by the notification model is based on message passing and retains the benefits of message queuing. In undirected communication the publisher does not necessarily know which parties receive the notification.

This also applies to message-oriented middleware such as JMS [7] that supports publish-subscribe type of communication. Undirected communication decouples producers and consumers from each other. In addition, many systems support filtering and pattern detection that are used to reduce the amount of transmitted information and to improve the accuracy of notifications.

Content-based routing is flexible because it does not require configuration information pertaining to channel names. Undirected communication may also be used to deliver the same set of information to a number of client devices. However, this requires associating user subscription information with a set of devices.

The main entities in a pub/sub system are the publishers and subscribers of information. A publisher publishes an event and a subscriber receives notifications of events that have occurred. There are many names for the entities in pub/sub or event systems, for example the terms subscriber, consumer, and sink are synonymous. Similarly, publisher, producer, source, and supplier are synonymous. The semantic meaning of an event and its notification is application and domain specific, and the communicating parties need to have a mutual agreement on the interpretation of a given notification.

An event system may be centralized or distributed in nature and the notification responsibility may be provided by different entities in the environment: publishers, a centralized router, or a sequence or a set of routers. In distributed environments a published event is communicated in an event message, also called a notification, using a message transport protocol. This is one of the defining characteristics of event and pub/sub systems – the use of asynchronous message passing. The entities may employ point-to-point messaging in communication, but communication may also be based on various multicast and broadcast technologies.

The *event router or broker* is a component that connects the publishers and subscribers and mediates event messages between them. Typically, an event router consists of two parts: a set of connections to neighbouring routers and a set of local clients. Both sets are associated with a *routing table* that contains information about which event messages should be forwarded to which neighbouring router or local client. Neighbouring routers may also be called interfaces or destinations. In filter-based routing the routing table contains a set of filters for each interface and local client. A router with only a single neighboring router is called an *edge router* or *border router*.

The distribution of event routers is necessary to achieve scalability, reliability, and high-availability. For example, if a publisher is responsible for directly notifying a set of subscribers, it is clear that the centralized nature of this kind of *direct notification* is limited in terms of scalability and performance. The scalability of direct notification may be improved by using intermediary components, but this is just a step towards a routing infrastructure. Indeed, many research projects have focused on infrastructure-based notification and investigated different distribution mechanisms for connecting publishers and subscribers efficiently.

In many cases, the subscriber is interested in a very specific event and if the event system does not provide any mechanism for defining interests, the subscriber will receive all event messages published by the producer or producers in question. This is called *flooding* and it is the trivial way to ensure that every subscriber will receive the correct notifications. A message that is sent to a client that does not match the client's interests is called a *false positive*. Similarly, a message that was not delivered, but should have been received by the client is called a *false negative*.

Flooding every event message everywhere is not a scalable solution, which has led to the development of various filtering languages and filter matching algorithms. The scalability limitation is obvious, because the forwarding of event messages requires processing time on various entities of the environment and message transmission uses network resources.

Table 7.1 Infrastructure interface operations

Operation	Description	Semantics
$Sub(X,F)$	X subscribes filter F	Sub/Adv
$Pub(X,n)$	X publishes notification n	Sub/Adv
$Notify(X,n)$	X is notified about notification n	Sub/Adv
$Unsub(X,F)$	X unsubscribes filter F	Sub/Adv
$Adv(X,F)$	X advertises filter F	Adv
$Unadv(X,F)$	X unadvertises filter F	Adv

Excess and uncontrolled messaging may lead to congestion. Congestion in turn may cause event messages to be dropped.

Filtering allows the subscribers to specify their interest beforehand and thus reduce the number of uninteresting event messages that they will receive. A filter or a set of filters that describes the desired content is included with the subscription message. Filters may also be used to *advertise* the future publication of events. This advertisement information given by publishers may be used to further optimize messaging and the processing overhead of routers. Many filtering languages have been developed, specified, and proposed. For example, the filtering language used by JMS is based on *Structured Query Language (SQL)*.

In order to support event filtering and event delivery, an event router needs to provide an interest-registration service and also have an interface for publishing events. Subscribers define their interests using this interest-registration service. Table 7.1 presents the pub/sub operations used by most event systems. The table presents the operations for two different semantics: *subscription semantics* and *advertisement semantics*. The advertisement semantics adds the operations for advertising and unadvertising a filter.

Depending on the expressiveness of the filtering language, a specific field, header, or the whole content of the event message may be filterable. In *content-based routing* the whole content of the event message is filterable. With the introduction of filtering we face the problem of how to propagate this filtering information in the distributed environment. It is not feasible to expect that a producer or a single router is capable of filtering event messages for a large number of subscribers.

The two important parts of a distributed pub/sub system are the router topology, by which we mean the exact nature of how the routers are connected with each other, and how routing information is propagated by the routers. By propagating routing information we mean how the interests, filters, of the subscribers are conveyed towards the publishers of that information. In essence, the routing information stored by a router must enable it to forward event messages either to other routers or to local clients that have previously subscribed to the event messages.

Expressiveness and scalability are important characteristics of an event system [7]. Expressiveness deals with how well the interests of the subscribers are captured by the notification service, and scalability deals with federation, resources and issues such as how many users can be supported and how many routers are required. In addition to expressiveness and scalability, an event system needs to be relatively simple to be manageable, implementable, and to be able to support rapid deployment. Moreover, the system needs to be extensible and interoperable. Other non-functional requirements are: timely delivery of notifications (bounded delivery time), support for Quality of Service (QoS), high availability and fault-tolerance. Event order is an important non-functional requirement and many applications require support for either causal order or total order.

7.2   Router Topologies

A number of different router topologies have been proposed in event literature. Well-known router topologies include: *centralized*, *hierarchical*, *acyclic*, *cyclic*, and *rendezvous point-based* topologies. Centralized routers represent the trivial case for distributed operation, in which subscribers and producers use a client-server protocol for sending and receiving event messages and invoke the interest-registration service provided by the router.

In hierarchical systems each router has a master and a number of slave routers. Notifications are always sent to the master. Notifications are also sent to slaves that have previously expressed interest in the notifications. The basic hierarchical design is limited in terms of scalability, because one master router is the root of the distribution tree and will receive all the notifications produced in the system.

For acyclic and cyclic topologies routers employ a different, peer-to-peer, protocol to exchange interest propagation information and control messages. In this context, the peer-to-peer protocol denotes that the topology is not hierarchical. Acyclic topologies allow more scalable configurations than hierarchical topologies, but they lack the redundancy of cyclic topologies. On the other hand, topologies based on cyclic graphs require techniques, such as the computation of minimum spanning trees, to prevent loops and unnecessary messaging.

The rendezvous point model differs from acyclic and cyclic topologies, because the routing of a specific type of event is constrained by a special router, the *Rendezvous Point (RP)*. The RP serves as a meeting point for advertisements and subscriptions and avoids the flooding of advertisements throughout the system. Rendezvous-based systems limit the propagation of messages using the RP and thus attempt to address scalability limitations presented by the flooding of subscriptions or advertisements. Typically, an RP is responsible for a pre-determined event type. RPs may be used to create a type hierarchy. In this case, a message needs to be sent to the proper RP and any super-type RPs, which may increase messaging cost and limit scalability

The hierarchical topology was used in the JEDI system [8, 9], and an acyclic topology with advertisements in Rebeca [10–12]. The SIENA project investigated and evaluated the topologies with different interest propagation mechanisms [13, 7]. In general, the acyclic and cyclic topologies have been found to be superior to hierarchical topologies [8, 13, 14]. The router topology in Gryphon [15, 16] is based on clusters called *cells* and redundant *link bundles* that connect cells. Most research has focused on static connections between routers. Dynamic connections between routers have been investigated in [17] and [18].

A number of *overlay*-based routing algorithms and router configurations have been proposed. An application layer overlay network is implemented on top of the network layer and typically overlays provide useful features such as fast deployment time, resilience and fault-tolerance. An overlay-routing algorithm leverages underlying packet-routing facilities and provides additional services on the higher level, such as searching, storage, and synchronization services.

Good overlay routing configuration follows the network level placement of routers. Many overlays are based on *Distributed Hash Tables (DHTs)*, which are typically used to implement distributed lookup structures. Many DHTs work by hashing data to routers/brokers and using a variant of *prefix-routing* to find the proper data broker for a given data item. Hermes [19] and Scribe [20] are examples of pub/sub systems implemented on top of an overlay network and are based on the rendezvous point routing model. The Hermes routing model is based on advertisement semantics and an overlay topology with rendezvous points. This model was found to compare favourably to the SIENA advertisement semantics using an acyclic topology [19].

7.3   Interest Propagation

The main functions of a router are to match notifications for local clients and to route notifications to neighboring routers that have previously expressed interest in the notifications. The interest propagation mechanism is an important part of the distributed system and at the heart of the routing algorithm. The desirable properties for an interest propagation mechanism are small routing table sizes and forwarding overhead [14], support for frequent updates, and high performance.

With subscription semantics the routers propagate subscriptions to other routers, and notifications are sent on the *reverse path* of subscriptions. In *simple routing* each router knows all active subscriptions in the distributed system, which is realized by flooding subscriptions. In *identity-based routing* a subscription message is not forwarded if an identical message was previously forwarded. This requires an identity test for subscriptions. Identity-based routing removes duplicate entries from routing tables and reduces unnecessary forwarding of subscriptions. In *covering-based routing* a covering test is used instead of an identity test. This results in the propagation of the most general filters that cover more specific filters. On the other hand, unsubscription becomes more complicated because previously covered subscriptions may become uncovered due to an unsubscription. *Merging-based routing* allows routers to merge exiting routing entries. Merging-based routing may be implemented in many ways and combined with covering-based routing [14]. Also, merging-based routing has more complex unsubscription processing when a part of a previously merged routing entry is removed.

With advertisement semantics the routers first propagate advertisements and then, on the reverse path of advertisements, the subscriptions. Notifications are forwarded on the reverse path of subscriptions in both semantics. Advertisements may be used with various routing mechanisms. Advertisements typically have their own routing table and they are managed using the same algorithms as subscriptions. The removal of an advertisement causes a router to drop all overlapping subscriptions for the neighbor that sent the unadvertisement message. Similarly, an incoming advertisement requires that overlapping subscriptions are forwarded to the neighbor that sent the advertisement message. The use of advertisements considerably improves the scalability of the event system [8, 13, 14, 22].

One of the first formulations of a wide-area pub/sub system based on these two semantics with optimizations was presented in the SIENA system, which used covering relations between filters to prevent unnecessary signaling. The SIENA system used the notion of covering for three different comparisons: matching a notification against a filter, covering relation between two subscription filters, and overlapping between an advertisement filter and a subscription filter. Covering and overlapping relations have been used in many later event systems, such as Rebeca [12] and Hermes [19, 21]. The combined broadcast and content-based (CBCB) routing scheme extends the SIENA routing protocols by combining higher-level routing using covering relations and lower-level broadcast delivery [22]. The protocol prunes the broadcast distribution paths using higher-level information exchanged by routers.

7.4 Routing Decision

Message routing systems may be classified into four categories: channel-based, subject-based, header-based, and content-based. Channel-based systems make the routing decision based on channel names that have been agreed by the communicating participants. Subject-based systems make the routing decision based on a single field. Header-based systems use a special header part of the message in order to make the routing decision. For example, SOAP [23] supports header-based routing of XML-messages. Finally, content-based systems use the whole content of the message in making the decision [24]. Next, we describe the four well-known categories of message routing systems.

Channel/topic-based. Routing decision is made based on the channel on which the event is published. A channel is a discrete communication line with a name. Named channels are also called topics, and they represent an abstraction of numeric network addressing mechanisms. Usually with channel-based messaging, new channels need to be added to the address space, because the producers and consumers must agree on a channel. Channel-based messaging allows the use of IP multicast groups [25]. The channels can be allocated to multicast addresses.

Subject-based. Routing decision is made based on the subject of the event. Subject-based routing is more expressive than channel-based routing. On the other hand, a single field may not be enough to properly describe the content of a message.

Header-based. Routing decision is made based on a number of fields in the message header. In header-based routing the message has two distinct parts: the header and the body. Only fields in the header are used for making routing decisions. Header-based routing is more expressive than subject-based and has performance advantage to content-based routing, because only the header of a message is inspected.

Content-based. Routing decision is made based on the whole content of a message, for example strongly typed fields in the event message. Content-based routing is the most expressive of the four types.

Content-based event routing has been proposed as one of the requirements for advanced applications, in particular for mobile users [26, 27] and context-sensitive messaging [28]. The latter mechanism formulates the current and future context of entities as event filters and subscribes to them. The Elvin event broker [29] is used to deliver messages to the recipients based on the subscribed context filters. Context-sensitive messaging may be used, for example, to control and monitor a set of mobile robots in a particular location [28].

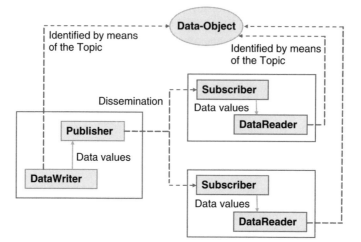

Figure 7.2   Overview of DDS

7.5   Standards

7.5.1   OMG Data Distribution Service (DDS)

The *Data Distribution Service for Real-Time Systems (DDS)* OMG specification defines an API for data-centric pub/sub communication for distributed real-time systems[30]. DDS is a middleware service that provides a global data space that is accessible to all interested applications. The specification describes the service using UML. The latest version is 1.2 and it was released in January 2007.

Figure 7.2 illustrates the key elements of the DDS model, namely the publisher, subscriber, and data-object. DDS uses the combination of a Topic object and a key to uniquely identify instances of data-objects. In this model, the subscriptions are decoupled from the publications. DDS creates a name space that allows participants to locate and share objects. In case a set of instances are under the same topic, these different instances must be distinguishable.

DDS uses a key to distinguish between these instances. The key consists of the values of some data fields. These fields need to be indicated to the middleware. Different data values with the same key value represent successive values for the same instance. Different data values with different key values represent different instances. If no key is provided, the data set associated with the Topic is restricted to a single instance.

A ContentFilteredTopic may be created for content-based subscriptions. In addition, the MultiTopic can be used to subscribe to multiple topics and combine/filter the received data. The filter language syntax is a subset of the SQL syntax.

The QoS usage follows the subscriber-requested publisher-offered pattern. In this pattern, the subscribers request desired QoS properties and these are matched against those offered by the producers.

DDS is suitable for signal, data, and event propagation. Signals represent continuously changing data, for example from a sensor. In this case, publishers may set the reliability property to best-effort and the history QoS property to retain the last signal (KEEP_LAST).

Data delivery, such as exchanging the state of a set of objects, can be realized by using reliable communication and requiring that the last data elements are stored by the system. Events are streams of values and publishers typically use reliable delivery and require that the system keeps a history of all messages (KEEP_ALL).

DDS complements CORBA, because it provides a service more suitable for asynchronous and dynamic operation. CORBA provides support for distributed objects in a client/server environment and supports remote method calls. DDS is more suitable for flexible QoS-aware data dissemination to many nodes in dynamic environments. Therefore, CORBA is object-centric, whereas DDS is data-centric.

The CORBA Event Service decouples producers and consumers, but it is not data-centric and does not offer QoS contracts. CORBA Notification Service offers a more data-centric approach with filters and QoS support. DDS differs from these two services, because it does not have to support the Common Data Representation or use the IIOP protocol. This means that DDS services do not interoperate unless an interoperability specification is adopted. In other words, a DDS implementation does not have to be CORBA based.

7.5.2   IP IMulticast

IP multicast is a simple, scalable and efficient mechanism to realize simple group-based communication. IP multicast routes IP packets from one sender to multiple receivers. Participants join and leave the group by sending a packet using the IGMP (RFCs 1112, 2236, 3376) protocol to a well-known group multicast address.

IP multicast maintains a group specific distribution tree (or forest), which is updated when nodes join and leave. The most widely used multicast protocol is the *Protocol Independent Multicast (PIM)*. PIM configures multicast distribution trees that deliver packets from senders to a multicast group to the receivers who have joined the group. PIM includes a number of specifications for different kinds of multicast algorithms, including *Sparse Mode(SM)* (RFC 2362), *Dense Mode (DM)* (RFC 3973), *Bidirectional PIM* (RFC 5015), and *Source Specific Mode (SSM)* (RFC 3569).

PIM-SM is suitable for inter-domain multicast with groups that have low numbers of subscribers. Clients join a PIM-SM multicast group by sending an IGMP Join message. This protocol works by constructing a unidirectional tree from each sender to the receivers in the multicast group by using a *rendezvous point (RP)*. PIM-SM supports both a shared RP-centered multicast tree and source-specific shortest path trees. The latter is typically useful for high data rate sources.

PIM-DM has been designed for groups where there are many subscribers. This high subscription density means that routers must process and forward packets published in the multicast groups. PIM-DM works by flooding packets to the network and then pruning the multicast forwarding tree. These prune messages are used to prevent future messages from propagating to routers that do not have subscribers for the group in question.

Bidirectional PIM is a variant of the PIM-SM protocol that builds bidirectional shared trees, by using the rendezvous point, connecting multicast sources and subscribers.

PIM-SSM creates a source specific delivery tree for each subscriber. The delivery tree is a (S, G) channel, with an IP unicast source address S and the multicast group address G as the IP destination address. The interdomain tree for IP packet forwarding is rooted at the source S, and is constructed using the PIM-SM protocol.

IP multicast groups are not very expressive. They partition the IP datagram address-space and each datagram belongs at most to one group. IP multicast faces also several deployment challenges that include multi-domain operation and RP discovery. Moreover, IP multicast is a best-effort unreliable service, and for many applications a reliable transport service is needed.

Event systems may use multicast to deliver notifications to appropriate event routers or servers. Not many event systems take advantage of network level IP multicast. An evaluation of different algorithms for mapping subscribers to multicast groups is presented in [25]. Multicast works well in closed networks, however, in large public networks multicast or broadcast may not be practical. In these environments universally adopted standards such as TCP/IP and HTTP may be better choices for all communication [15].

7.5.3   RSS and Atom

Really Simple Syndication (RSS) is a family of specifications for the definition of web-based information feeds using XML. RSS is essentially a simple pub/sub system that is based on polling the URL that identifies a feed and then determining if information has changed. RSS builds on existing web standards, namely HTTP and XML, and it has become ubiquitous. RSS is used to disseminate updates, for example, pertaining to blog entries, news, video and audio resources.

A RSS document defines a feed that includes textual description of the elements and related metadata such as URIs and author information. RSS feeds complement the current websites, because they allow a standardized way to quickly disseminate updates and aggregate them. RSS is supported by a number of web browsers and custom feed reader programs. The client program then checks the RSS feed URL regularly for any new information.

Atom is an alternative syndication protocol defined in RFCs 4287 and 5023. Atom is similar to RSS and uses HTTP and XML to realize web feeds. Atom specifications emphasize internationalization support, modularity, and security features.

7.5.4   Java Distributed Event Model

The Distributed Event Model of Java is based on Java Remote Method Invocation (RMI) that enables the invocation of methods in remote objects. This model is used in Sun's Jini architecture[31]. The architecture of the Distributed Event Model is similar to the architecture of the Delegation Model with some differences. The model is based on the Remote Event Listener, which is an event consumer that registers to receive certain types of events in other objects. The specification provides an example of an interest registration interface, but does not specify such.

The Remote Event is the event object that is returned from an event source (generator) to a remote listener. Remote events contain information about the occurred event, a reference to the event generator, a handback object that was supplied by the listener, and a unique sequence number to distinguish the event globally. The model supports temporal event registrations with the notion of a lease (Distributed Leasing Specification). The event generators inform the listeners by calling the listeners' notify method.

The specification supports Distributed Event Adaptors that may be used to implement various QoS policies and filtering. The handback object is the only attribute of the Remote

Event that may grow to unbounded size. It is a serialized object that the caller provides to the event source; the programmer may set the field to null. Since the handback object carries both state and behavior it can be used in many ways, for example to implement an event filter at a more powerful host than the event source.

7.5.5 SIP Event Framework

Asynchronous notifications are useful in implementing many types of SIP services, such as automatic callback services, buddy lists, and message waiting indications. The SIP event framework (RFC 3265) defines an extensible pub/sub system. The SIP event framework allows SIP user agent's to create, modify, and remove subscriptions. The event framework can be extended by using Event Packages, which are additional specifications that define a set of state information pertaining to notifications. Event packages define additional syntax and semantics for the framework.

The SIP event framework defines two new methods for pub/sub, namely the SUB-SCRIBE and NOTIFY methods. The former is used to request current state and state updates from a remote node.

The latter message method and message type is used to inform subscribers about changes in states. NOTIFY messages are sent only when there is a prior subscription associated with the state change. When a new SUBSCRIBE is processed, a NOTIFY is sent that informs the subscriber of the current state of the subscribed resource.

A subscription can be removed or modified by using the SUBSCRIBE method and by indicating the desired outcome using header parameters, such as the 'Expires' header. When a subscription is removed, a final NOTIFY message will be sent regarding the resource's state.

SIP events are identified by using three fields, namely the Request URI, Event Type, and optionally message body. The Request URI of a SUBSCRIBE request must contain enough information to route the request to the proper entity. It must also identify the resource associated with the subscription.

Figure 7.3 illustrates the use of SIP events for presence subscriptions and notifications across multiple domains. The SIP presence event package is defined in RFC 3856. According to RFC 2778, presence service is a system that accepts, stores, and distributes presence information to interested parties (called watchers). A presence protocol is a protocol for providing a presence service over any IP network.

A presentity is typically identified in a general way using a presence URI, which has the form pres:user@domain. These URIs are resolved to protocol specific URIs, such as the SIP or SIPS URI, using domain-specific mappings maintained on a server. A presence server is a physical entity that can act as either a presence agent or as a proxy server for SUBSCRIBE requests. A presentity publishes information to a presence agent, which is then responsible for notifying any subscribed watchers.

Subscribing to presence information of a presentity proceeds as follows. The subscriber sends a SUBSCRIBE request that defines the presentity in the Request-URI using SIP URI, SIPS URI, or a presentity URI. This SUBSCRIBE request is then routed by SIP proxies to a presence server or an edge presence server. In the latter case, the edge presence server acts as the presence agent for the presentity. The presence agent is responsible for authenticating and authorizing the subscription and subsequently sending notifications to the subscriber.

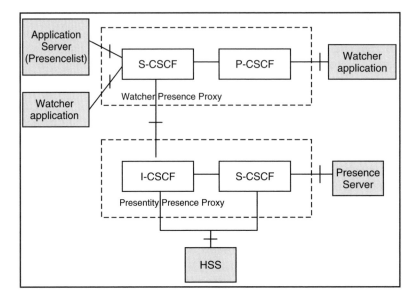

Figure 7.3   Example of IMS and SIP Events

In the figure, the home domain hosts the presence server and HSS. A number of watcher applications are monitoring user presence in the other domain. The presentity presence proxy and watcher presence proxy mediate subscriptions and notifications across the domains.

7.5.6   CORBA Notification Service and Channel Management

The CORBA Notification Service [5] extends the functionality and interfaces of the older Event Service [4] specification. The Event Service specification defines the event channel object that provides interfaces for interest registration and event notification. One of the most significant additions to the Notification Service is event filtering. Filters allow consumers to receive particular events that match certain constraint expressions. Filtering reduces the number of events sent to the consumers and improves the scalability of the event handling system.

An event is created by the event supplier and is transferred to all relevant event consumers. The set of suppliers is decoupled from the set of consumers, and the supplier has no knowledge of the number or identity of the consumers. The consumers have no knowledge of which supplier generated the event. The Event Service and Notification Service are based on the event channel pattern. The event channel is responsible for asynchronously transferring events between suppliers and consumers. Suppliers and consumers connect to the event channel using the interfaces supported by the channel. An event is a successful completion of a sequence of operation calls made on consumers, suppliers, and the event channel.

Internally, both Event Service and Notification Service utilize the proxy pattern heavily. For each external supplier they create an internal proxy consumer, and for each external consumer they create an internal proxy supplier. In the Notification Service the proxies can have their own QoS and filtering rules derived from a common administration object that was used to create the proxies.

CORBA Event Service and Notification Service do not specify an event discovery service or a mechanism to federate event channels. Moreover, the procedure for connecting event channels is complex. The OMG Telecommunications Domain Task Force addresses these issues in the CORBA Management of Event Domains Specification, which specifies an architecture and interfaces for managing event domains. An event domain is a set of one or more event channels grouped together for management, and for improved scalability.

The specification defines two generic domain interfaces for managing generic typed and untyped channels. Moreover, a specialized domain for both channels and logs is defined by the OMG Telecom Log Service specification.

The specification addresses:

- connection management of clients to the domain;
- topology management;
- sharing the subscription and advertisement information in an event domain, even when connections between event channels change at runtime;
- event forwarding within a channel topology; and
- connections between event channels.

The specification supports the creation of channel topologies of arbitrary complexity, allowing cycles and diamond shapes in the graph of interconnected channels. However, if events may reach a point in the graph by more than one route duplicate events need to be detected and removed. Moreover, if no timeouts are specified, events in a cycle will propagate infinitely. Therefore, the specification defines mechanisms that are used to detect cycles or diamonds in the network topology. Graph topology enforcement is done at channel connection time, and the domain management refuses illegal connections. Event suppliers inform the proxy consumers of event type changes using the offer_change callback. The channel is responsible for sharing this information with the consumers by executing offer_change on them. The consumer may be another channel and thus the change may propagate throughout the channel topology. Subscription changes work similarly, and the channel is responsible for invoking the subscription_change operation on all suppliers.

Event suppliers attached to the channel can obtain the types of subscriptions of event channels anywhere downstream by invoking obtain_subscription_types on the proxy consumers. Similarly an event consumer can obtain the event types offered by suppliers on any event channel downstream by invoking obtain_offered_types on its supplier channels.

7.5.7   WS-Eventing and WS-Notification

Two key mechanisms have been proposed for realizing pub/sub for web services, namely WS-Eventing and WS-Notification. The former was submitted to W3C in 2006 as member submission and the latter was standardized by OASIS in 2006.

The Web Services Eventing (WS-Eventing) specification describes a protocol that allows web services to subscribe or to accept subscriptions for event notifications. An interest registration mechanism is specified using XML Schema and WSDL. The specification supports both SOAP 1.1 and SOAP 1.2 Envelopes.

The key aims of the specification are to specify the means to create and delete event subscriptions, to define expiration for subscriptions, and to allow them to be renewed. The specification relies on other specifications for secure, reliable, and/or transacted messaging. The specification supports filters by specifying an abstract filter element that supports different filtering languages and mechanisms through the Dialect attribute.

WS-Notification version 1.3 was standardized by OASIS in 2006. These specifications provide a way for a web service to disseminate information to a set of other web services. WS-Notification consists of the following specifications[32]:

- The WS-Base Notification specification defines the pub/sub interfaces for web services.
- The WS-Topics specification defines a mechanism to organize items of interest, called topics. The specification defines an XML model for meta-data and several topic expression dialects for filtering.
- The WS-Brokered Notification specification defines the web services interfaces for notification brokers. This specification supports the creation of distributed pub/sub topologies based on WS-Notification.
- The WS-Notification Policy specification defines a set of policy statements that can be used with the other specifications.

7.6   Research Systems

7.6.1   Scalable Internet Event Notification Architecture (SIENA)

Scalable Internet Event Notification Service (SIENA) is an Internet-scale event notification service developed at the University of Colorado. SIENA balances expressiveness with scalability and explores content-based routing in a wide-area network. The basic pub/sub mechanism is extended with advertisements that are used to optimize the routing of subscriptions [7].

Several network topologies are supported in the architecture, including hierarchical, acyclic peer-to-peer, and general peer-to-peer topologies. Servers only know about their neighbors, which helps to minimize routing table management overhead. Servers employ a server-server protocol to communicate with their peers and a client-server protocol to communicate with the clients that subscribe to notifications. It is also possible to create hybrid network topologies.

SIENA is similar to IP multicast; however, the two mechanisms differ in the way they support groups of subscribers. IP groups are not very expressive. They partition the IP datagram address space and each datagram can belong to at most one group. Clearly, this creates problems if an event that spans several groups of subscribers is to be delivered.

SIENA is implemented with a flat event namespace, i.e. event names have no structural correlation with each other. An event consists of a set of attribute-value pairs. Each attribute has a name and a value. SIENA supports the types null, string, long, integer, double, and boolean. A filter consists of an attribute name, a constraint operator, and a constraint. SIENA does not support wildcards in the attribute name so the attribute names must match exactly to the names in the published event. A filter may include several filtering clauses, which are ANDed together. Thus every filtering clause or component must return true in order for the filter to pass the event. SIENA supports the operators equal, less than, greater than, greater than or equal to, less than or equal to, string prefix, string suffix, always matches, not equal, and substring.

SIENA supports patterns, which are based on event attribute values and event combinations. A pattern is a sequence of filters that is matched to a temporally ordered sequence of notifications. Network latencies may cause some events to arrive in the wrong order, and these are ignored by the SIENA solution.

7.6.1.1 Routing

In SIENA, each event consists of a set of attribute-value pairs that are matched with filters. Each server on the event system routes events to other servers based on subscription information, advertisement information, and filters. Each subscriber may specify a filter to constrain the subscription. In the same fashion, each advertisement may also include a filter. SIENA evaluates the filters and follows a policy where events are replicated downstream and filtered upstream. This means that events are replicated to the clients at the last possible moment, thus reducing the bandwidth needed to transmit the events. Upstream filtering means that events are filtered as close to the sources as possible in order to reduce the number of uninteresting events transmitted over the network. The simple filter syntax allows the decomposition of a complex filter into several more general filters, which can be evaluated upstream. A filter is only applied if it is less general than the one used in upstream.

The same principle of upstream filtering also applies to event patterns. Patterns are decomposed (factored) into elementary filters that are delegated to other servers. In the delegation process a server tries to assemble sub patterns that are delegable to other servers. SIENA uses covering relations to determine when a filter covers a notification, a subscription covers a notification, an advertisement covers a notification, or an advertisement covers a subscription. For example, subscription S1 covers S2 if it evaluates to true in every instance where S2 is true. Servers propagate the most generic subscription that covers a given set of subscriptions. This minimizes the downstream data structures, however, the complex computation cost is paid closer to the subscriber, because the subscriptions need to be matched and evaluated. The results of SIENA indicate that the covering relations exhibit a complexity that is quite reasonable for a scalable service.

The SIENA system supports two different notification semantics: subscription-based and advertisement-based. In subscription-based semantics subscriptions are introduced at every node of the event service and a notification is routed if it covers a subscription. In advertisement-based routing servers use the information provided by event producers to route incoming subscriptions. A subscription is only forwarded if it is related with an active advertisement. The system uses a *partially ordered set (poset)* as the routing table data structure. The complex routing table update processing restricts the scalability of the system especially in environments in which the routing tables are updated frequently.

7.6.1.2 Forwarding

The forwarding algorithm that was developed in conjunction with the SIENA project consists of a forwarding table and a set of processing functions. Conceptually the forwarding table is a mapping between predicates (sets of filters) and interfaces to neighboring nodes.

Each predicate is a disjunction of filters, where each filter is a conjunction of elementary conditions. Each elementary conjunction must return true in order for a filter (and predicate) to map to an interface. Each filter may map to several interfaces.

The forwarding algorithm iterates over the event attributes. It searches for a partial match from the set of filters, where a constraint belonging to a filter is matched by the given attribute. If the filter (with the partial match) is not yet associated with an interface, the algorithm increases a counter to keep track of matched constraints for the given filter. If the counter size is equal to the number of constraints in the filter, the filter is said to match. After processing one filter the algorithm checks if all filters are matched. The algorithm stops if either all attributes of the notification or all filters are processed. The number of interfaces thus imposes an upper bound on the processing along with the number of attributes and filters. The forwarding algorithm is optimized using binary trees and lookup indices for attributes used in the filters.

The filter-matching algorithm has been extended with several optimizations [33]. The algorithm uses a matching structure based on an index and selections over attribute filters. The article proposes several enhancements, namely the selectivity table that is used to prune those predicates that cannot be matched.

7.6.1.3 Mobility Support

SIENA has been extended to support mobility and wireless clients. The generic mobility support involves a handover protocol, which uses existing pub/sub primitives: publish and subscribe [34]. The benefits of a generic protocol are that it may work on top of various pub/sub systems and requires no changes to the system API. On the other hand, the performance of the mobility support decreases, because mobility-specific optimizations are difficult to realize when the underlying topology is hidden by the API.

The SIENA generic mobility support service, the *ping/pong protocol*, is implemented by proxy objects that reside on access routers. The handover involves the following phases:

- The client arrives to access point *B* from *A* and sends the *move-in* request to the new local proxy.
- A ping request is sent and a response will be eventually received from the old proxy. The response can also be called a *pong*. The pong message ensures that subscriptions are fully propagated from *B* to *A*.
- The client sends a *download* request for buffered events.
- The buffered events are sent to the proxy.
- Finally, in the client receives the messages and duplicates are removed.

7.6.2 Elvin

Elvin is a general event notification service and a content-based routing service. Elvin uses a client-server architecture in notification delivery. Clients establish sessions with Elvin servers and subscribe and publish notifications.

An Elvin notification is a list of name-value pairs, similarly to that of SIENA. Basic primitives are a 32- and 64-bit integer, a 64-bit double precision floating point, an internationalized string (UTF-8 encoded), and an array of bytes. Subscription expressions are defined using logical expressions with a C-like syntax: stock = = 'abc' && value > 80.

The expressions are evaluated with Lukasiewiczs tri-state logic that uses an additional value of indefinite (i.e. true, false, indefinite).

Elvin has language bindings for C, C++, Java, Python, Smalltalk, Emacs Lisp, and Tcl. Elvin is content-based, because it allows routing decisions to be made based on the whole message. Elvin features a decoupled security model, in contrast with the traditional point-to-point model, in which communication between publishers and subscribers is authenticated with keys. Producers and consumers can have overlapping key sets. This supports multi-party authorization.

Service discovery is done using a lightweight protocol that is based on multicast. Once a server has been deployed on the network, clients use the protocol to discover the server and dynamically register. Clients also listen to router advertisements, which are also distributed using multicast.

7.6.2.1   Clustering

Elvin supports local clustering of servers that improves scalability and distributes the local load. Clustering is used to implement a distributed, but single-subscription, address space. Routers within a cluster communicate using a reliable multicast protocol over an IP network. An Elvin router may force a client to reconnect to another server in order to reduce load. The Elvin cluster is similar in functionality to a web farm. An Elvin router is a daemon process that runs on a single server and distributes Elvin messages.

Each router in an Elvin cluster shares client subscription information with every other node. Not all subscription information is shared, but only sufficiently in order for a router to decide if a given notification has any subscribers at any server. The initial forwarding decision in server-server communication is done based on a list of terms. Messages are first analyzed at a local router and then multicast to the cluster. The set of destination routers is determined before multicasting by matching the message against the term list. Each packet contains the unique identifiers of the routers that have matching terms. This hasty approach results on a number of unnecessary notifications at the router level.

The Elvin cluster topology consists of a single master router and a number of slave routers. The master router maintains management data. All slave routers listen to management traffic within the cluster and keep information about every node. Routers also keep information about subscription terms of other servers, current states, the list of URLs offered by a router for client connection, and current router load and statistics. Master servers listen for join packets and keep track of the cluster as a whole. A new master router is elected using an election protocol if the old one fails. Communication between clients and routers employs RPC-style communication with positive and negative acknowledgments. Delivery has best effort, at-most-once semantics. In the client-server protocol the server may drop notifications, but is obliged to warn the client that it has done so.

7.6.2.2   Federation

There is a different protocol for linking distributed clusters of servers to a federated system. The Elvin federation protocol assumes that the federated topology forms a spanning tree. Moreover, the linking protocol supports the definition of pull filters that constrain the notifications sent to other clusters.

7.6.2.3 Quench

In Elvin terminology quench means an operation supported by all event producers that gives the producers the possibility to evaluate a subscription expression to cease producing events that are no longer needed.

Quench is also used to determine which notifications should be produced. The quench is a semantic extension of the subscribe mechanism. In Elvin quench is implemented in the client-server protocol. Any client may request to be notified when the subscription information of the server changes. The client may request information on named attributes in subscriptions. The requested information is sent as an abstract syntax tree. There is also support for an automatic quench, which is implemented in the client library.

7.6.2.4 Mobile Support

Elvin has been extended to support mobile users. One of the requirements was persistence in order to keep undelivered notifications. Elvin is non-persistent by design so a prototype proxy was designed to store notifications. The proxy model extends the client-server architecture of Elvin by introducing the proxy as a third component. Proxies act as normal clients to servers, but as proxy servers to clients. In this design, clients connect to these proxies, which mediate the Elvin service.

The proxy is able to handle multiple clients with separate sets of subscriptions. Elvin did not support subscription grouping by the client, so support for this was added to the system (the concept of a session). These sessions need not be client-specific, but may rather span multiple clients or applications. This stems from the observation that many people have several devices, but may wish to receive the same set of information regardless of the medium.

In order to manage the storage space for undelivered notifications, the proxy supports the definition of a time-to-live (TTL) for each subscription. In addition, clients may specify the maximum number of notifications to keep. In the current prototype clients explicitly connect to the proxy, and they must connect to the same proxy to retrieve notifications. Proxy discovery and roaming between proxies is not supported. The Elvin proxy service is proposed as a solution to proxy roaming and client migration between networks. However, the difficulty lies in that the proxy is a stateful entity, whereas normal Elvin servers are stateless.

7.6.2.5 Non-destructive Notification Receipt

For users who use many different devices and wish to share notifications, Elvin supports non-destructive notification receipt. This means that the proxy does not destroy a notification upon its successful delivery. Elvin ensures that notifications are never delivered to the same client more than once. Because sessions may contain a number of clients, Elvin supports additional management functionality regarding the set of subscription set by clients. Each client is informed of the current subscription status. There may also be a number of sessions per client, in which case only one notification is sent even if there are multiple matches.

7.6.3 JEDI

Java Event-based Distributed Infrastructure (JEDI) [9] is a distributed event system developed at Cefriel at Politecnico di Milano. In JEDI the distributed architecture consists of a set of *dispatching servers (DS)* that are connected in a tree structure. Each DS is located on a node of the tree and all nodes except the root node are connected to one parent DS. Each node has zero or more descendants.

Event subscription and unsubscription requests are propagated by each DS upwards towards the root. Event notifications are processed similarly and forwarded by the local DS to its parent. Upon receiving an event, each DS checks its descendants if they have an interest in the event, and, if required, forwards the event down the tree. This strategy requires that a given DS knows the event requests of its descendants in order to make the forwarding decision. Moreover, since all requests and notifications are propagated up the tree, the communication and processing overhead of the nodes near the root may become a bottleneck. If any of the nodes near the root become disabled, parts of the tree become isolated. In this case the system needs to deal with segmentation and to be able to mend the tree or negotiate a new root and a new tree.

A JEDI event is an ordered set of strings, the first string being the name of the event followed by event parameters. An Event Dispatcher can subscribe to a single event or an event pattern. Event patterns are used to filter events based on parameter matching, for example foo(aa*,bb) matches all events named foo that have exactly two parameters and the first parameter starts with aa and the second parameter is exactly bb. JEDI preserves causal ordering of messages, that is, if event e1 caused the firing of event e2, e1 must be delivered first to all interested subscribers. This mechanism allows a pair of components to synchronize through the generation of events.

The dispatching servers in the JEDI architecture support mobility by allowing clients to disconnect, move to a new dispatching server, and connect while retaining all the notifications. This is supported by two new operations, namely the *move-out* and *move-in* commands. The dispatching servers manage temporary storage for notifications. They also coordinate that no duplicates are received and that the notifications are causally ordered. The new dispatching server contacts the old one directly in order to receive the accumulated notifications. The old DS notifies its parent-dispatching server to route any further notifications for this client to the new DS. Notifications are routed in the JEDI dispatching tree from producers to consumers and there is no possibility for adapting the routing strategy to reflect changes in the pattern of communication.

The system offers good performance if the tree is organized in a good way that minimizes network traffic. In essence, when clients migrate from one dispatching server to another the load placed on the servers changes. It may be necessary to recreate the dispatching topology to reflect these changes. JEDI approaches the adaptation of pub/sub systems to more dynamic environments by extending the event routing mechanism with the addition of a new spanning tree routing algorithm. Now, a delegate leader is responsible for each subscription. The delegate accepts subscriptions of similar type and becomes the leader of the subscribers. It also manages the distribution of the group in the tree. Each dispatcher knows the group leaders for all subscriptions.

7.6.4   Rebeca

Rebeca is a distributed event system that supports mobile users and context-aware sub-scriptions [10, 35]. The system supports both logical and physical mobility. The basic system is an acyclic routed event network using advertisement semantics. The mobility protocol uses an intermediate node between the source and target of mobility, called Junction, for synchronizing the servers.

If the brokers keep track of every subscription, the Junction is the first node with a subscription that matches the relocated subscription propagated from the target broker. If covering relations or merging are used this information is lost, and the Junction needs to use content-based flooding to locate the source broker.

7.6.5   STEAM

The *STEAM (Scalable Timed Events and Mobility)* event system is specifically designed for wireless ad-hoc networks [36]. The system uses three different filters to address the problems related to dynamic reconfiguration of the network topology. Specifically, the STEAM system is intended for WLANs using the ad hoc network model, and the main application domain is traffic management. The system uses an implicit event model in which entities subscribe to interesting event types locally, and not by using a centralized broker. STEAM exploits a group communication service for notifying interested entities. Groups are geographically bound and nodes are identified using beacons.

The three filter types supported by STEAM are subject, proximity, and content filters. Events consist of a name and a set of typed parameters. The name also determines the structure of the event. A subject filter is matched against the event and mapped onto a proximity group. A proximity filter corresponds to the geographical aspect of the proximity group. A proximity filter specifies the scope in which events are disseminated. A proximity filter applies to an event type and is established when the type is deployed. In essence, upon publication of an event the source matches the subject and proximity, and the subscribers match the content. This requires that the proximity filter at the producer must have location information from the subscriber. The paper does not explain how this information is acquired, how often it is updated, and how the security implications are handled. In essence the protocol is a wireless application-level broadcast protocol with subject-based filtering at the source and content-based filtering at the client.

Producers announce the event types they intend to raise (publish) with the geographical area, called the proximity, within which events of this type are to be disseminated. The proximities may be defined independently of the physical range of the communication system. The routing layer may support multi-hop communication.

STEAM uses a *Proximity-based Group Communication Service (PGCS)*. In this service, groups are assigned certain geographical areas. A node that wants to join a group needs to be located in the groups area. STEAM provides a *Proximity Discovery Service (PDS)* that uses beacons to discover proximities. Once a proximity is discovered the associated events are delivered to the client if it has a matching subscription. PDS causes the middleware to join a group if either a subscription or an announcement matches the group. The proximities are static but clients may move.

An experimental scenario is presented that pertains to traffic lights at an intersection with and without filtering. The results suggest that distributed filtering, although simple in this case, is beneficial in ad-hoc environments and may reduce the amount of transmitted traffic significantly.

7.6.6 Fuego Event Service

The Fuego event service was developed in the Fuego Core 2002/2004 project at the Helsinki Institute for Information Technology HIIT. The event service addresses the challenges in the mobile computing environment by providing an asynchronous content-based pub/sub system that supports client mobility [37].

The event service for mobile computing consists of two parts: the client-side API, and the server-side system. The client-side API is similar in functionality to JMS and offers the basic pub/sub functionality and session management. The server-side provides an extensible framework for content-based routing with optimizations and mobility support. The generic router implementation allows pluggable routing algorithms and routing table user interfaces.

A key component of the architecture is the Fuego event router. The event router is a component that connects the publishers and subscribers and mediates event messages between them.

Events are represented according to the XML Infoset specification. All remote API calls use the SOAP request-response protocol, and the notification of events uses asynchronous SOAP messaging. The event router is implemented as an Apache Axis web service. The client side API implementation uses a SOAP protocol adapted for wireless by default; however, a lightweight version of the API was created for Java ME end systems that uses HTTP 1.1 and a proprietary binary message format.

Filtering is a central core functionality for realizing event-based systems and accurate content-delivery. Filtering may be optimized by using covering relations or filter merging. The event service supports both through new generic data structures for content-based routing. Any objects that implement methods for covering and merging are automatically optimized. The system has a default filtering language, which is based on typed tuples and disjunctive normal form based attribute-filters. Typically, events and filters are represented as lists of typed tuples. In this case, an attribute filter is a 3-tuple <name, type, constraint>.

The router toolkit does not specify any particular routing topology, but rather allows the developers to use the generic API methods available, leverage the efficient data structures for various configurations, and develop various mobility protocols. As an example, the Fuego router toolkit has been used to implement an event channel-based configuration.

The server-side system consists of a set of event routers. Each router has two components: a local routing table and a remote routing table. The local routing table stores filters set by local clients and provides queue management for mobile and wireless clients. Disconnected clients may retrieve queued events upon reconnection using push or pull semantics. The remote routing table is responsible for communicating with other routers and forwarding events in the distributed system. In order to support extensibility, the local and remote routing tables and algorithms are separate objects, which may be changed if necessary.

This modularity allows the implementation of various distributed event routing semantics and router topologies. Subscription semantics and advertisement semantics are examples of two different interest propagation mechanisms. The former propagates subscription messages throughout the system and events are routed on the reverse-path of subscriptions. In the latter, advertisements are propagated throughout the system and subscriptions are routed on the reverse-path of advertisements. Supported routing topologies include hierarchical, event channel, and peer-to-peer topologies. The system also has separate user interface modules for the routing tables.

The client-side API supports expressive operation with three mechanisms: multiple sessions, expressive pull functionality, and fast subscriptions. The first approach allows clients to create multiple sessions at different access servers for subscriptions with different maximum queue sizes and delivery options. The client may have several sessions at different access servers, for example to support different modes of operation. The second approach is realized by pulling the notifications that match subscription identifiers or arbitrary filters. Thus the client may subscribe different events, which are stored in the session and when running on a small client only the essential events may be retrieved using the pull operation.

7.7 Advanced Topics

Mobility support is a relatively new research topic in for pub/sub. Mobility is an important requirement for many application domains, where entities change their physical or logical location. In this book we presented a comparison of mobility support protocols in Chapter 3. We observed that mobility protocols can be characterized based on the nature and number of indirection points and the target entity. For example, Mobile IP is a protocol that supports the mobility of hosts. Wireless CORBA was presented as an example of an object level mobility support protocol. The SIP framework supports the mobility of sessions. We also discussed pub/sub mobility protocols and observed that they require their own solutions in order to update the event-routing topology and optimize event flow.

In this chapter, we have outlined several pub/sub mobility support technologies, namely SIENA, Elvin, JEDI, Rebeca, and the Fuego event system. JEDI was one of the early systems to incorporate support for mobile clients with the move-in and move-out commands. JEDI maintains causal ordering of events and is based on a tree-topology, which has a potential performance bottleneck at the root of the tree with subscription semantics. Elvin is an event system that supports disconnected operation using a centralized proxy, but does not support mobility between proxies.

SIENA, Rebeca, and Hermes[19] support content-based routing of events using covering relations. To our knowledge, covering relations were first introduced in the SIENA project and they support the optimization of event-based communication. Optimizations that aggregate content-based routing state, such as covering relations, are problematic for mobility support because they lose information pertaining to the original subscribers and advertisers. This can be solved by using techniques that partially flood control messages on the reverse path of messages that were used to build the routing table [38].

A number of overlay-based pub/sub routing algorithms and router configurations have been proposed. An application layer overlay network is implemented on top of the network layer. Typically overlays provide desirable features such as fast deployment time, resilience and fault-tolerance.

An overlay-routing algorithm leverages underlying packet-routing facilities and provides additional services on the higher level. These services include searching, storage, synchronization, and multicast services.

Good overlay routing configuration follows the network level placement of routers. Many overlays are based on Distributed Hash Tables (DHTs), which are typically used to implement distributed lookup structures. Many DHTs work by hashing data to routers/brokers and using a variant of prefix-routing to find the proper data broker for a given data item. Hermes [19] and Scribe [20] are examples of pub/sub systems implemented on top of an overlay network and are based on the rendezvous point routing model. The Hermes routing model is based on advertisement semantics and an overlay topology with rendezvous points.

Typical fixed-network pub/sub routing algorithms are deterministic in nature. Basic routing algorithms do not cope with topology changes. The routing algorithms should be able to adapt to the network environment and at the same time take the supply and demand of information into account. Several solutions have been presented for on the fly configuration of the pub/sub network, for example the GREEN system [39]. In addition, probabilistic algorithms have also been proposed for better routing support in peer-to-peer and ad hoc environments.

A topic-based multicast algorithm for peer-to-peer event dissemination is presented in [40]. The algorithm is data-aware in the sense that it exploits information about process subscriptions and topic inclusion relationships. This data-awareness is used to limit the membership information that each process needs to maintain.

Gossip-based broadcast algorithms form a family of broadcast algorithms that contrast reliability guarantees with scalability properties. A lightweight probabilistic broadcast, lpbcast, is a decentralized gossip-based broadcast algorithm that offers scalability in terms of throughput and memory-management [41]. Decentralization means that the algorithm is based only on local information.

The construction of an optimal pub/sub dissemination tree for routing information from source to interested recipients was analyzed in [42]. A greedy algorithm was proposed as a solution to the tree building problem that builds the tree in a fully distributed fashion.

Bibliography

[1] Eugster PT, Felber PA, Guerraoui R and Kermarrec AM 2003 The many faces of publish/subscribe. *ACM Comput. Surv.* **35**(2), 114–131.

[2] Gupta S, Hartkopf J and Ramaswamy S 1998 Event notifier, a pattern for event notification. *Java Report* **3**(7), 19–36.

[3] Yu H, Estrin D and Govindan R 1999 A hierarchical proxy architecture for internet-scale event services *Proceedings of 8th International Workshop on Enabling Technologies: Infrastructure for Collaborative Enterprises (WETICE '99)*, pp. 78–83, Palo Alto, CA, USA.

[4] Obj 2001a *CORBA Event Service Specification v.1.1*.

[5] Obj 2001b *CORBA Notification Service Specification v.1.0*.

[6] Obj 2001c *Management of Event Domains Specification*. http://www.omg.org/cgi-bin/doc?formal/2001-06-03.

[7] Carzaniga A, Rosenblum DS and Wolf AL 2001 Design and evaluation of a wide-area event notification service. *ACM Transactions on Computer Systems* **19**(3), 332–383.

[8] Bricconi G, Tracanella E, Nitto ED and Fuggetta A 2000 Analyzing the behavior of event dispatching systems through simulation. In *HiPC* (ed. Valero M, Prasanna VK and Vajapeyam S), vol. **1970** of *Lecture Notes in Computer Science*, pp. 131–140. Springer.

[9] Cugola G, Di Nitto E and Fuggetta A 1998 Exploiting an event-based infrastructure to develop complex distributed systems *Proceedings of the 20th international conference on Software engineering*, pp. 261–270. IEEE Computer Society.

[10] Fiege L, Gärtner FC, Kasten O and Zeidler A 2003 Supporting mobility in content-based publish/subscribe middleware. In *Middleware* (ed. Endler M and Schmidt DC), vol. **2672** of *Lecture Notes in Computer Science*, pp. 103–122. Springer.

[11] Mühl G 2002 *Large-Scale Content-Based Publish/Subscribe Systems* PhD thesis Darmstadt University of Technology.

[12] Mühl G, Ulbrich A, Herrmann K and Weis T 2004 Disseminating information to mobile clients using publish/subscribe. *IEEE Internet Computing* **8**(1), 46–53.

[13] Carzaniga A 1998 *Architectures for an Event Notification Service Scalable to Wide-area Networks* PhD thesis Politecnico di Milano Milano, Italy.

[14] Mühl G, Fiege L, Gärtner FC and Buchmann AP 2002 Evaluating advanced routing algorithms for content-based publish/subscribe systems In *The Tenth IEEE/ACM International Symposium on Modeling, Analysis and Simulation of Computer and Telecommunication Systems (MASCOTS 2002)* (ed. Boukerche A, Das SK and Majumdar S), pp. 167–176. IEEE Press, Fort Worth, TX, USA.

[15] IBM 2002 *Gryphon: Publish/subscribe over public networks*. (White paper) `http://researchweb.watson.ibm.com/distributedmessaging/gryphon.html`.

[16] Strom RE, Banavar G, Chandra TD, Kaplan M, Miller K, Mukherjee B, Sturman DC and Ward M 1998 Gryphon: An information flow based approach to message brokering. *CoRR*.

[17] Cugola G, Frey D, Murphy AL and Picco GP 2004 Minimizing the reconfiguration overhead in content-based publish-subscribe *SAC '04: Proceedings of the 2004 ACM Symposium on Applied Computing*, pp. 1134–1140. ACM Press.

[18] Virgillito A, Beraldi R and Baldoni R 2003 On event routing in content-based publish/subscribe through dynamic networks *Proceedings of the Ninth IEEE Workshop on Future Trends of Distributed Computing Systems (FTDCS 2003)*, pp. 322–328. IEEE.

[19] Pietzuch PR 2004 *Hermes: A Scalable Event-Based Middleware* PhD thesis Computer Laboratory, Queens' College, University of Cambridge.

[20] Rowstron AIT, Kermarrec AM, Castro M and Druschel P 2001 Scribe: The design of a large-scale event notification infrastructure. In *Networked Group Communication* (ed. Crowcroft J and Hofmann M), vol. **2233** of *Lecture Notes in Computer Science*, pp. 30–43. Springer.

[21] Pietzuch P and Bacon J 2002 *Hermes: A distributed event-based middleware architecture Proceedings of the 1st International Workshop on Distributed Event-Based Systems (DEBS'02)*.

[22] Carzaniga A, Rutherford MJ and Wolf AL 2004 A routing scheme for content-based networking *Proceedings of IEEE INFOCOM 2004*. IEEE, Hong Kong, China.

[23] W3C 2003 *SOAP Version 1.2*. W3C Recommendation.

[24] Carzaniga A, Rosenblum DS and Wolf AL 2000 Content-based addressing and routing: A general model and its application. Technical Report CU-CS-902-00, Department of Computer Science, University of Colorado.

[25] Opyrchal L, Astley M, Auerbach J, Banavar G, Strom R and Sturman D 2000 Exploiting IP multicast in content-based publish-subscribe systems *Middleware '00: IFIP/ACM International Conference on Distributed systems platforms*, pp. 185–207. Springer-Verlag New York, Inc., Secaucus, NJ, USA.

[26] Carzaniga A and Wolf AL 2001 Content-based networking: A new communication infrastructure. In *Infrastructure for Mobile and Wireless Systems* (ed. König-Ries B, Makki K, Makki SAM, Pissinou N and Scheuermann P), vol. **2538** of *Lecture Notes in Computer Science*, pp. 59–68. Springer.

[27] Cugola G, Di Nitto E and Picco GP 2000 *Content-based dispatching in a mobile environment Workshop su Sistemi Distribuiti: Algorithmi, Architectture e Linguaggi*.

[28] Loke SW, Padovitz A and Zaslavsky AB 2003 Context-based addressing: The concept and an implementation for large-scale mobile agent systems. In *DAIS* (ed. Stefani JB, Demeure IM and Hagimont D), vol. **2893** of *Lecture Notes in Computer Science*, pp. 274–284. Springer.

[29] Segall W and Arnold D 1997 Elvin has left the building: A publish/subscribe notification service with quenching *Proceedings of the 1997 Australian UNIX Users Group, Brisbane, Australia*. `http://elvin.dstc.edu.au/doc/papers/auug97/AUUG97.html`.

[30] Obj 2007 *Data Distribution Services, V1.2*.

[31] Waldo J 2000 *The Jini Specifications*. Addison-Wesley Longman Publishing Co., Inc., Boston, MA, USA.

[32] Niblett P and Graham S 2005 Events and service-oriented architecture: the OASIS web services notification specifications. *IBM Syst. J.* **44**(4), 869–886.

[33] Carzaniga A and Wolf AL 2003 Forwarding in a content-based network *Proceedings of ACM SIGCOMM 2003*, pp. 163–174, Karlsruhe, Germany.

[34] Caporuscio M, Carzaniga A and Wolf AL 2003 Design and evaluation of a support service for mobile, wireless publish/subscribe applications. *IEEE Transactions on Software Engineering* **29**(12), 1059–1071.

[35] Zeidler A and Fiege L 2003 Mobility support with REBECA. *ICDCS Workshops*. IEEE Computer Society.

[36] Meier R and Cahill V 2003 Exploiting proximity in event-based middleware for collaborative mobile applications. *DAIS*, pp. 285–296.

[37] Tarkoma S, Kangasharju J, Lindholm T and Raatikainen K 2006 Fuego: Experiences with mobile data communication and synchronization *17th Annual IEEE International Symposium on Personal, Indoor and Mobile Radio Communications (PIMRC)*.

[38] Tarkoma S and Kangasharju J 2007 On the cost and safety of handoffs in content-based routing systems. *Computer Networks*.

[39] Sivaharan T, Blair G and Coulson G 2005 GREEN: A Configurable and Re-configurable Publish-Subscribe Middleware for Pervasive Computing *Proceedings of DOA 2005*.

[40] Baehni S, Eugster PT and Guerraoui R 2004 Data-aware multicast. *Proceedings of the 2004 International Conference on Dependable Systems and Networks (DSN 2004)*, pp. 233–242.

[41] Eugster P, Handurukande S, Guerraoui R, Kermarrec A and Kouznetsov P July 2001 Lightweight probabilistic broadcast *Proceedings of The International Conference on Dependable Systems and Networks (DSN 2001)*.

[42] Huang Y and Garcia-Molina H 2003 Publish/subscribe tree construction in wireless ad-hoc networks. *4th International Conference on Mobile Data Management (MDM 2003)*, pp. 122–140.

[43] Sun 2001 *Java Message Service Specification*.

8

Data Synchronization

A common theme throughout this book has been device mobility, but user mobility has not been quite as prominent. By user mobility, we mean users moving around, using different devices depending on their location. After all, while a mobile phone or a PDA is handy to have while traveling, it is hardly the pinnacle of computing devices, so a user probably prefers something a bit more powerful and comfortable to use when device form factor and portability are not issues.

The concept of using multiple devices brings with it the problem of data management. A user is hardly going to think of each of their devices existing as a bubble of its own, processing data of its own without concern for any other devices. No, the user thinks of all the data on the devices combined as being their data, and wants this to be consistent across all the devices. That is, there is a need to keep the data on all devices in synchrony.

There is a simple answer to the synchronization problem, and one that has recently gained ground under the terms 'Software as a Service' and 'cloud computing'. That answer is keeping all the data on a centralized server and always accessing it through the network. While this answer is attractive in many ways, it cannot be considered acceptable when network connectivity is slow, expensive, and not always available, as is the case with mobile devices. Therefore, we must accept the fact that the data is distributed among a number of different devices, without constant connectivity to each other.

8.1 Synchronization Models

Our basic model of data synchronization is as follows: We assume there to exist a user who possesses a collection of data. For the moment, we shall assume nothing about the syntax or semantics of the data, but treat it simply as a set of opaque binary files (we will consider later what happens when this assumption is relaxed). The user has a number of devices, each with its own storage, and wishes to be able to access and modify the data on any of them.

If there were no need to modify the data, synchronization would be simple: just copy all the data to each device. The possibility of modification means that any updates to the data need to be propagated to all devices. Furthermore, it is common that a user modifies their

Mobile Middleware Sasu Tarkoma
© 2009 John Wiley & Sons, Ltd

data on two separate devices, so that there are two sets of changes, possibly incompatible, that both need to be applied. Above all, synchronization should ensure that no data is lost.[1]

8.1.1 Basics of Synchronization

Like with all systems, there are some choices to be made when designing a synchronizer. In synchronization, the high-level choices are mostly *when* to synchronize and *how* to do it.

There are essentially two ways of determining when to synchronize. First, the process can be started manually by the user activating some trigger. Second, it can be performed automatically by the synchronizer when an opportunity is detected. While the automatic way provides a more seamless illusion of the user having just a single collection of data, the manual way is much easier to get right, since it just punts the detection of appropriate conditions to the user.

Considering the automatic process in a mobile environment a bit further: when should a synchronization be triggered? It is not usually feasible to start using a network connection immediately as one comes into range, as mobility might often mean that available network access points flicker into and out of view frequently. Therefore the system would need to include some detection of, say, whether the user is mobile or has become stationary, and in the latter case, also to predict how long the user is going to be stationary, as synchronization can consume some time. Thus, it is probably not feasible to use automatic synchronization in the near future on mobile devices.

Data synchronization has much in common with *data replication*. In both, the aim is to store the same data in a number of locations. Usually, the purpose of data replication is to provide fault tolerance in the case of storage location unavailability. Stretching the concept slightly, we could say that this is also the purpose of synchronization, except that in this case we expect most copies of the data to actually be unavailable most of the time.

Consideration of the data replication field allows us an axis on which to divide the overall style of synchronization:

Pessimistic While there can be several copies of the data, permit only one modifiable copy. Synchronization of the modifications is simply copying the modified copy. Synchronization may also transfer the modifiability status.

Optimistic Explicitly permit there to be multiple copies that can be independently modified. Requires a way to reconcile concurrent modifications.

A pessimistic replication system needs more control than an optimistic one, since there needs to be some way of identifying the modifiable copy, in effect a lock on the data. The model of a pessimistic system is usually a hub-and-spoke model, with the hub serving as a central clearing house that gives locks to the modifiers. In contrast, an optimistic system can in theory permit completely unconstrained modification and synchronization.

We shall assume that the underlying system is based on optimistic replication. This is not a limitation, since a pessimistic system is essentially a special case of an optimistic one, so our coverage can be more general. Also, we feel that the optimistic model is

---

[1] A tall order, as sometimes the user will make mistakes, and there is rarely a way to reliably identify these as mistakes.

more 'correct' for the use cases relevant to the scenario outlined in the beginning, but this does not mean that pessimistic replication is an outdated idea. A system designer should be familiar with many different architectural choices to enable the most suitable design decisions to be made.

It is often helpful to think of the data as *versioned*. That is, each state is assigned a version number that increases as modifications are made. In fact, many synchronization or synchronization-like systems work by assigning explicit version numbers to the data. Designing a version numbering system for an optimistic replication system is a challenging problem, and there exist many different solutions.

For simplicity, we shall assume there to be some form of versioning that obeys the following rules:

1. If file B is the result of making a modification to file A, then the version number of B compares higher than that of A.
2. If two files have been independently modified from some base at different locations, their version numbers are not comparable.

This ensures that it is possible to detect when a file has been modified concurrently while keeping a linear sequence of versions at each individual location.

8.1.2 Synchronization Process

Synchronization itself can be divided into two phases [1]:

Update detection Recognition of an update that should trigger synchronization of changes.
Reconciliation Combination of the updates to build a synchronized version.

Neither is a trivial problem, and a number of solutions exist for each.

In an actual synchronizer implementation, there is a third phase between these two: *update propagation*. When changes have been made, they need to be transmitted to wherever they are going to be reconciled. For this purpose, a communication protocol will need to be designed. Update propagation is somewhat peripheral to the current discussion, though, so we will expand on it only later.

In the update detection phase, files are identified as either *clean* or *dirty*, in analogy to caches or virtual memory. A dirty file is one that has potentially been changed and must therefore be synchronized. It is essential that the identification of a file as dirty is conservative, in that it always identifies a modified file as dirty. It is a matter of efficiency to identify unmodified files as clean, but accuracy is still of large importance; after all, one would not like the whole file system to be synchronized every time independently of whether changes were made.

A fairly simple solution to the update detection problem is to use modification timestamps. If the data consists of files, these should be available, as most common file systems provide this information. Thus, the synchronizer only needs to store the time of the most recent synchronization and synchronize exactly those files that have been modified since.

There are a couple of problems with this approach. The first is that it requires going through the whole file system, looking at modification timestamps, which can be an expensive operation, especially on a large file system. Modern operating systems include ways to monitor a file system, so if the synchronizer uses such a mechanism, it will be informed precisely of any changes. An innovative way that does not require such a monitor would be to make a directory's modification time mean the last time when any file in that directory or its subdirectories was changed. This way the scanning for modified files could prune all the branches of the directory tree that do not contain modified files, significantly cutting the number of files that need to be examined.

The second problem is the semantics of the modification time, which requires care in designing the synchronizer. For instance, on Unix file systems, the file modification time refers to the last time the file contents were modified, which does not include moving, or renaming, files. Thus, if an unmodified file is moved on top of another file, this will not be noticed by looking at file timestamps. Rather, what is needed is the full chain of modification timestamps of directories all the way up to the root directory. Still, despite these problems, using modification times is a reasonable strategy.

Modification times, however, are not completely accurate. After all, it might happen that one modifies the file but leaves it in the same state. To get fully accurate change detection, one must resort to a comparison of the file contents with the previous ones. This, of course, means storing the old contents from the previous synchronization run for comparison purposes, essentially (at least) doubling the storage requirements.

After updates have been detected, they will need to be synchronized, and, if concurrent modifications have been made, those will need to be reconciled. At this point we must, by necessity, consider the syntax and semantics of the data. Namely, if we were to treat the data as opaque, the only thing to do in reconciliation would be to present the two modifications to the user and ask them to create a new version that integrates the changes.

Reconciling concurrent modifications requires some way to determine which modifications have been made that need to be reconciled. There are, again, two approaches here:

Edit logs This approach requires cooperation from any application that edits the data. For each piece of data, a log of all the edits made to it is stored. When reconciling, these logs are composed together to form a joint edit log starting from the latest common ancestor version.

State comparison Take the two modified files and merge their contents by some appropriate algorithm. Most often, this process is aided by comparing against the latest common ancestor, using a process called *three-way merge*.

Again, the state comparison model is more general than the edit log model, since it is applicable in more situations. Considering applications that users normally use to modify their data, such as text editors or word processors, it is rare to have a complete edit log available, with semantically meaningful information. An edit log could perhaps be faked somewhat by making comparisons whenever the user saves a file, creating a sequence of edits, but there is no way to ensure the kind of fine granularity that a pure edit log provides.

One clear benefit of edit logs compared to any state-based approaches is that they are better able to provide more precision to the modifications. Namely, modifications can be divided into four categories:

- insertion of a piece of structure;
- deletion of a piece of structure;

- moving a piece of structure from one place to another; and
- changing the content of some piece of structure.

We use the term 'piece of structure' to evoke the concept of a structured file. This is not too limiting, as many types of data can be considered to have some structure, even if they are not rigidly defined as structured.

Of these four types of operations, it is rare for a state-based approach to consider the move operation. Rather, it is commonly represented as a deletion followed by an insertion. However, consider the case where one location has moved a piece of structure and another location has changed its content. If moves are recognized, these do not cause any conflict as both operations can be performed. But if the move is represented as a deletion–insertion pair, the change made at the other location will either cause a conflict or, in the worst case, be completely lost.

Attitudes toward implementing the move operation are changing, though, and it is increasingly seen as a useful primitive operation. As a state-based approach cannot support it directly, there are a few different ways to implement it. One is to embed unique identifiers inside the files themselves. While accurate, this is not applicable to every kind of file, but it works well enough for editable text files, which are pretty common. Another method is to examine deletion–insertion pairs and compare files to determine if they are 'close enough' to be a potential move. This is a heuristic, so it cannot capture every move and can potentially denote non-moves as moves as well, but it has the benefit of working on all kinds of files without the need to embed extraneous information inside them.

8.1.3 Architectures for Synchronization

We touched on the architectures for synchronization briefly above when considering optimistic and pessimistic replication. There are a few ways of constructing an architecture for synchronization, and as time has passed, the trends have moved toward more and more flexible arrangements. Still, there are also benefits to older, more rigid styles.

In a synchronization architecture, the design starts by considering a single user's data, as the data of different users can be seen to be independent of each other. Any device where the data is kept is a component in the architecture. Connections are the synchronization connections. That is, components A and B are connected if they ever synchronize the data between them.

A useful way to think about synchronization is as a kind of sequence diagram as seen in Figure 8.1. Here, each device is thought of as a running process and a synchronization between two devices is shown as a double-arrow connection between them. Note that synchronization always involves only two devices, not more.

The sequence diagram view of synchronization denotes the actual actions in a system. In contrast, the architecture of synchronization denotes, with its connections, all possible synchronizations. For instance, allowing the synchronizations that happen in Figure 8.1 actually requires a pretty advanced system.

The simplest architecture of all is the centralized model, demonstrated in Figure 8.2. In the centralized model, there is one distinguished device, with which all other devices must synchronize. The data at the distinguished device is also often called 'master copy' to reflect its special role.

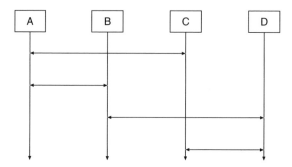

Figure 8.1   Synchronization sequence

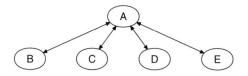

Figure 8.2   Centralized synchronization

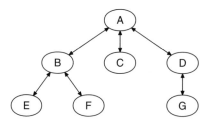

Figure 8.3   Tree synchronization

The centralized model appears pretty rigid, but it often works quite well. It is normal for the distinguished device to be some sort of a storage facility that is highly available, like a file server, but it can also be a regular workstation. If there is no need to modify the data on multiple devices between synchronization opportunities, the centralized model can be used without any ill effects.

A development of the centralized architecture is the tree (or star) model, shown in Figure 8.3. Here, instead of there being a central repository, each piece of data is *branched* from some other piece of data.

The figure is intentionally drawn to evoke parallels with the centralized model, with one node 'at the top'. In both the centralized and tree models, it is common to imagine that the child nodes in the tree are always the active synchronizers. That is, each device initiates synchronization only with its parent.

Limiting synchronization to just the tree model has advantages in version numbering. Namely, since each piece of data has a parent, its version number can incorporate the parent version number. This way, it is simple to recognize whether there is any need

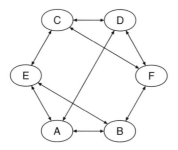

Figure 8.4 General synchronization

for synchronization, since the embedded parent version number tells the highest parent version with which the data is synchronized.

The tree architecture also helps with merging of changes. It is necessary for each device to store only the contents of the data when it was last synchronized with the parent and the version number it had at the time. If the parent also stores the versions that its children were synchronized with, the communication can be optimized, but merging will still work even if the parent does not do this.

The tree architecture can also be constraining. If there are a large number of devices, with a complex synchronization tree, propagating the modifications from one device to another can be very complex, since it needs to follow the tree connections. Relaxing all conditions on the architecture we arrive at the general architecture shown in Figure 8.4.

While the general architecture allows maximal freedom, it can be extremely problematic. Since any two devices can synchronize at any time, it is difficult to locate common ancestors of the two file versions being synchronized. Common ancestors are needed for both the edit-log and state-based merging methods, so the general model requires solving this problem.

One solution is to assign version numbers at each device and have the version number of each piece of data be the composite of all device-assigned version numbers of the pieces of data that have been synchronized into it. So, if we assume that all the potential synchronizations of Figure 8.4 have happened, a piece of data at the E device would have version numbers consisting of components from all the devices A, B, C, and E.

Assigning version numbers is not the only problem, though. For instance, it may not even be possible to locate the closest common ancestor of two versions, since that can be a version produced somewhere that is neither of the two synchronizing devices, so the synchronization process will be suboptimal.

Finally, there is the question of storage space. As noted, having common ancestor versions available is a prerequisite for merging the changes. With general synchronization, there is no way to know which older versions will be needed in synchronization, since there is no control of the synchronization history. Thus, it may be necessary to store a number of older versions, and even to transmit these during synchronization.

To summarize, a centralized architecture is rigid, but it is also efficient. Storage of old versions can be left to the distinguished device, which can also perform any necessary merges, saving space on the other devices. The tree architecture relaxes some of the assumptions of the centralized architecture, leading to more flexibility with a minor

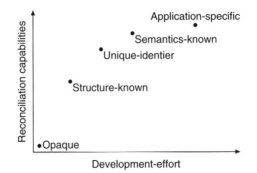

Figure 8.5   Effect of file type information on synchronization

decrease in efficiency due to the lack of a central controller. Finally, architectures more general than the tree suffer from efficiency issues in reconciliation, and are rarely seen in current systems that are not designed to store all the old versions.

8.1.4   Type-Specific Synchronization

Different types of files require different synchronization and especially reconciliation algorithms. The more we know about a file's type, the more 'intelligent' we can make the synchronization. Namely, most files have some structure in them, and knowing this structure is pretty much essential to being able to merge any concurrent changes. Furthermore, sometimes even semantic knowledge of the file is required, since different modifications, affecting different parts of the structure, can still have conflicting semantics.

In general, we can view the knowledge of a file's syntax and semantics affecting synchronization roughly as shown in Figure 8.5 (adapted from [2]). In the figure, development effort is contrasted with reconciliation capabilities. The figure demonstrates how increased development effort brings with it improved reconciliation capabilities, but also the diminishing returns of increased effort. We note that reconciliation capabilities essentially measure the amount of information that the synchronizer has about the file type, and, in a sense, the development effort axis also measures the specificity of the reconciler, in that the points towards the increasing direction also indicate less general reconciliation.

An opaque file does not give us any capabilities, so user intervention is always required. Superficial knowledge of the file's structure actually helps quite a lot, and if there are unique identifiers associated with the atomic pieces of structure, it is possible to go even further by being able to easily recognize moves and not be dazzled by deletion–insertion pairs that superficially appear as moves.

Going even further, including semantic knowledge in the synchronization permits even better reconciliation support, and also conflict recognition. For example, in a file containing markup, knowing that two different kinds of markup are semantically equivalent helps to avoid conflicts, and on the other hand, being aware that a file system requires files in a directory to have unique names allows identifying conflicts. Finally, building a fully application-specific synchronizer gives the best capabilities, but also increases development effort significantly as well as taking synchronization out of the domain of middleware, which is focused on generic solutions that apply to many different applications.

For text files, structure-aware synchronization can be built on the venerable line-based Unix tools `diff` and `patch`. `diff` computes the difference between two text files, expressed as a sequence of inserted and deleted lines, and `patch` is capable of applying this difference to an existing file. Furthermore, `patch` includes some heuristics so that it can apply the difference even if the file it is applied to is not an exact replica of the original target of `diff`.[2]

There exists much less work on synchronization of non-line-based data. Recently, however, with the increased interest in XML, tree synchronization has become an item of interest. Synchronization of trees is naturally present in every file system synchronizer, as current common file systems are invariably organized in a tree structure. However, file systems also include much metadata, which will usually not be available in an XML document, necessitating other kinds of techniques.

XML synchronization, on its own, already occupies the 'structure known' point in Figure 8.5. Furthermore, XML applications are typically lax in allowing especially attributes from foreign namespaces to appear on elements. This permits immediate increase to the 'unique identifier' point, since the `xml:id` attribute is specified to have exactly the required semantics.

Getting to the 'semantics known' point easily and generically would require further work, though. An approach to try here would be to go for schema-aware synchronization, i.e., synchronization that understands the schema that the documents are supposed to follow and ensures that synchronized versions also follow the schema. This is not an easy problem, though it may be possible to handle some special cases and punt the others to be handled by the user.

Still, just the immediately available techniques are sufficient to implement equivalents of the `diff` and `patch` tools for XML documents, and several such tools already exist. One difference between these XML tools and the line-based `patch` is the general lack of heuristics to get inexact patching functionality. But if the unique identifiers are included, inexact patching loses its appeal, as the change points can be precisely determined.

An interesting synchronization method is provided by the `rsync` utility. It has the advantage of not requiring any knowledge of the data syntax or semantics, and can thus work on any type of data. `rsync` does not do reconciliation but only synchronization of single-side changes. However, it does not perform a full copy but instead computes *rolling checksums* that can be efficiently computed for all blocks of a certain size in a file, including overlapping ones. The use of rolling checksums permits the detection of the same piece of data even if it has moved in the file itself, allowing `rsync` to send a much smaller amount of changes than copying the file or even copying it from the first change onwards would require.

8.2 File Systems and Version Control

We have considered the situation as being that a user possesses a collection of data but has not imposed any form of structure on that collection. Of course, in real life, many devices, including desktop computers and high-end mobile devices, arrange a user's data into a

---

[2] These heuristics are not always reliable, though, as the writer of this found out when a program debugging went awry due to a difference having been applied in a wrong location.

file system of some sort. The common arrangement is a hierarchical tree consisting of files containing the actual data and directories containing both files and other directories.

With this practical consideration in mind, it appears obvious that a synchronizer needs to support synchronizing file systems, or, perhaps more properly, directory trees, since it is rarer for complete file systems to be synchronized than just a part of them. This leads into the concept of *distributed file systems* that are, essentially, synchronization built into a replicated file system.

Note especially that a distributed file system is not even close to the same thing as a networked file system.[3] A networked file system is still a single data repository, even if it can be accessed over the network from multiple devices. In particular, networked file systems do not have to deal with synchronization issues while synchronization is a crucial concept in distributed file systems.

The primary concern of a distributed file system is synchronizing the changes made to the directory tree, and not really synchronizing the contents of the files. While the file system will notice that the contents of a file have changed and it needs to be synchronized, the actual synchronization is typically left to an external plug-in of some sort. Commonly, users may define their own synchronizer plug-ins for different types of files, but there is usually some form of fallback when nothing more specific applies. Of course, sometimes this fallback is simply informing the user of a conflict, giving them two versions of the file and the changes made in each, and letting them ensure that all the changes have been accounted for.

There is much in common with file system synchronization and XML synchronization covered above, as both cases synchronize tree-structured data and have similar concerns regarding structure preservation. However, in certain cases, file systems have things much easier.

For one, in most modern file systems, files are not identified by their names but by unique identifiers. This means that file moves can be easily supported, since the identifier remains the same even if the file is given a new name. This also permits supporting hard links, that is, alternate names for the same file. Thus, a distributed file system functions at least at the 'unique identifier' point of Figure 8.5.

Second, if a distributed file system is implemented in the same way as a regular local file system, it will have access to every modification made to the file system and can therefore use the edit-log method to synchronize changes. As we noted, edit logs can be more accurate than state-based methods, since they are aware of the actual edits made.

Version control systems are also synchronizers at heart. A version control system is designed to allow multiple people to make modifications concurrently to a shared set of data, which is exactly the model of synchronization we have been covering. The intent of version control is slightly different, with a focus on different people each having a single copy instead of a single person having multiple copies, but in practice this makes little difference.[4]

Traditional version control systems use the centralized architecture and are based on pessimistic replication. That is, there is a version control server and when a user wishes

---

[3] Well, terminology differs. Some sources consider these two to be synonymous, but we feel there is value in having a term for both kinds.

[4] In fact, version control systems can be seen as a bit more general, since a single person can be expected to possess more coordination when making concurrent changes.

to modify some file, she 'checks it out', taking a lock on it that prevents others from checking the same file out. This model makes synchronization trivial because there is not even a need for reconciliation.

The Concurrent Versions System (CVS) was a major step forward in version control systems. Instead of requiring locking of data, CVS uses optimistic replication (but still the centralized architecture). Synchronization with the server is handled by CVS using a three-way merge. If any conflicts happen here, the user is informed and is expected to correct them.

In recent years, especially in the Open Source community, there has been widespread interest in distributed version control. Distributed systems abandon the centralized architecture of CVS, which is especially attractive in Open Source where everyone is a potential contributor, since it is not feasible to give everyone who asks rights to access a software project's central version control server. With distributed version control, contributors can develop their own work without needing such access, which increases the pool of contributors.

Typically, a distributed version control system uses *change sets*, intended to be small atomic changes. Each change set is given a unique name, so the system can recognize which changes have already been applied to the data. The synchronization process is then an attempt to apply any missing change sets to create the synchronized version. In a way, this approach more resembles the edit log method, with change sets functioning in the role of individual edits, though the modifications inside a single change set are by necessity handled through the state-based approach.

8.3   Synchronization in Middleware

As noted in the beginning of this chapter, data synchronization is becoming (or has already become) an important issue, especially for heavy mobile-device users (see also [3]). As most of the synchronization procedure is generic, it should therefore be a part of a mobile middleware platform, with hooks for applications, for instance, to provide specific reconciliation functionality.

Despite its superficial dissimilarity to the other middleware topics covered previously, synchronization can also use both messaging and publish/subscribe functionality. It is not a full dependency, such as publish/subscribe has on messaging (see Figure 8.6), but usage of the other components can ease the design of the synchronizer and provide opportunities for optimization.

Where the other components can help is in the update propagation phase, as update detection and reconciliation are more local activities. Like in many other protocols, update

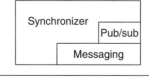

Figure 8.6   Three middleware components

propagation can be seen as having two communication channels: a control channel for establishing which updates need to be propagated and a data channel for actually transmitting the updates.

In most cases, the protocol used on the data channel should not be based on messaging but instead be explicitly designed for synchronization. This is because the needs of the data channel are not met well by a messaging system and the data channel is, after all, the main consumer of resources in synchronization. For example, basing the data channel protocol on HTTP is a reasonable choice: HTTP is designed for transferring data and is so prevalent that the design and implementations have been optimized well for its purpose.

One case where publish/subscribe can be useful even for the data channel is in online editing where the application is built on top of this middleware (see [4] for an example). Here, each edit gets published as an event immediately as it is made, leading to a form of continuous reconciliation, as multiple editors are each making their own modifications. This view is somewhat different from the view we have focused on, namely a single user with the same data on multiple devices, but it is still well within the application area of middleware synchronization.

The control channel protocol, on the other hand, is typically exactly what a messaging system is designed for: small structured messages with some interaction pattern that can, depending on the synchronizer, be more complex than simple request-response.

Publish/subscribe can also be used in update detection. If the synchronization architecture is constrained, it limits the places where updates can be detected. Still, it is beneficial to know that updates have happened elsewhere even if the synchronizer does not allow immediate synchronization, since they can be evaluated and propagated through the system in case they are significant.[5]

8.4   Case Studies

As in previous chapters, we shall next turn to some examples of actual synchronization implementations and see how they fit into the model we have built above. The most important of these in practice in our context is Synchronization Markup Language, or SyncML, from the OMA. We also look at one of the earliest distributed file systems, Coda, and the Unison file synchronizer that embodies many of the principles we have advocated above.

8.4.1   OMA Synchronization (SyncML)

The SyncML Initiative produced a language for synchronization called Synchronization Markup Language (SyncML) [5, 6]. The intent was to unify the synchronization landscape that was littered with a number of proprietary protocols, making interoperability difficult and user experience confusing.

SyncML is not a complete synchronizer, and is not intended to be. The purpose of SyncML is only to specify an interoperable protocol for synchronizing data. That is, it does not care about update detection or reconciliation, only about update propagation, the transfer of updates from one device to another.

---

[5] It has undoubtedly happened to many that important modifications made to some document were forgotten as it was being edited in some other location.

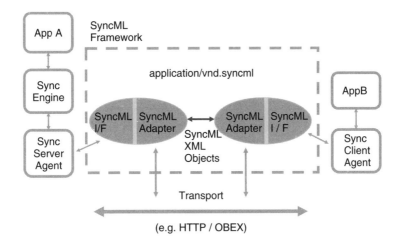

Figure 8.7   SyncML Overview

SyncML update semantics is based on the edit log model. Possible edits include insertion, deletion, replacement, and copying of objects, the atomic units of SyncML. The
architecture is centralized. The synchronization process is initiated by the client, which
sends any edits it has made to the server. The server is then supposed to integrate these
edits and send its own updates back to the client. The server-sent updates also include
any modifications made to resolve conflicts.

Figure 8.7 gives an overview of the SyncML framework. The framework consists of
several components, such as the Representation Protocol, the Synchronization Protocol,
the SyncML API and adapter, and bindings for HTTP, *Object Exchange Protocol (OBEX)*,
and WAP WSP. The framework does not include applications, synchronization engines,
and synchronization agents. The App A in the figure synchronizes its data with the App B
running on the server using the synchronization engine. The engine on the client monitors
changes to local data, and the engine on the server monitors changes made by all clients
to the data and has an understanding of the application semantics. Thus the server keeps
track of the versions of the modified data items. The server can also detect conflicts
between changes; however, conflict resolution is not part of the SyncML specification.

Figure 8.8 illustrates the usage of the SyncML protocol. The client asks the server for
synchronization and provides information for the server that can be used to check the
synchronization state. The client and server communicate using messages, exchanged for
example with HTTP POST. These messages are called SyncML Packages. The server
then sends an initialization message to the client that informs the client to gather changes
and then send those back to the server. The server then can apply changes and merge data.
The server then sends the updates to the client and the client sends an acknowledgment
back to the server.

Nowadays, the SyncML Initiative, like many other similar groups, has joined the Open
Mobile Alliance (OMA). The OMA has adopted SyncML and is responsible for its further
development. Formally, SyncML is now OMA Data Synchronization and Device Management, but is still called SyncML practically everywhere, including device manufacturer
data sheets.

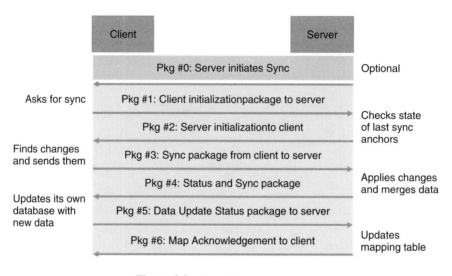

Figure 8.8 SyncML protocol steps

8.4.2 The Coda File System

Coda [7] is one of the earliest distributed file systems, as we use the term. In many ways, Coda is 'just' a networked file system with features that allow it to function when good network connectivity cannot be assured, but these features really make it a distributed file system that has a need for synchronization.

Coda uses a centralized architecture with somewhat optimistic replication. The view of Coda is that there exists a server that contains the file system and clients access it over the network. Where the synchronization comes in is that Coda is designed to function even when disconnected from the network, so a client can 'hoard' files when connected, use them like normal files when disconnected, and reintegrates any changes when connected again.

The intent behind Coda is that disconnections are not the usual case, so the full file system is not always available on the client. Rather, the client is expected to know which files are important and ensure that they are present by making them 'sticky'. If an attempt is made to access a file that is not present on the client while disconnected, this will fail since the file is not there. This is why we called Coda's replication 'somewhat' optimistic above.

Reconciliation in Coda is based on the edit log method. Since Coda is a file system, it naturally has access to all the modifications made to the structure and can store this log for reconciliation purposes. Coda considers a file to be an atomic unit, which means that concurrent modifications to the same file always result in a conflict. Coda supports application-defined conflict resolvers that can be used to resolve the conflicts automatically, but if the conflict cannot be resolved by the Coda server, the user must do it manually.

8.4.3 The Unison Synchronizer

The Unison file synchronizer [8] is a user-level implementation of file synchronization. This means that it lacks the full access to modification information that a file system has, but it also means that it is closer to what synchronization in middleware could be, since a middleware platform also sits on top of an operating system, lacking the complete view of the system that an OS component has. This means that Unison is a state-based synchronizer.

Unison has a formal specification, unlike many other synchronizers, which limits its defined properties. For instance, it only supports synchronization between two locations, not arbitrary many. However, this limitation is not as grave as it seems to be, since the tree-based model can function with this limitation. After all, each edge in the tree splits the tree into two pieces, and any synchronization activity happening in the other piece can be treated as changes directly on the file at the other endpoint of the edge.

One case where the user-level nature of Unison most clearly manifests itself is in metadata synchronization. In real file systems, files possess some metadata, for instance, permissions or file types, and a synchronizer should synchronize any changes in this data too. However, while this is easy in a distributed file system, a user-level synchronizer is usually used on a number of different file systems, making this problem somewhat harder. Unison does have a solution to this, based on treating the metadata specially but still as an integral part of the file. Synchronization in mobile middleware will most definitely have to handle this issue due to the widely varying nature of mobile device storage systems.

Bibliography

[1] Balasubramaniam S and Pierce BC 1998 What is a file synchronizer? *Proceedings of the Fourth Annual ACM/IEEE International Conference on Mobile Computing and Networking*, pp. 98–108.

[2] Lindholm T 2003 XML three-way merge as a reconciliation engine for mobile data *Proceedings of the 3rd ACM international workshop on Data engineering for wireless and mobile access*, pp. 93–97. ACM Press.

[3] Raatikainen K, Christensen HB and Nakajima T 2002 Application requirements for middleware for mobile and pervasive systems. *ACM SIGMOBILE Mobile Computing and Communications Review* **6**(4), 16–24.

[4] Kangasharju J, Lindholm T, Ramya SK, Tarkoma S and Raatikainen K 2007 Collaborative XML editing on small devices: An application of mobility middleware *Pervasive 2007 Demo Session*.

[5] Syn 2002a *SyncML Representation Protocol, version 1.1*.

[6] Syn 2002b *SyncML Sync Protocol, version 1.1*.

[7] Satyanaraynan M et al. 1990 Coda: A highly available file system for a distributed workstation environment. *IEEE Transactions on Computers* **39**(4), 447–459.

[8] Pierce B and Vouillon J 2004 What's in Unison? A formal specification and reference implementation of a file synchronizer. Technical Report MS-CIS-03-36, Dept. of Computer and Information Science, University of Pennsylvania.

9

Security

This chapter examines security in the mobile service access setting. We discuss potential security threats to mobile communications both between the client and the fixed-network, and then within a mobile service platform. Then, we outline solutions for security threats. We consider transport layer security protocols, XML security, including WS-Security and SAML 2.0. We also examine identity management by discussing Liberty Alliance, *Generic Bootstrapping Architecture (GBA)* from 3GPP, *Trusted Platform Module (TPM)*, and the OpenID and OAuth specifications. Finally, we briefly consider issues related with spam and downloaded code.

9.1 Basic Principles

The three core security principles are *confidentiality*, *integrity*, and *availability*. Confidentiality is the property of preventing disclosure of information to unauthorized individuals or systems. For example, a credit card transaction on the Internet requires the credit card number to be transmitted from the buyer to the merchant and from the merchant to a transaction processing network. Data integrity means that data cannot be modified without first obtaining authorization to do so. Moreover, for any information processing system to work properly, it must ensure that information is available when it is required. Availability requires support from the whole architecture, including storage, processing, and communications components.

We summarize the eight well-known security principles including the three mentioned above.

- *Authentication.* Determine identity of an entity.
- *Authorization.* Determine what the entity is allowed to do.
- *Integrity.* Ensure that a message or data element was not altered in transit.
- *Confidentiality.* Ensure that only proper entities can access the data.
- *Availability.* Ensure that service is available for users.
- *Non-repudiation.* Verify the identity of senders or data sources using signatures that are analogous to handwritten signatures. Non-repudiation requires that any events are logged and can be audited. Non-repudiation implies that an entity cannot deny having participate in a transaction.

Mobile Middleware   Sasu Tarkoma
© 2009 John Wiley & Sons, Ltd

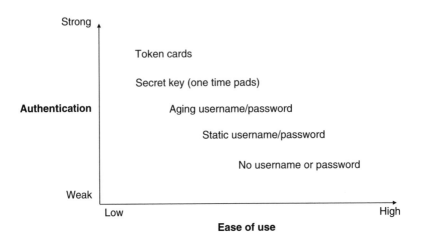

Figure 9.1 Many levels of authentication

- *Privacy.* Ensure that access to data and enable control of information that can identify an individual.
- *Digital Rights Management.* Limit what users can do with the data.

Authentication is a basic building block and necessary for authorization and non-repudiation. Figure 9.1 illustrates the many levels of authentication. Authentication can be characterized by its ease of use and level of security. Simply not using any form of authentication is the easiest choice; however, rife with security problems. A commonly used solution is to use a static username and password. This solution is usable if the password does not change frequently and can be chosen by the user; however, it is subject to brute force attacks and in many cases passwords are much too simple and can be guessed by the attacker. Therefore, aging username and password is more secure albeit less user-friendly solution. This is the current best practice in companies. A more secure approach is to use secret keys known only to the user and the service and agreed beforehand using an out-of-band channel. Typically one time pads are used and they are delivered using postal mail. This is the common solution used by banks in order to authenticate users. A more secure variant of this is based on a hardware-based crypto module that is time-synchronized with an authentication server. The module generates one-time pads for the user.

9.2 Cryptography

Information security is based on cryptographic techniques, which are used to protect information from unauthorized disclosure and tampering while in transit or in storage. Typical mechanisms include authentication methods, digital signatures, message digests, non-repudiation, and encryption. Cryptographic techniques are subject to various kinds of attacks and care must be taken to ensure that proper techniques with reasonable parameters are used. For example, the key size in encryption is an important parameter. A short (weak) key will result in weak encryption that may be broken by unauthorized parties. The storage and dissemination of keys is also important and prone to various kinds of problems. Many *Public Key Infrastructure (PKI)* solutions have been developed to address issues in key management.

The two crucial mechanisms for modern data communications security are data encryption and digital signatures. The former transforms a given plaintext using a cipher algorithm into an encrypted ciphertext. Only those entities that have the correct key can decrypt the information. Keys are either *symmetric* or *asymmetric*.

Symmetric keys are shared by the communicating parties and they result in a more straightforward ciphering mechanism of the two, but this approach requires that a secure mechanism is used to derive the key. Asymmetric keys come in two parts, namely public and private keys. The public keys are shared and the private keys are never shared. Using cryptographic techniques the holder of a private key can prove its ownership of a public key. The public key can then be used to encrypt data sent to the holder of the private key.

Encryption protects the confidentiality of messages; however, encryption alone is not sufficient to determine the authenticity of the message. The former mechanism, namely signatures, are fundamental in supporting message authenticity verification and authentication of communicating entities. The analogue of digital signatures is the handwritten signature on a piece of paper. The aim of the signature is to allow recipients of a message to be able to decide based on the signature that the message is correct and proper.

Digital signatures are a form of asymmetric cryptography and typically three algorithms are required for realizing a signature scheme:

- *A key generation algorithm.* This algorithm selects a private key using random uniform distribution from a set of possible private keys. The algorithm outputs the private key and a corresponding public key.
- *A signing algorithm.* This algorithm produces a signature given a message and a private key.
- *A signature verification algorithm.* This algorithm either accepts or reject the input message given the message, a public key, and a signature over the whole message or parts of the message.

Two main properties are required, namely that a signature generated for given message using a private key, should verify the same message using the corresponding public key. Moreover, it should be computationally infeasible to construct a valid signature verifiable with a public key without having the corresponding private key.

Typically, instead of signing all the bits of a message, a hash value or *message digest* is computed from the bits, and this value is then signed. The main reasons for signing this hash value are as follows. Signing a hash value is more efficient than signing the whole message. Standard hash functions, such as SHA-1, are used which supports compatibility and arbitrary input can be converted to a common format that is easily verifiable.

Checksums are often used as a basic integrity checking mechanism. They are transmitted with the data and the receiver applies a formula to the data and then compares the obtained value with the integrity code that was received with the data. Examples of checksum methods include parity bits and *Cyclic Redundancy Checks (CRC)*. A hash function is a function that takes an arbitrary input and then produces a fixed-size output, the hash value. A one-way hash function is such that it is very difficult to invert. A cryptographically secure hash function, for example SHA-1, is such that a small change in the input results in a very large change in the output.

A *Message Authentication Code (MAC)* is a short bitstring that is used to authenticate a message and verify its integrity. A MAC is a one-way hash function combined with a secret key. A MAC code can be computed in several ways, for example by taking a hash of the message and then encrypting it with the secret key. Another approach is to concatenate the message with the secret key and then hashing. MACs are widely used to protect the integrity of messages.

A *keyed-Hash Message Authentication Code (HMAC)* is a type of MAC that (RFC 2104) utilizes an iterative hash function over blocks of fixed size. HMAC is more secure than plain MAC. HMAC-SHA-1 and HMAC-MD5 are used by the IPsec and TLS protocols.

9.3   Public Key Infrastructure

Public Key Infrastructure (PKI) is a system that binds public keys with associated user identities using a *certificate authority (CA)*. A registration and issuance process is used to establish the binding for a public key and a user identity. As a result of the registration and issuance, the CA issues an unforgeable certificate that contains information about the identity, public key, validity conditions, and other relevant attributes. The certificate is signed by the private key of the CA and can be verified by using the public key of the CA. The CA can also be called as a *trusted third party (TTP)* or an *Identity Provider (IdP)*.

The main benefit of a PKI system is that it allows users to authenticate each other without prior trust relation. This is possible because the CA is trusted. PKI thus supports the communicating parties to establish confidentiality, message integrity, and authentication without prior shared secret information.

The challenges with PKI systems pertain to certificate revocation, conditions for certificate issuance, cost of certificate issuance, heterogeneous regulations and jurisdictions, and scalability of the system.

9.4   Network Security

A distributed system must be able to cope with a number of security issues. Table 9.1 illustrates the main security threats and issues and typical solutions. Namely, DoS attacks, eavesdropping, message spoofing, message replaying, and message integrity compromises.

To overcome these different security challenges, a number of protocols have been developed. Figure 9.2 presents a view of the protocol layers and identifies some common security related protocols used in today's Internet. On the link layer, we have authentication protocols such as *Password Authentication Protocol (PAP)*, *Challenge-Handshake Authentication Protocol (CHAP)*, and link encryption protocols such as *Wi-Fi Protected Access (WPA and WPA2)* and *Wired Equivalent Privacy (WEP)*. WPA is a certification program administered by the Wi-Fi Alliance to indicate compliance with the security protocol created by same alliance. This protocol was developed in response to several serious weaknesses in the previous system, WEP. WEP is known to have a number of security problems. Therefore WPA or WPA2 should be used instead.

Extensible Authentication Protocol (EAP) is a set of guidelines that describe different authentication message formats (RFC 3748). EAP based protocols define how messages

Table 9.1   Security threats and issues

Issue	Description
Denial-of-Service (DoS) attacks	Firewalls/NATs/packet filtering offer some resistance. Computational puzzles. Minimizing state kept by nodes. Efficient message authentication prevents spoofed messages.
Traffic monitoring / eavesdropping	Message data encryption.
Man-in-the-middle attacks	Encryption and mutual authentication.
Message/packet spoofing	Sender verification and authentication
Replay attacks	Message encryption. Message Sequence numbering.
Message tampering	Message authentication and integrity checking.
Identity masquerading	Identity verification.
Loss/theft of device	Secure storage and encrypted media. Passwords and other credentials that are not stored on the device.
Wireless signal jamming	Physical measures are needed to avoid this attack, for example frequency hopping.
Viruses and trojans	System scanning and monitoring for suspicious activities and program signatures.
Wireless spam (SMS, MMS, SIP, ...)	Various methods have been proposed including sender authentication and human interactive proofs.

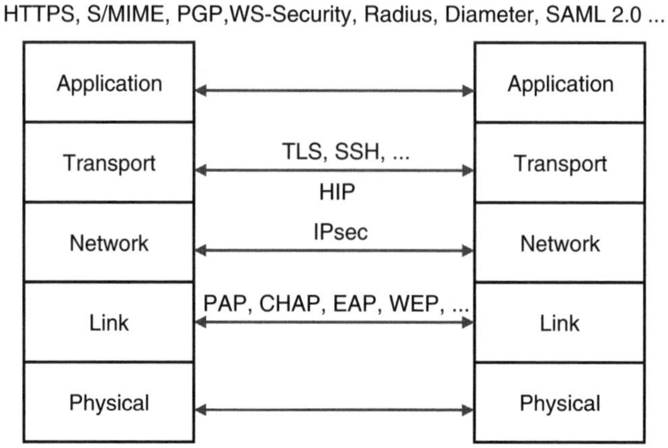

HTTPS, S/MIME, PGP,WS-Security, Radius, Diameter, SAML 2.0 ...

Figure 9.2   Layering and protocols

are encapsulated and communicated between two hosts. EAP is a generic protocol; however, the typical usage scenario is communication with a wireless access point. In general, EAP is used by an 802.1X enabled *Network Access Server (NAS)* device. For example, EAP is used by an 802.11 a/b/g Wireless Access Point in order to authenticate a user and negotiate a secure *Pair-wise Master Key (PMK)*.

Commonly used authentication methods in wireless networks include EAP-TLS, EAP-SIM, EAP-AKA, PEAP, LEAP and EAP-TTLS. *EAP-SIM (EAP for GSM Subscriber Identity* is used for authentication and session key negotiation based on the *Subscriber Identity Module (SIM)* of GSM. EAP for UMTS *Authentication and Key Agreement (EAP-AKA)* is used for authentication and session key distribution using *Universal Subscriber Identity Module (USIM)* in UMTS networks.

On the network layer, we have IPSec for securing IP traffic. IPSec is implemented by two communicating hosts and traffic is secured end-to-end. IPSec has two modes, namely the transport and tunnel modes. The transport mode secures the payload of IP packets, and the tunnel mode is able to secure whole IP packet by encapsulating them into new packets. In this mode, the IP addresses of the outer IP header are the tunnel endpoints, and the IP addresses of the encapsulated IP header are the actual source and destination addresses. Transport mode is the default mode and used for client-server communications. Tunnel mode is typically used by gateways to traverse unsecure networks.

Above the network layer, we have the *Transport Layer Security (TLS)* (RFC 5246), which is the de facto standard for securing Web interactions. Other security protocols include *datagram TLS (dTLS)* (RFC 4347), *Secure Shell (SSH)*, and more recently, the *Host Identity Protocol*.

On the application layer, we find the ubiquitous HTTP and its secure variant HTTPS that uses TLS, *Pretty Good Privacy (PGP)*, WS-Security for web service security, *Remote Authentication Dial-in User Service (RADIUS)* and *Diameter* for managing authentication, and SAML 2.0 for representing security assertions, to name some well-known solutions.

RADIUS is an *Authentication, Authorization, and Accounting (AAA)* protocol specified most recently in RFC 2865. RADIUS operates using UDP and is used to transport EAP messages, usually between a NAS and AAA server. It relies on the combination of *Message Digest 5 (MD5)* hashing and shared secrets to secure its transmissions. Despite its popularity and industry support, RADIUS has several limitations such as security vulnerabilities and unreliability due to the use of UDP. RADIUS is usually applied in trusted network segments or within encrypted *Virtual Private Networking (VPN)* tunnels. Its successor, known as Diameter, is specified in RFC 3588 and offers improved operational security features and better scalability.

From Figure 9.2 we can observe that a number of security techniques and protocols are needed, and they are applied on different layers of the protocol stack. In many cases, multiple solutions are needed to be applied either in sequence or concurrently. For example, link layer security solutions, transport layer solutions, and application layer solutions are often used simultaneously. This stems from the layered nature of the protocol stack and that there are no standard and deployed APIs for determining what security solutions are being used by lower layers.

The commonly used URL syntax allows applications to specify a secure version of a protocol, for example HTTPS, however this is limited to those parts of the protocol stack that process URLs. URL-based security pertains to some transaction and further mechanisms are needed to protect parts of a web page or message.

Table 9.2 presents an overview of security mechanisms for SIP. S/MIME is used to encrypt and sign message payloads. It is useful for end-to-end secrecy. RFC 3261 [1] defines the *Secure SIP (SIPS)* URI that forces the communication over a set of TLS connections. This provides end-to-end secrecy given that the intermediaries are trusted.

Table 9.2 Security techniques

Name	Description
S/MIME	For encrypting and signing message payloads.
	The public key of the recipient must be known.
SIPS URI (TLS)	Tight coupling between SIP and TLS.
	Applications and proxies need to be TLS aware.
	Integrates well with current browsing technologies.
	Must be applied on a hop-by-hop basis with SIP intermediaries.
IPsec	Tight coupling with applications not required.
	IKE key agreement protocol is heavy.
	Must be applied on a hop-by-hop basis with SIP intermediaries.
	NAT/firewall traversal issues.

For this security mechanism, the applications need to be aware of the SIPS URI. According to RFC 3261, in a SIPS session, the SIP user agent contacts the SIP proxy server and requests a TLS session. The proxy server then responds with a public certificate. The user agent and the proxy exchange session keys. If there are multiple hops, the proxy then contacts the next hop until the final destination is reached.

IPSec does not require a tight coupling between applications and the security solution; however, in order to support SIP message forwarding, it must be applied on a hop-by-hop basis rather than end-to-end. IPSec can be seen as a heavier protocol than the session layer TLS due to complex key agreement and there are also a number of NAT and firewall traversal issues. It is expected that the Host Identity Protocol would also be useful in this case due to its simple authenticated key agreement scheme and mobility support

TESLA is an interesting protocol outlined in RFC 4082. TESLA provides delays per-packet data authentication and integrity checking for unicast and multicast flows. The protocol uses low-cost operations per packet, can tolerate packet losses, and does not require per-receiver state at the sender. The protocol is based on delayed disclosure of keys. Source authentication is realized by using MAC chaining. This results in some delay in authentication, typically on the order of one round-trip-time. TESLA has been applied for securing communications in sensor and satellite networks, as well as for general multicast streams.

9.5 802.11X

IEEE 802.1X is an IEEE standard for port-based Network Access Control and part of the IEEE 802 family of standards. 802.1X is implemented by many wireless access points to increase security. This protocol enables the authentication of devices that are attached to a LAN port. Upon successful authentication, the protocol establishes a point-to-point connection for communications. This protocol is useful for closed wireless access points and it is based on EAP.

The main entities in the protocol are the supplicant, authenticator, and an authentication server. Figure 9.3 illustrates the protocol workflow. The supplicant is typically a software module executed on a client device such as a mobile phone or a laptop. The authenticator

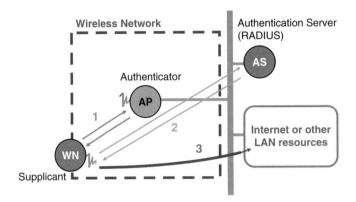

Figure 9.3 Overview of 802.1X

is a wireless access point or a wired Ethernet switch. The authentication server is usually a RADIUS server and database.

In this model, the authenticator is a security checkpoint that guards access to the protected network. The supplicant cannot communicate beyond the authenticator unless proper credentials are presented and the supplicant is authenticated. In this 'unauthorized' mode the supplicant can only send 802.1X traffic and other types of traffic are blocked in the data link layer.

The authenticator sends out an EAP-Request to the supplicant and the supplicant responds with the EAP-response packet. This packet is then forwarded by the authenticator to the authentication server. Typical credentials include user name and password and digital certificates. Upon receiving the credentials from the supplicant, the authenticator then forwards the credentials to the authentication server for verification. If the credentials are valid the supplicant is authorized to access resources beyond the authenticator. In this case, the authenticator sets the port associated with the supplicant to 'authorized' mode and normal traffic is allowed to the protected domain. When the supplicant leaves the network, it sends an EAP-logoff message to the authenticator. The port state is then reset to 'unauthorized'.

9.6 AAA, RADIUS, Diameter

Authentication, Authorization, and Accounting (AAA) (pronounced triple-A) is a central security architecture for distributed systems. The three security functions enable networks and services to control which users are allowed access, what functions they are allowed to use and how much resource they have used. AAA can be extended with auditing resulting in the acronym AAAA.

RFC 2903 gives an outline of the generic AAA architecture. The three key elements in the AAA model are:

- Authentication, which establishes the user's identity and verifies its authenticity. This is performed before granting access to resources and performing operations on behalf of the user. The identity is established by using different credentials, such as passwords, one-time tokens, digital certificates, and phone numbers.

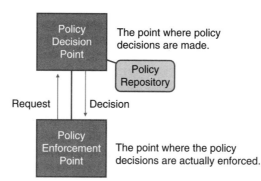

Figure 9.4   AAA policy management

- Authorization is responsible for granting the user the privilege to access networks and resources, and perform commands. Authorization may be based on restrictions, for example access location, time-of-day restrictions, or restrictions against multiple logins by the same user. Authorization decision may also be based on the service type and access method.
- Accounting pertains to the process of collecting data about usage patterns by individual users. This information may be used for planning, billing, management, or other purposes. Accounting can happen in real-time or in a delayed fashion (batch accounting).

AAA servers typically take part in implementing an organization's security policy. A security policy is a set of principles and rules used by an enterprise to protect system resources. In the AAA Authorization Framework (RFC 2904) policy management functionality is divided between two entities the *Policy Decision Point (PDP)* and *Policy Enforcement Point (PEP)*.

Figure 9.4 illustrates the policy management components. A policy is retrieved by a PDP from a policy repository and it is evaluated by the PDP and enforced at a PEP. Any of the AAA servers involved in an authorization transaction may retrieve a policy, evaluate it, and then enforce the policy. The elements are logical and they can all reside on a single AAA server or be distributed in the environment. As a general requirement, an AAA protocol needs to be able to transport both policy definitions and the information needed to evaluate policies.

The five authentication methods used with AAA systems are:

- *ASCII.* This form of authentication requires the user to type in a username and password that are sent in clear text. The credentials are then matched with those in the user database stored in ASCII format.
- *Password Authentication Protocol (PAP).* This protocol is used to authenticate PPP connections. PAP passes passwords and other user information in clear text.
- *Challenge-Handshake Authentication Protocol (CHAP).* This protocol provides the same functionality of PAP, but it is more secure because it avoids sending the password and other user information over the network.
- *Token-Card.* This authentication technique uses one-time passwords. Token-card authentication systems provide strong access security.

- *Extensible Authentication Protocol (EAP).* EAP is similar to CHAP, but it offers more flexibility. It allows authentication by encapsulating various types of authentication exchanges. EAP messages can be encapsulated in the packets of other protocols, such as RADIUS.

Figure 9.6 presents an example of AAA in a mobile dial-in environment. A user connects with a NAS device in the visited network. The NAS device consults the local AAA server for authentication and authorization. Since the user is roaming from another network, the AAA contacts the home network AAA for the security decisions. This example illustrates a crucial requirement for AAA, a distributed network of servers is needed that may span multiple organizations and domains.

Figure 9.5 illustrates the three key interactions between the client, AAA, and server (RFC 2904). The first interaction involves the *push sequence*, in which a client first obtains security tokens from the AAA server and then presents them to a server. The second interaction is the *pull sequence*, in which the client invokes the service, which then pulls the token from the AAA server. The final interaction is the *agent sequence* where the client contacts the AAA server, which acts as an intermediary. These three basic interactions can be combined in order to create more complex sequences.

The *Remote Authentication Dial In User Service (RADIUS)* is a popular AAA protocol defined in RFC 2865. RADIUS has been designed to authenticate dial-in-access customers and it is extensively used in dial-in lines and 3G networks. The system is based on a centralized user database for passwords and other information pertaining to users. RADIUS is based on the client-server model. A central security server holds the user profile information used by client access devices to enforce security. When a user makes a request to connect to a network or service, the client device queries the server to check if the request is valid.

The RADIUS transport is implemented using UDP and it does not have keep-alive signaling support, which makes it prone to packet losses and server failures. RADIUS

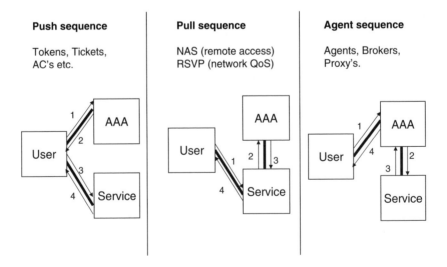

Figure 9.5   AAA sequences

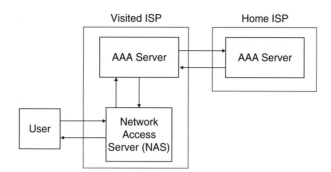

Figure 9.6   Example of AAA in mobile dial-in

offers also limited support for scalability, because it does not support proxies and requires manual configuration of shared secrets. From the security viewpoint, RADIUS is assumed to be used in trusted network segments so it does not offer per packet confidentiality.

The RADIUS AAA server receives user requests for access from a client, attempts to authenticate the user, and returns the configuration information and polices to the client. The RADIUS AAA server may be configured to authenticate an Access-Request locally or to act as a proxy client and forward a request to another AAA server. After forwarding a request, it handles the message exchanges between the NAS and the remote server.

If the RADIUS AAA server successfully authenticates a client, it will send an Access-Accept packet to the client. Otherwise it will send an Access-Reject. An Access-Accept data packet may include authorization information that specifies what services the user can access and other session information. Upon receiving an Access-Accept packet, the client will generate and send an Accounting-Request to start the session. The Accounting-Request packet describes the type of service being delivered and the user that will use the service. The server will respond with an Accounting-Response to acknowledge that the request was successfully received and recorded.

Diameter is a newer AAA protocol that is a replacement for RADIUS (base protocol defined in RFC 3588). Diameter is a network protocol for providing AAA services to roaming users. The protocol addresses some of the limitations of RADIUS, for example security, reliability, and support for proxies and peer-to-peer operation between servers. Diameter is based on TCP or SCTP and can be secured using IPSec or TLS. In addition to RADIUS and Diameter, *TACACS+ (Terminal Access Controller Access Control System)* is a proprietary Cisco protocol that is largely the equivalent of RADIUS, but with some improved features.

9.7   Transport-layer Security

The most common way to provide security for websites is to use *Transport Layer Security (TLS)*. TLS provides server authentication and optionally client authentication. To avoid user interaction, the server certificate needs to be signed by a known trusted *Certificate Authority (CA)*. TLS does not solve security problems encountered with multiple intermediaries, which is becoming common with single sign-on and web services. Therefore message-based security techniques are also typically needed for many services.

The TLS protocol allows applications to communicate across a network in a way designed to prevent eavesdropping, tampering, and message forgery. TLS provides end-point authentication and communications privacy over the Internet using cryptography. Typically, only the server is authenticated while the client remains unauthenticated. TLS also supports *mutual authentication*; however, both communicating endpoints need to have certificates. Mutual authentication requires a PKI deployment to clients unless TLS-PSK or the *Secure Remote Password (SRP)* protocol are used.

TLS involves three basic phases illustrated by Figure 9.7:

- peer negotiation for algorithm support;
- key exchange and authentication;
- symmetric cipher encryption and message authentication.

During the first phase, the client and server negotiate cipher suites, which determine the ciphers to be used, the key exchange and authentication algorithms, as well as the *message authentication codes (MACs)*. The key exchange and authentication algorithms are typically public key algorithms, or as in TLS-PSK preshared keys can be used. The message authentication codes are made up from cryptographic hash functions using the HMAC construction.

Typical algorithms for TLS include:

- For key exchange: RSA, Diffie-Hellman, ECDH, SRP, PSK.
- For authentication: RSA, DSA, ECDSA.
- Symmetric ciphers: RC4, Triple DES, AES, IDEA, DES, or Camellia.
- For cryptographic hash function: HMAC-MD5 or HMAC-SHA.

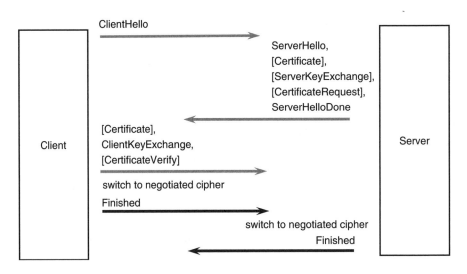

Figure 9.7 Overview of TLS handshake

9.8   Web Services Security

The web services security architecture consists of a number of specifications, which are illustrated by Figure 9.8. The web services specifications have been developed by numerous industry and university partners and they have been standardized by W3C or OASIS, or are in the process of being standardized. The key specifications include *WS-Security (Web Services Security, WSS)*, WS-SecureConversation, WS-Trust, WS-Policy, WS-Privacy, WS-Federation, and WS-Authorization. These specifications work on top of the SOAP stack and they are exclusive to the application layer.

The Web Services *Interoperability Organization (WS-I)* has produced a document describing SOAPs security challenges, which include the following [2]:

- SOAP does not provide a mechanism for identification and authentication. This means that there is no way to verify the origin of a SOAP message.
- SOAP does not perform any authentication between SOAP endpoints or intermediaries.
- SOAP does not provide a mechanism for ensuring data integrity or confidentiality.
- SOAP does not provide a mechanism for preventing SOAP messages or parts of SOAP message from being replayed.

As mentioned already in this chapter, although transport layer security, namely TLS, can be used to protect end-to-end communications, additional security solutions are needed for message specific security. End-to-end communications can be compromised if one of the intermediaries is not trustworthy. Any point in the communications where the message is processed as plaintext is a potential security vulnerability point. Figure 9.9 illustrates a communications environment that has multiple security contexts. End-to-end transport security is not sufficient to guarantee security when the security context changes. Moreover, any messages that are stored in a queue or database should be protected. Transport layer security does not help here if a message has been received in a secure way, then decrypted, and stored in an unencrypted form. At this point a malicious program could modify the message, remove it, or simply copy the message.

We conclude that XML message specific security measures are needed. The key specification for confidentiality is the XML Encryption specification from W3C and defines a mechanism to encrypt XML documents. The XML Signature specification defined jointly by the W3C and IETF defines a mechanism for selectively signing parts of an XML document. The XML Encryption and XML Signatures specifications define XML vocabularies

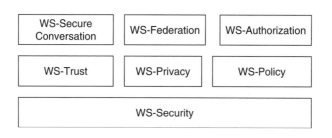

Figure 9.8   Web services security architecture

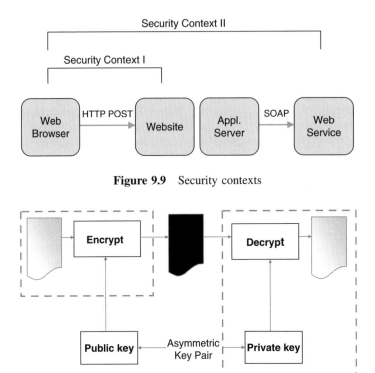

Figure 9.9   Security contexts

Figure 9.10   Encrypting a message

and processing rules enabling confidentiality and integrity to be applied to a variety of content.

The *XML Key Management Specification (XKMS)* from W3C defines protocols for public key management services. Public key management includes the creation of a public and private key pair, the binding of this key pair with an identity and other attributes, and how to present key pairs in different formats.

XML Encryption serves the purpose of maintaining confidentiality of information, both while in transit as well as when stored. Figure 9.10 illustrates how XML encryption works with public key cryptography. The sender selectively signs the parts of the XML document using the receiver's public key. The receiver can then decrypt these parts using its corresponding private key. The message includes details about the encryption algorithm used and optionally about the key. The keys can be either symmetric or asymmetric.

XML Signatures maintain integrity and authenticity of information both during transit and in storage. Figure 9.11 illustrates XML document signing using public key cryptography. First, a message digest is computed from the selected parts of the message. This digest is then signed using the sender's private key. The message contains the signed digest and details about the algorithms used and optionally about the key that was used to perform the signature. The receiver can then compute a new digest upon receiving the message and verify that the digest is the same that the sender signed. If the digest is not the same, then the message has been changed during transit.

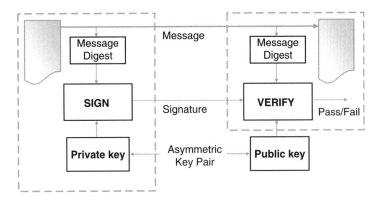

Figure 9.11   Signing a message

Signing XML is more challenging than encrypting XML, because the digest needs to be calculated from exactly the same data. Unfortunately, XML processing steps and serialization may slightly change the XML, for example, by removing whitespaces or by changing the order of element attributes. This means that a special process called *canonicalization* is needed to preprocess the parts of the XML document that are used in computing the digest.

There are two types of canonicalization, namely inclusive and exclusive. Inclusive canonicalization copies all namespace declarations that are currently in force. This is useful in the typical case of signing part or all of the SOAP body. This approach may have problems when the context changes later.

Exclusive canonicalization tries to understand what namespaces are actually used and copies those. This is useful when an XML document is inserted into another document, for example a signed SAML assertion. This should be used with WS-Security. The XML Signature specification defines this process in detail.

Figure 9.12 presents an example SOAP message that has both encrypted content and a signature. The WS-Security SOAP header security block includes a binary security token (certificate) that is used to sign parts of the message (1). The block also contains a signature, which has information necessary to perform canonicalization, calculate digest, and then sign the digest with the identified method (2). The header contains also an EncryptedKey element which contains the encryption key or certificate. The body of the message contains encrypted XML content in the EncryptedData element (3). In this case the whole body element has been encrypted.

From the mobile middleware viewpoint, WS-Security and WS-SecureConversation are the key specifications and we will focus on these. We briefly review the other specifications. The WS-Trust provides a framework for defining and maintaining trust relationships between web services. Public key security relies on trust in certificate authorities. Enterprises can use this specification to build trust relations between their services and identity providers. WS-Policy allows web services to describe their policies regarding security, Quality of Service, etc. using XML. Web service consumers can specify their policy requirements in a similar fashion. WS-Privacy is a specification that addresses how privacy practices can be stated and implemented by web services.

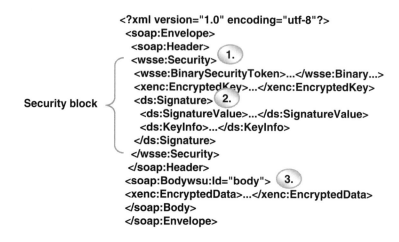

```
<?xml version="1.0" encoding="utf-8"?>
 <soap:Envelope>
  <soap:Header>
  <wsse:Security> 1.
    <wsse:BinarySecurityToken>...</wsse:Binary...>
    <xenc:EncryptedKey>...</xenc:EncryptedKey>
    <ds:Signature> 2.
     <ds:SignatureValue>...</ds:SignatureValue>
     <ds:KeyInfo>...</ds:KeyInfo>
    </ds:Signature>
   </wsse:Security>
  </soap:Header>
  <soap:Bodywsu:Id="body"> 3.
  <xenc:EncryptedData>...</xenc:EncryptedData>
  </soap:Body>
  </soap:Envelope>
```

Security block

Figure 9.12 Example of XML Encryption and Signature

WS-Authorization specification defines rules for authorization data and policy management for web services. The WS-Federation specification defines how various identities are federated across administrative domains.

WS-Security provides the necessary facilities for securing web services communication. The latest version of this specification is 1.1 that was released in February 2006. WSS contains specifications on how message integrity and confidentiality can be achieved. Specifically, WSS supports message source authentication, integrity, and confidentiality. The first is responsible for ensuring that the sender of a message can be accurately identified. The second is responsible for ensuring that messages are not tampered in transit. Finally, the third is responsible for ensuring that only the ultimate target of a message can read the message.

WS-SecureConversation specification is used together with WS-Security, WS-Trust and WS-Policy to allow sharing of security contexts. The aim is to allow secure conversations between sites using web services for communication. Conversations persist for the duration of several interactions and can thus be used to optimize communications. WS-Secure Conversation describes how message exchanges can be securely managed. It also deals with security context exchange and establishing and deriving session keys.

XML gateways also support schema validation and some support for SOAP intrusion prevention against the following attacks that target vulnerabilities native to XML and XML based services. A number of attacks can be devised against SOAP-based services. XML gateways and middleboxes can be used to detect and mitigate these attacks. Attacks against XML-based services include the following:

- *WSDL scanning.* an potential attacker may attempt to obtain the WSDL of a web service. The WSDL may be useful in preparing the attack and finding vulnerabilities.
- *Parameter tampering.* an attacker may modify parameters and arguments that a web service expects to receive. The aim of these modifications is to gain access to unauthorized features or otherwise disrupt the service.
- *Replay attacks.* an attacker may resend SOAP requests or parts of the requests.

- *XML parser attacks and buffer overflows.* the attacker may send large XML documents in order to attempt to overload the XML parser. Buffer overflows may result in arbitrary code being executed thus compromising the system.
- *External reference attacks.* the attacker may attempt to bypass security measures by using external references that are downloaded after the XML has been validated.
- *Schema poisoning.* the attacker may attempt to use a schema that the XML validator will accept, but that will still allow harmful content to pass the validation.
- *SQL injection.* the attacker may inject harmful program codes or statements into the XML content, for example SQL statements.

9.9 Security Tokens

Security tokens are the building blocks for secure distributed systems. Security tokens typically contain information pertaining to an entity, for example authentication or authorization decisions. Tokens can be realized using symmetric or asymmetric cryptographic techniques. Kerberos (RFC 4120) tickets are examples of the former, and various public key-based tickets are examples of the latter.

Kerberos is a secret key-based distributed authentication system. The *Key Distribution Center (KDC)* is the core element of this system, which consists of the *Authentication Server* and a *Ticket Granting Server (TGS)*. The AS knows the users secret keys. Clients first authenticate with AS using a secret key and a challenge-response protocol. After this the AS generates a *Ticket Granting Ticket (TGT)*. TGT consists of a session key encrypted using the client's secret key and a temporal credential secured using a secret key established between AS and the TGS. When the client then presents the TGT with an authenticated (client's identity secured using the session key) to the TGS, TGS can verify that the client is authenticated and can issue further tickets for other services. Tickets granted by TGS for a service are secured using a symmetric key shared by TGS and the service in question. A service receiving a request with the TGT and a ticket granted by TGS can then verify that they are valid.

Many current web applications use security tokens for authentication and authorization. Figure 9.13 illustrates the key entities in web services security. The requestor accesses

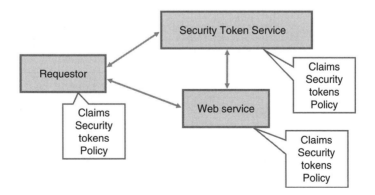

Figure 9.13 Security tokens in web services

services by first contacting a *Security Token Service (STS)* and then utilizes any security tokens given by the STS to use a web service. There is a trust relation between the web service and the STS.

Web services security provides extensive support for including various security tokens in messages. The WS-Security specification provides details how to use various kinds of security tokens, such as SAML, Kerberos, and X.509 certificates. WS-Security is tightly coupled with the SOAP protocol and it utilizes the SOAP header system for storing security tokens.

A security token can be passed using any protocol supported by the communicating peers. HTTP POST is commonly used to exchange tokens. In the REST model, a token can be passed within a name-value pair in the URL QueryString. The token may indicate access limitations, for example 1000 calls per day.

We illustrate tokens in current websites with an example. A token-based login is used by Amazon to support third party applications. First, a developer registers their application with a portal and receives a shared key from Amazon. Then, the developer forwards customers from their application to the portal login page, which redirects users back to the application with an application specific token. The token is encrypted with the shared key and the application grants access based on this portal issued token.

Amazon's developer tokens use HMAC for authentication at the message level. Request authentication is the process of verifying the identity of the sender of a request. Each account is assigned an Access Key ID and a corresponding Secret Access Key. The Secret Access Key is used with HMAC-SHA1 authentication. The HMAC signature is calculated over service, operation, and timestamp. The timestamp prevents replay attacks. Alternatively, an X.509 certificate can be used. The certificate can be self-signed and it is registered with the portal. A request signature that is an HMAC-SHA1 hash calculated using the Secret Access Key. A Secret Access Key is a 40-character alphanumeric sequence.

We present another example of security tokens by considering how the YouTube account authentication works when uploading a video using third-party application (or website). This is illustrated in Figure 9.14. First, the user clicks a link to upload video and then is redirected to the YouTube authentication proxy service. The authentication proxy requests user to login and then if the authentication is successful the user is directed to the third

Figure 9.14   Overview of YouTube upload authentication

party application with an authentication token. By default the token is for single-use only; however, using an optional procedure it can be exchanged for a session token that does not expire. The authentication token can then be used by the third party application to receive a single-use upload token. The user's browser can then upload the video using the upload token. Alternatively, the third-party application can directly upload the user's video with the authentication token.

9.10 SAML

Security Assertion Markup Language (SAML) is an XML-based standard for exchanging authentication and authorization assertions between security domains. The SAML assertions are created by identity providers and consumed by service providers and clients. SAML has been standardized by the OASIS Security Services Technical Committee.

The single most important problem that SAML is trying to solve is the Web Browser *Single Sign-On (SSO)* problem. Single sign-on solutions are popular in intranets and typically implemented using web cookies. Extending these SSO solutions beyond intranets has proven to be challenging, and many non-interoperable standards have been developed for SSO. The SSO specifications from the Liberty Alliance are the most deployed solutions currently. In recent years, SAML has become the standard that underlies many of the SSO solutions in the enterprise domain, such as the more recent Liberty specifications.

SAML assumes the principal, typically the end user, has registered with at least one identity provider. The identity provider is responsible for locally authenticating the principal. The implementation of the local authentication is beyond the scope of SAML. The service provider relies on the identity provider to properly authenticate the principal. When requested by the principal, the identity provider issues a SAML assertion that is passed to the service provider. This assertion includes statements regarding the principal that the identity provider believes to be true. The service provider can then make an access control decision based on this assertion. The decision will be also based on a prior trust relation between the service provider and the identity provider.

9.11 XACML

XML Access Control Markup Language (XACML) is an initiative to develop a standard for XML-based access control and authorization. A typical access control and authorization scenario features three key entities, namely a subject, a resource, and an action. Each of the entities can have their own attributes describing their properties and state. A subject issues a request for permission to perform an action on a resource.

The goals of the XACML specification standardized by OASIS include the following:

- Creation of a standard way of describing access control entities and their attributes.
- Providing a mechanism that offers finer grained access control.

Typically, proprietary access control systems utilize *Access Control Lists (ACLs)* in representing information about entities and their attributes. XACML differs from proprietary systems in that it is XML-based and standardized. XACML utilizes the SAML specification to represent various assertions on entities and their attributes. XACML and SAML complement each other. SAML provides the mechanism for exchanging security

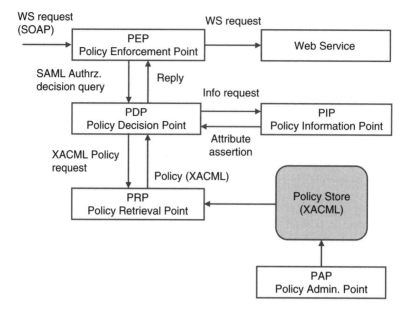

Figure 9.15   Overview of XACML

assertions, XACML is a mechanism for making decisions that result in those assertions.

Figure 9.15 presents the XACML model. XACML follows the PDP and PEP model
outline already in this chapter. A request is intercepted by a PEP. PEP then creates an
XACML request and sends it to a PDP. PDP is responsible for evaluating the request
and then sending back a response. Typically, the response signifies either to grant access
or deny access. PDP makes this decision based on relevant policies and rules specified
by the policies. The PDP uses a *Policy Access Point (PAP)* in order to retrieve relevant
policies. PDP may also contact a *Policy Information Point (PIP)*, which provides values for
attributes pertaining to the subject, resource, or environment. Finally, when the decision
made by PDP is received by PEP, the PEP can implement the policy-based decision.
Typically, PEP either grants or denies access.

In summary XACML defines the following:

- An XML vocabulary for expressing authorization rules.
- An XML vocabulary for expressing a variety of conditions to be used in creating rules.
- A means for creating policy statements, a collection of rules applicable to a subject.
- Mechanisms for combining and evaluating rules.
- A means for evaluating requests against rules.

9.12   Single Sign-On (SSO)

Security tokens and web services security solutions are instrumental in enabling various
login functions to services. Typically, users have to have separate accounts for services
provided by different companies. A number of *Single Sign-On (SSO)* solutions have been

proposed to alleviate this issue and enable flexible authentication to multiple services provided by different companies.

SSO enables users to access multiple services with a single login. Technically, it is a specialized form of software authentication that enables a user to authenticate once and gain access to the resources of multiple software systems. SSO systems require that a special trusted entity, called the *Identity Provider (IdP)*, coordinates and manages the underlying authentication session.

Windows CardSpace (also called InfoCard) is Microsoft's client software for realizing the Identity Metasystem [3]. The system is illustrated in Figure 9.16. The idea in CardSpace is to store references to users' digital identities on behalf of the users, and then allow the users to select what identities use for services. The identity selection is presented through a card abstraction. The motivation is to provide a consistent user experience for selecting the identity to be used.

When an Information Card enabled application or website wishes to obtain information about the user, the application or website requests a particular set of claims from the user. The CardSpace UI then appears, switching the display to the CardSpace service, which displays the user's stored identities as visual Information Cards. The user selects the InfoCard to use and the CardSpace software contacts the issuer of the identity to obtain a digitally signed XML token that contains the requested information.

CardSpace supports both user created and managed Information Cards. The cards contain one or more fields of identity information. Managed InfoCards are issued by a third party identity provider. InfoCards could be issued by a bank, employer, or a government agency.

The CardSpace architecture is based on the web services protocol stack and it utilizes heavily the web services security (WS-*) specifications. This makes CardSpace interoperable and a platform that supports the WS-* stack can be made to integrate with CardSpace.

In order for an identity provider to issue InfoCards, they need to provide interfaces through which consumers can obtain the cards, and have a *Security Token Service (STS)* that accepts WS-Trust requests and can return secure tokens. The STS can be implemented by the provider or it can be provided by a third party. From the developer viewpoint, CardSpace is used in websites using the HTML $< OBJECT >$ tag. Using this tag, the developer needs to specify what credentials the website requests from the user and then implement the necessary code to decrypt and validate the token returned by the STS.

CardSpace is an abstract Identity Metasystem and supports different security token formats. As such, it accommodates other Internet identity systems, such as OpenID and SAML.

9.13 Generic Bootstrapping Architecture (GBA)

The *Generic Authentication Architecture (GAA)* was created by 3GPP to ensure a secure environment for accessing service from mobile phones. GAA includes many solutions and is based on both asymmetric and symmetric cryptographic techniques. GAA consists of three parts:

- TS 33.220 *Generic Bootstrapping Architecture (GBA)*. Subscriber authentication is based on HTTP Digest AKA (RFC 3310).

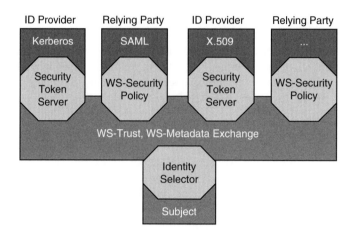

Figure 9.16   Microsoft CardSpace

- TS 33.221 *Support of Subscriber Certificates (SCC)*. In this part, a PKI Portal issues subscriber certificates for UEs and delivers an operator CA certificate. The issuing procedure is secured by using shared keys derived using GBA. This offer an increased level of security compared to basic GBA; however, a PKI infrastructure is required. Typical usage scenario is HTTP over TLS.
- TS 33.222 Access to Network Application Function using HTTPS. This part defines how HTTPS can be used to access services.

GAA and GBA involve of the following elements:

- *User Equipment (UE)*.
- Application server (called *Network Application Function (NAF)*, which provides the service that requires authentication.
- *Bootstrapping Server Function (BSF)*, which is responsible for authenticating the user to the service.
- Mobile network operator's *Home Subscriber Server (HSS)*, which stores user profiles.

Figure 9.17 illustrates the entities of the GAA architecture. GBA is responsible for negotiating a shared session key between. The BSF is used to perform mutual authentication. If SCC is used, then a PKI portal is used for obtaining certificates. UE can then use the obtained certificate to authenticate with service providers.

GBA enables the authentication of a user based on the user's *Subscriber Identity Module (SIM)* card[4]. The protocol outlined in Figure 9.18 is based on HTTP and can be used to realize converged services without relying on SIP. The authentication utilizes a shared secret stored in a mobile phone's SIM card or other tamper resistant module and the *Home Location Register (HLR)* or *Home Subscriber Server (HSS)* maintained by the SIM card's provider. The motivation behind GBA is to enable authenticated service access using the existing provisioned SIM cards. The technology can be used to authenticate users with any service, for example web services, that have a contract with the mobile operator. The GBA protocol enables IMS operators to provide single sign-on for users.

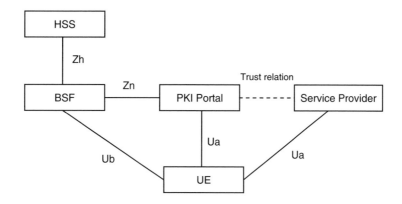

Figure 9.17   Overview of GAA

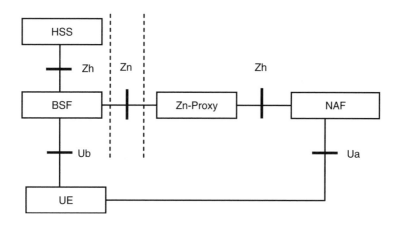

Figure 9.18   Overview of GBA

The BSF is a key component of the architecture that provides mutual authentication of the user equipment and servers. The usage of the protocol can be divided into two procedures: the bootstrapping authentication procedure and the bootstrapping usage procedure.

The former protocol proceeds as follows. First, when an unauthenticated UE contacts a service provided by the NAF, the NAF refers this request to the BSF. At this point, UE and BSF authenticate each other using the *Authentication and Key Agreement (AKA)* protocol standardized by 3GPP and used in today's 3G networks. The BSF also contacts the HSS in order to verify the UE's authenticity. After this phase is complete, the UE and BSF derive a session key that is used to encrypt data exchange with service provided by the NAF.

The latter part involves the UE contacting the service and NAF again. Now, the communications can be secured using the session key. The service provided by NAF can also access the user's profile and other information at HSS. The session key is created for a limited time and for a specific domain. After the bootstrapping, the UE and NAF can run some application specific protocol which is secured using the session keys.

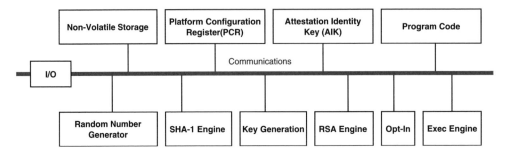

Figure 9.19 Overview of TPM

A special variant of the GBA can be used to allow the authentication of 2G UEs. This is needed, because GBA has been developed for 3G AKA authentication. The 2G authentication is done using a TLS tunnel with the BSF. This tunnel is used to exchange the required messages.

9.14   Trusted Platform Module

The *Trusted Platform Module (TPM)* is a specification that defines a secure hardware module that offers certain cryptographic functions [5]. TPM has been standardized by the Trusted Computing Group. Thus the TPM pushes some security features into hardware rather than doing all security operations in software. The motivation is that such a design can result in more secure systems. Many laptops are currently sold with TPM technology.

The TPM module offers several functions to the operating system and applications, including (Figure 9.19):

- generation of cryptographic keys;
- hardware pseudo-random number generator;
- remote attestation support, which allows the creation of unforgeable hash key summary of hardware and software configuration;
- sealed storage, which encrypts data in such a way that TPM controls the release of the decryption key;
- a unique RSA key included on the chip, which effectively allows authentication of the device.

9.15   OpenID, OAuth, MicroID

OpenID is a simple mechanism for proving ownership of a URL based on a verified identifier, such as an email address. OpenID is a basis for decentralized single sign-on (SSO) on the web [6]. The OpenID specification does not specify how different sites and domains should be federated. It simply provides a mechanism for authenticating the OpenID identifier provided by a user with the respective identity provider. Figure 9.20 presents the key interactions in the OpenID user authentication process. The relaying party and an OpenID provider may generate a shared key using Diffie-Hellman key derivation in an optional part of the specification called association.

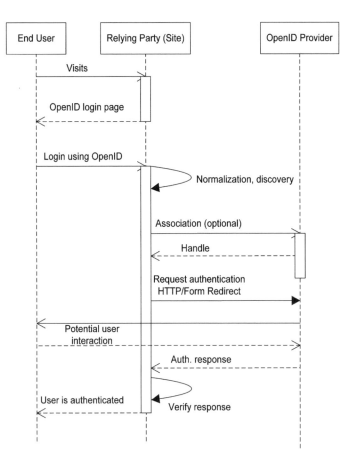

Figure 9.20 Overview of OpenID

Recently, Information Card and CardSpace support has been added to the
`MyOpenID.com` maintained by JanRain. Users can sign-in to the site using self-issued
information cards without entering passwords. Information Card support is available for
Windows platforms MacOS X, and Linux.

Information Cards are based on open standards. Anyone can issue, implement, and
accept Information Cards. The system uses WS-* specifications instead of HTTP redirect.
This means that the implementation is more complicated than OpenID and that changes
are required in the web browsing stack.

Support for OpenID Provider Authentication Policy Extension (PAPE) is an OpenID
extension that allows the Relying Party (RP) to specify an authentication policy. It allows
the IdP to inform the RP what authentication policy was used. `MyOpenID.com` recently
added support for logging in with client certificates. This avoids password prompts, but
requires that the client-side certificate is transferred to all machines that are used by a
particular user.

The OAuth protocol enables websites or applications (Consumers) to access Protected
Resources from a web service (Service Provider) via an API, without requiring Users to

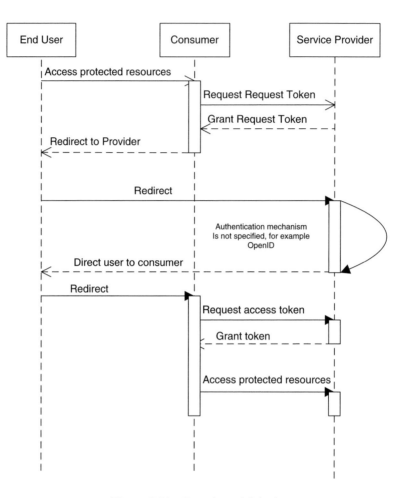

Figure 9.21 Overview of OAuth

disclose their Service Provider credentials to the Consumers [7]. Figure 9.21 presents an outline of the protocol.

An example use case is allowing printing service `printer.example.com` (the Consumer), to access private photos stored on `photos.example.net` (the Service Provider) without requiring Users to provide their `photos.example.net` credentials to `printer.example.com`.

OAuth does not require a specific user interface or interaction pattern, nor does it specify how Service Providers authenticate Users, making the protocol ideally suited for cases where authentication credentials are unavailable to the Consumer, such as with OpenID.

OAuth aims to unify the experience and implementation of delegated web service authentication into a single, community-driven protocol. OAuth builds on existing protocols and best practices that have been independently implemented by various websites. An open standard, supported by large and small providers alike, promotes a consistent and trusted experience for both application developers and the users of those applications.

MicroID [8] is a way to create a hashed identifier for a resource based on two URIs, namely the communication URI and the content URI. MicroID works as follows:

1. Hash a communication URI (email address).
2. Hash the URL of the size where content is hosted.
3. Append two hashes, hash concatenation.
4. Prepend resulting hash with the inputted URI schemes and the hashing algorithm using the following format: uri+uri:algo:hash.

Only identity issuers can create MicroIDs for individuals. Individual entities cannot create them. The issuer must verify that the individual or service provider has control over the service URI that is being included in the identifier. The MicroID system does not bind the identity with content. This means that anyone who can edit a page can change the MicroID values on that page. This can be avoided by signatures and integrity checking.

9.16   Spam

Spam is a world-scale problem and recently many solutions have been proposed [9, 10]. These solutions form the basis also for wireless spam prevention. On the other hand, mobile devices differ from the traditional desktop computer in that they have less computational resources and memory to support heavy anti-spam measures.

The main factor facilitating spam distribution is the nearly zero cost of sending it. Existing proposals and techniques vary widely in cost of deployment, effectiveness, scalability, and user annoyance. Spam prevention can also be applied on different levels, for example, on the network-level, on the level of SMTP servers, and on the level of web-based email services, such as HotMail and Google Mail. In addition, spammers may be able to circumvent network and SMTP server level countermeasures by using web-based services, that need to be considered as well.

Existing mechanisms to stop inbound spam utilize mechanisms of *source identification*, *source analysis*, and *distributed key mechanisms*. Source identification is performed using lists, such as *Real-time Blackhole List* supported by SendMail. Source analysis is performed by SpamAssassin [11] and analysis services. Tools, such as SpamAssassin, perform various checks to detect spam including string matching, known sources, resolvable names in headers, black hole lists, and header integrity. In distributed key methods, SMTP server peers exchange email summary information. This information can then be used to block spam, but the limitation is that it can also block legitimate bulk email.

Outbound spam prevention techniques include SMTP *access control*, reverse DNS lookups, *rate and connection limiting* and forcing end subscribers to access a POP server before allowing outbound email (*POP before SMTP*). Relays typically have access control lists of IP addresses that are used to determine which hosts are allowed to relay email.

Microsoft's Penny Black project [12] has investigated several techniques to reduce spam by introducing a cost to the sender. They considered several different forms of payment including CPU cycles, memory cycles, and *human interactive proofs*. Typically, senders compute a time or memory consuming function per message or per challenge. In a variant, senders acquire tickets from ticket granting servers. The tickets are used to authorize messages.

Several anti-spam mechanisms, namely filtering and human interactive proofs, have been proposed to be used in combination by ISPs in order to stop outbound spam. An interactive proof could be used to prove that a customer is not a spambot. The SmartProof system developed at Microsoft Research combines several techniques in an anti-spam challenge response system [13]. The combined techniques are tickets (or postage), machine learning, computational challenge, and interactive proofs. This system requires a challenge server for computing challenges and storing micro-payments and email clients need to support the challenge response mechanism.

One frequently proposed solution for spam prevention is the *electronic coin method*. The basic idea is that in order for a client to send email, it needs to spend some computation time to create a *coin*. Each coin allows the client to send an email. For example, coins in HashCash are based on partial hashes that the client needs to solve. To prevent spammers from leveraging the computational power of zombie networks, the computation of a partial hash should be difficult to parallelize.

The Sender ID Framework was developed by Microsoft and industry partners to create an email authentication system that prevents spoofing and phishing. *Email spoofing* means that the sender's address is forged. *Phishing* means that the spammer poses as a trusted party, such as an ISP or a bank. The system prevents these ills by verifying that the domain name from which email is sent is correct. The address of the sending server is checked against a registered list of servers that the domain owner has authorized to send email.

SenderID requires that the recipient email server performs the verification. Email senders have to add an *Sender Policy Framework (SPF)* record to the DNS of their domain. The SPF record is used to prevent return-path address forgery by allowing domain owners to identify sending email servers using DNS. SenderID creates additional load for DNS servers, because each message has to be checked separately.

DomainKeys proposed by Yahoo is a system for protecting email sender identity. Outbound and inbound mail servers are modified to sign and verify email messages. Other crypto-based authentication mechanisms are similar to DomainKeys, but the signed part varies. Google Mail (gmail.com) has adopted several sender authentication technologies. They publish SPF records and sign messages using DomainKeys.

The problem of IP-based authentication schemes is forwarding, which requires the rewriting of the return-path. The return-path is used to verify the sending domain and this verification fails unless the return-path is updated when forwarding a message. The problem of crypto-based schemes are content changes and mailing lists. This requires that mailing list managers re-sign messages. This is expected by DomainKeys.

9.17 Downloaded Code

Today many mobile applications are distributed over the air. Software running on the mobile device, for example a browser or a native service, downloads the application binary from the Internet, installs it, and then executes the program. It is clear that downloading arbitrary code from the Internet is rife with security problems. The commonly used solutions for making application downloads safe for consumers include:

- *Sand-boxed virtual machine for executing the program code.* The virtual machine has built in safeguards that prevent malicious activity. This offers a good level of protection;

however, it may significantly limit what the application can do. From the developer point of view this may pose severe restrictions for application development, for example Java ME does not by default have file access APIs.

- *Application certification program.* Most native applications today require a digital certificate if the application performs non-trivial activities, such as accessing file system or communications APIs. For example, Symbian devices and Apple's iPhone applications are certified.

Both forms of protection are commonly used to protect, for example, Java and .NET applications.

The Java language is compiled into a platform independent bytecode format. This means that the binary bytecode retains properties of the original source code and with today's tools it is not difficult to reverse engineer Java bytecode. One commonly used technique to prevent reverse engineering of applications is to utilize bytecode obfuscation techniques. Obfuscation transforms the bytecode into a more difficult format to reverse engineer although retaining the same functionality as the original code. This is a countermeasure against unwanted disclosure of information and tampering with the program code.

A Java virtual machine performs bytecode verification before execution. This gives some protection against malicious code. This protection can be improved by digitally signing the application as mentioned above.

.NET applications are also executed in a secure environment. The Common Language Runtime is run in a sandbox environment. To alleviate the concerns with software downloads, code access security has been integrated with the .NET framework. Code access security permissions determine what applications can do in the sandbox. This security mechanism provides varying levels of trust for applications depending on user given parameters and the origin of the application.

Bibliography

[1] Rosenberg J, Schulzrinne H, Camarillo G, Johnston A, Peterson J, Sparks R, Handley M and Schooler E 2002 SIP: Session initiation protocol. RFC 3261, IETF.
[2] Schwarz J, Hartman B, Nadalin A, Kaler C, Davis M, Hirsch F and Morrison KS 2005 Security Challenges, Threats, and Countermeasures Version 1.0. Technical report, Web Services Interoperability Organization.
[3] Chappel D 2006 Understanding Windows CardSpace http://msdn2.microsoft.com/en-gb/library/aa480189.aspx.
[4] 3GP 2007 *3GPP TS 33.220: Generic Authentication Architecture (GAA); Generic Bootstrapping Architecture.*
[5] Tru 2008 *Trusted Platform Module (TPM) Specifications.*
[6] ope 2006 OpenID specification http://openid.net/specs.bml.
[7] oau 2007 OAuth specification 1.0 http://oauth.net/documentation/spec.
[8] Miller J n. d. MicroID - small, decentralized, and verifiable identity http://microid.org/.
[9] Hulten G, Goodman J and Rounthwaite R 2004 Filtering spam e-mail on a global scale *WWW Alt. '04: Proceedings of the 13th international World Wide Web conference on Alternate track papers & posters*, pp. 366–367. ACM Press, New York, NY, USA.
[10] Yerazunis WS, Chhabra S, Siefkes C, Assis F and Gunopulos D 2005 A unified model of spam filtration *In the proceedings of the MIT Spam Conference 2005.*
[11] Apa 2005 *The Apache SpamAssassin Project.*
[12] Mic 2005a *The Penny Black Project.*
[13] Goodman J and Rounthwaite R 2005 *SmartProof.*

10

Application and Service Case Studies

This chapter presents examples in how to apply the presented architectural elements and technological solutions in building, deploying, and maintaining mobile services and applications. We focus on the emerging converged communications environment and consider applications and services such as VoIP, location-based services, Mobile Web Servers, mobile advertisement, mobile video, and push email.

10.1 Mobile Services

Today's mobile telecommunications business model has been based on so called walled gardens, which are essentially closed business ecosystems that are tightly controlled by one or more telecom operators.

The security and user experience of these closed ecosystems can be guaranteed; however, they lack incentives for third party service provisioning and creation. Indeed, the lack of incentive for smaller companies to launch their services for the mobile environment slows down the ecosystem and its evolution.

The customers of mobile applications and services, developers, and network providers create a value chain. In order for the value chain and service ecosystem to work, it needs to be profitable for the stakeholders.

This model of doing business is contrasted with service ecosystem on the Internet, which is more open and based on inherently interoperable protocols. The development pace of Internet and web services is fast. New services, and also protocols for supporting them in an interoperable fashion, are rolled out much faster than what is typical in the telecommunications world. Indeed, the aim of many current mobile development efforts is to bridge the telco world with the Internet and web world. This trend is called converged communications, and it is happening around SIP and web protocols.

Similar concerns pertain also to services provided by the mobile operator. Figure 10.1. illustrates the stovepipe service model and the IMS service model. The stovepipe model builds a service stack for each possible service tailoring for the task at hand. The stacks

Mobile Middleware   Sasu Tarkoma
© 2009 John Wiley & Sons, Ltd

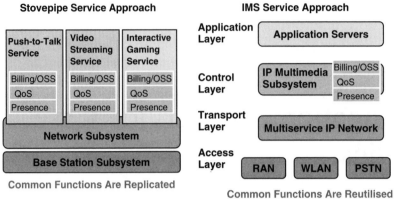

Stovepipe Service Approach   **IMS Service Approach**

Common Functions Are Replicated

Common Functions Are Reutilised
Access Independence

Figure 10.1   IMS service approach

are run on top of a network subsystem. This approach is criticized for the overhead and management complexity it introduces.

In addition to the limitation of having somewhat closed service environments, the mobile developer faces a number of challenges due to the heterogeneous nature of devices and platforms. This is where mobile middleware is especially needed – to support software development and enable faster life-cycles and less testing.

Figure 10.2 illustrates the differences of the desktop world and the mobile world. In traditional software development, a sequence of steps can be identified, which include having a product idea, doing software design, implementing the design, building the product,

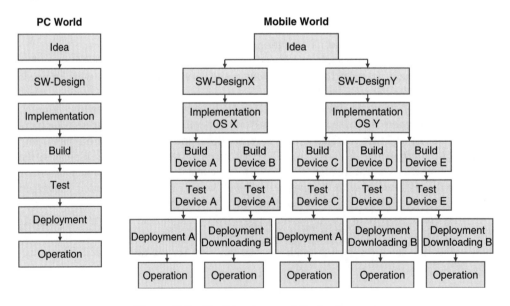

Figure 10.2   Traditional vs. mobile development

testing the product, deploying the product, and then finally maintaining the product. Of course, many software products are developed to support multiple operating systems; however, middleware solutions can be used to alleviate these problems and the typical end systems are usually similar.

Therefore, the IMS service approach introduces common layers for services. The Access and Transport Layers provide the basic access for service usage. The Control Layer is responsible for ensuring that the plumbing works, namely that billing, quality of service, and other signaling related functions are properly executed and maintained. Finally, the Application Layer provides the services by hosting the on application servers.

As mentioned above, the mobile and the IT and web worlds are converging, and going further towards the web would be to simply use web technologies in providing service to mobile customers. Indeed, this is something that is currently happening with mobile products. The mobile web browser is becoming increasingly sophisticated and able to execute Javascript and Flash Lite.

We observe that although web technologies are becoming increasingly popular for the development of mobile services, utilizing web requests and client-side scripting are not enough for many application domains. Moreover, there is significant overhead in utilizing HTTP in the mobile environment and more efficient protocols are needed.

One key observation is that HTTP works rather well even in the mobile environment when the application has been developed for browsing. Things become more tricky when real-time communications are needed or information needs to be pushed to users. Moreover, it is expected that in the future the mobile phone interfaces with a number of other devices in the environment. This calls for whole new APIs and ways of accessing that volatile data.

On the service provider side, the billing models for web services and mobile services have been very different. Web users usually do not pay for web services, but rather the revenue for the service providers come from advertisement or simply sales of products. The Internet service provider, on the other hand, gets payment for the raw IP connectivity either on per-use or flat rate basis.

10.1.1 Accountability

Interestingly, one of the main differences in these models is accountability. The Internet way of doing business does not have a very flexible system for accountability. The system is limited by how the routing system works, namely there are network prefixes and domain-level paths across the global network. It is possible to track a particular network prefix to a company or an individual, but it requires the cooperation of many different parties, possibly in different jurisdictions.

In practice, IP address offers a very rudimentary level of accountability. To contrast this, the mobile telecommunication systems have a very strong level of accountability. This is realized through the SIM card or similar hardware module and a mobile operator maintains a binding between the secret stored by the SIM card and the user records stored at a *Home Subscriber Server (HSS)*. Through a vast number of roaming contracts, the SIM card will then work all over the world.

The mobile operators therefore have a very good basis for expanding their product base from basic data transport and voice service to interactive services of the modern era. The billing and charging instrument is an immensely powerful enabler and with recent standards such as the *Generic Bootstrapping Architecture (GBA)*[1], technology is becoming mature enough to fully utilize the possibilities of credentials stored in mobile devices.

The possible downside of having strong accountability is the potential vendor lock-in and privacy concerns. Mobile-based service credentials do not necessarily introduce any new limitations that are not already in place. For example, with the advent of flat rate access, many people use their 3G connection as the primary means to access the Internet. The SIM-based credential is used to authenticate their access to the network. Of course, any credentials derived from the SIM-based credential must not be linkable to true identity of the user. This is needed to guarantee user privacy.

10.1.2 Identity Management

Lightweight identities have become popular on the web. The commonly known standards, such as the Liberty specifications for single-sign on, have been complemented by more bottom-up specifications such as OpenID and OAuth. The aim of OpenID and OAuth is to take web industry practices in service security and produce specifications of them in order to ensure interoperability. Indeed, OpenID has become a success story with hundreds of millions of active accounts.

The interesting point in OpenID is its support for decentralized operation. This radically contrasts the strong accountability that we have in today's mobile access networks. With OpenID, anyone can become an identity provider and vouch for some identities in a URL space. OpenID hence, at least in theory, prevents lock-in and supports competition between identity providers. Essentially, users have the freedom to choose which identity providers to use and trust. On the other hand, the accountability offered depends solely on how trustworthy an identity provider is from the viewpoint of the customers.

For example, if a bank or a mobile operator rolls out an *Identity Provider (IdP)*, then this service would be able to offer a very good level of accountability given that proper security protocols (TLS) are used. On the other hand, if a computer hobbyist rolls out an IdP, only a minimum notion of accountability can be sustained. Given that IdPs proliferate, the problem is then for a user to be able to assess IdPs and choose proper ones for the required trust level. This is something that is in the background of the IdentityCard and CardSpace proposals from Microsoft.

It is therefore necessary that the mobile platform supports billing regarding service usage. Therefore, a uniform billing and charging system is needed, which is not typically provided by ISPs.

With the mobile world, on the other hand, the product development process is more complicated. Since the operating systems can be widely different and require different kinds of optimizations, the software product needs to be tailored for the particular OS. Mobile middleware, such as Java, is now becoming more platformized to better abstract the OS; however, much work still needs to be done in this area. For instance, Java ME is still a fragmented platform and many features may not be available on a given system. The fact that there are typically a multitude of devices running a given operating system makes the life of a developer more difficult. Each product needs to be built and tested for each possible device, which is a daunting task given the many different mobile device types available today.

10.1.3 Taxonomy of Services

Figure 10.3 presents an overview of mobile services. The top of the figure presents consumer services and the bottom business services. For the consumer segment, we have four important subcategories, namely information, communication, entertainment, and transaction. The first includes various kinds of static and dynamic content, such as news, weather updates, etc. The second includes various messaging and advertisement solutions. Moreover, emergency and security services are also included in this category. Third, mobile entertainment is a growing market with the advent of flat rate 3G contracts. Ringtones and more recently music are thriving businesses. The fourth category pertains to transactions in mobile business, such as sales, ticketing, banking, and so on.

For the business segment, *Mobile Supply Chain Management Systems (M-SCM)* includes fleet management and various tracking functions needed in logistics. *Mobile Customer Relations Management (M-CRM)* on the other hand includes services pertaining to sales and customer management. These two examples of the business segment can be thought of as supporting cooperation with external parties. M-workforce services support the internal management through mobile services such as calendar, email, and groupware.

Voice services are the traditional services offered by mobile telecommunications systems. Until recently, voice services have been based on circuit-switched technology, which can support strict QoS requirements with the price of flexibility. It is commonly agreed that circuit-switched technology does not support general data traffic as well as packet switched technology. This has motivated the current trend to implement all communications services using the IP protocol. Indeed, all-IP networking looms in the near future.

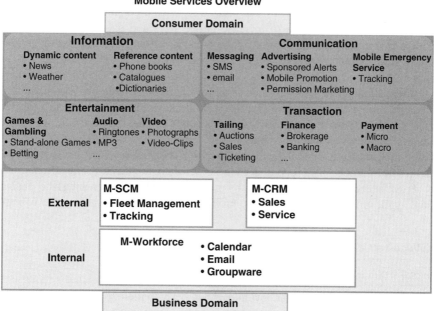

Figure 10.3 Overview of mobile services

Packet switching technology offers more flexibility for services with the price of not being able to guarantee strict QoS.

Voice services include the following subcateogries:

- Telephone calls.
- Voice mail, which supports voice messages left for a subscriber that can be retrieved later from a voice mailbox.
- Voice conference.
- *Private Branch Exchange (PBX)*, which allows connections among the internal telephones of a private organization, with the possibility of connecting the system to the public switched telephone network.

Voice-over-Internet Protocol (VoIP) is an Internet protocol for the transmission of voice. There are many concrete protocols that can be used to implement VoIP. The main benefit of VoIP is its easy and low-cost deployment over packet-switched networks. Compared to the more traditional circuit-switched telephone systems, VoIP can offer additional features at a very low cost for consumers.

VoIP can be seen as an integral part of a mobile service platform. In order to support mobile devices, a number of solutions are needed. The SIP architecture can be seen as a basic enabler for VoIP with its extensive signaling capabilities and NAT traversal extensions. As an alternative, Skype has emerged as a very popular proprietary alternative technology.

A number of mobile application types have been proposed that are outlined in Table 10.1 [2]. The table also presents the key challenges and commonly used optimization techniques.

Streaming media applications do not cope well with high jitter and low throughput. The commonly used solution is to increase buffer sizes and use layered encoding. The buffer size is a crucial parameter for application performance. A too small buffer size may result in interrupted playback and a too large buffer size may delay the start of the playback. Layered encoding allows the playback of the media even if some packets are lost. The decoding process of multiple layers is more computationally demanding than the decoding of a single layer; however, the computational capabilities of mobile devices is increasing every year making this feasible on high-end devices. Layered media allows applications to adapt to changes in bandwidth.

In mobile commerce, high latency and security issues are central challenges. High latency results in poor usability and in time sensitive applications result in financial losses. High latency can be mitigated by good protocol and user interface design. In addition, applications can reduce latency problems by minimizing the number of network

Table 10.1 Network centric mobile application types

Application type	Challenges	Optimizations
Streaming Media	high jitter, low throughput	buffering, layered encoding
Mobile Commerce	high latency, security	adaptive design, minimized comms.
Pervasive Gaming	latency variations	system specific timeout values
Web Browsing	low throughput, high load	phone caching, backoff algorithm

round-trips and by only transferring required data. Security and privacy are also vital for mobile commerce applications. These can be supported by utilizing well-known security techniques, namely encryption and digital signatures. These security mechanisms add to the overhead of the application and special solutions may be needed for very resource constrained devices.

Pervasive gaming is an emerging application area. A crucial challenge for this application type is to be able to synchronize game states across devices in real-time or near real-time. In this domain, key challenges includes latency variations.

Web browsing is one of the killer applications of the Internet and mobile web browsing is a key selling point for the latest generation of mobile phones. Web browsing is hindered by low throughput and high latencies, especially if regular web pages are requested over a wireless access network. The WAP model discussed previously utilized a special markup language for WAP pages; however, this has now been replaced by regular HTML and XHTML. For best usability and response times, web pages need still to be adapted for the mobile devices. The Opera mini is a server-centric solution for mobile browsing, in which a server farm preprocesses web pages for easier usability on the devices.

Mobile application developers need to be aware of the environment for whom they are creating their applications and understand a number of metrics including latency, throughput, jitter, and network timeouts. This environment is difficult, because these values are not constant, but they vary depending on the network provider and the device type.

Figure 10.4 presents a categorization of mobile applications based on their characteristics, namely requirement for real-time communications, and requirement for sessions. The figure also highlights the role the SIP IMS for the different application types. Web browsing, messaging, and e-commerce are examples of non-real-time communications in which sessions are not used. Instant messaging and group chat are examples of real-time communications that do not necessarily require a session. Video-on-Demand, streaming video, peer-to-peer, IPTV, and VPNs require a session but their real-time requirements

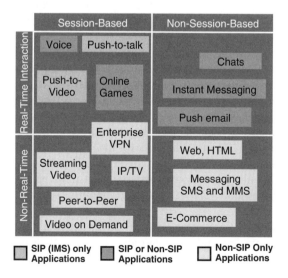

Figure 10.4 Mobile applications

vary. For example, streaming and IPTV use buffering to cope with network latency. These are contrasted with real-time session-based applications, such as voice, push-to-talk, push-to-video, and online games. These applications require real-time interaction.

10.1.4 Location Awareness

Location awareness is an important feature of mobile systems. Location awareness is fundamentally the knowledge of an entity of its location. Location can be understood in various ways, for example location can be described in terms of physical coordinates, or relations to other objects. The former is a form of absolute location awareness, and in these systems position is defined in a coordinate system. Today the two dominant location techniques are GPS and signal triangulation. The key metric for these systems is precision, namely how accurate the location estimates are. Moreover, some techniques, such as GPS, only work outdoors. Signal triangulation or simply the access point identifier are common techniques for indoor positioning. In systems based on relations between spaces, typically *landmarks* play a crucial role. Landmarks can be used to position entities relative to each other.

In addition to absolute and relative location awareness, transitions between locations are important. Indeed, transition awareness means that application behavior depends on some degree on the knowledge of transitions between spaces [3]. Events play a crucial role in signaling changes between application states. Indeed, as outlined in Chapter 7, publish/subscribe is a frequently used technique for creating adaptive mobile applications that react to their environment.

A middleware system thus needs to be able to support various notions of location awareness. This implies also some level of context awareness. APIs are needed for obtaining location information, sharing this information, reasoning about the locations, and also ways to react to the environment.

Location can be seen to be one of the most utilized context attributes. For example, in mobile emergency services location is typically crucial. In addition, navigation services have become very popular in recent years with the advent of GPS enabled phones and dedicated navigation instruments. Various tracking services rely on location information sent by wireless devices. For example in mobile supply chain management (M-SCM) goods are tracked and monitored when they are delivered and processed. Location is also expected to be utilized in mobile advertisement systems.

GPS is the commonly used global positioning solution. The GPS system consists of 24 satellites equally spaced in six orbital planes 20 200 kilometers above the Earth. GPS receivers can determine their 3D position, latitude, longitude, and altitude, within an accuracy of approximately 10 meters. In order for the receivers to be able to establish their position, they need to have a clear view of the skies and at least three or four satellites. *Assisted-GPS (A-GPS)* utilizes a mobile network or a third party service provider that provides additional information that can be used to look for specific satellites. A-GPS can be more accurate than GPS.

Indoor Global Positioning System (Indoor GPS) exploits the advantages of GPS and extends the system for indoor environments. This is achieved by introducing pseudo-satellites that generate the GPS signals indoors.

Cell Identification (Cell-ID) is a technique that uses the base station identifier to derive an approximate location of the mobile terminal. The system is based on the assumption

Table 10.2 Positioning for business-to-consumer mobile services

Example application	Environment	Accuracy req.
Emergency calls	Outdoor	Medium to high
Automotive assistance	Outdoor	Medium
Travel services	Outdoor	medium to high
m-yellow pages	Outdoor	Medium
Banners, alerts, marketing	Outdoor	Medium to high
People tracking	Indoor/Outdoor	High
Indoor routing	Indoor	High

Table 10.3 Positioning for business-to-business mobile services

Example application	Environment	Accuracy req.
Vehicle tracking	Outdoor	Medium
Product tracking	Indoor/Outdoor	Medium to high
Traffic management	Outdoor	Medium
Product replenishment	Indoor	High
Mobile sales	Outdoor	Medium to high
m-customer support	Outdoor	Medium
Field personnel support	Outdoor/Indoor	Medium to high

that the base stations are stationary and their location is known. The accuracy of this technique is generally low, typically several hundreds of meters.

Table 10.2 presents different positioning requirements for business-to-consumer mobile services. Table 10.3 presents positioning requirements for business-to-business services, respectively. The positioning methods have different target environments, accuracies, and they are based on different technologies [4].

Figure 10.5 illustrates the implementation of a mobile map service [5]. A remote facade component implementing the pattern is used to support mobile devices. The remote facade pattern provides a unified interface to a set of interfaces in a subsystem. In this example, mobile devices send requests with addresses and receive as responses directions and maps. The remote facade is responsible for processing the input addresses and querying a map web service for the coordinates, routes, route segments, directions, and maps. This simplifies the development of mobile map applications, because the developer does not have to know about the specific web services being used.

10.1.5 Pervasive Services

We outline a taxonomy for pervasive and mobile services after the work presented in [3]. The following features can be used to classify these services.

- Transitionality, which means the degree to which the application is used in different environments. Today we are seeing the emergence of an increasingly heterogeneous operating environment, which requires that the middleware supports awareness of the environment and changes in it.

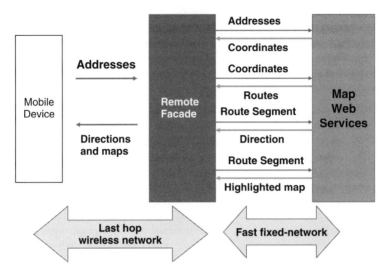

Figure 10.5 Mobile map applications

- Time Constraints, which can be understood in the applications timeliness requirements. These typically vary between delay-tolerant operation to real-time operation. Some applications can tolerate disconnections and delays, such as web browsing. Other application classes have stricter requirements regarding the communications environments, such as voice services, which do not cope well with jitter and delays.
- Goal, which is the type of benefit that can be expected to be achieved from the application. The goal can be improved access to information, timeliness of information, or simply entertainment.
- Collaboration, which denotes the degree to which multiple entities collaborate in realizing the application and work to meet its goals. Collaboration entails that content and context is shared and manipulated by the entities.
- Lifetime, which signifies the length of time an application is executed. This may be from several seconds to days or even years. The required longevity has implications for the middleware and operating environment, for example, battery consumption needs to be understood to ensure longevity.
- Centricity. An application can be data or action centric. A data-centric application deals with information gathering and processing. An action centric application also performs actions in response to changes in the environment. Typically, these are combined in an application. The centricity has implications for storage requirements, computation requirements, and communication.

Many pervasive middleware systems have been proposed in the research community. We briefly outline their key features and properties. We observe that in order to cope with the notions of transitionality and centricity, the proposed middleware systems typically offer either an event-based API or a tuple space API, or a combination of them.

- *Ambient Networks [6]*. The EU 6th Framework Programme Ambients Networks project investigated how to enable heterogeneous networks to cooperate at runtime. As a result, an architecture called the *Node ID architecture* was developed that presents multiple technologies, address domains, and various middleboxes as first-order components. The node ID architecture uses cryptographic node identifiers, gateways, distributed hash tables, and specific protocol constructs to the issue of cooperation among heterogeneous networks.

- *Aura [7].* This system provides a distraction-free pervasive computing environment. Aura develops the system architecture, algorithms, interfaces and evaluation techniques to meet the goal of pervasive computing.
- *Gaia [8].* Gaia supports dynamic adaptation to the context of mobile applications. This system supports the development and execution of portable applications in active spaces.
- *Lime [9].* Lime is a mobility-aware tuple space. It defines programming constructs, which are sensitive to the mobility constraints. Lime explores the idea by providing developers with a global virtual data structure and a tuple space. The content of the tuple space is determined by the connectivity among mobile hosts.
- *MobiLife [10].* This system emphasizes self-awareness in order to support automatic configuration arrangement of devices, services, and local connectivity in the user's local environment. It also enables automatic and multi-modal interfaces that enhance the user experience and minimize the active user effort needed in managing the local environment.
- *Rover [11].* Rover is a software toolkit that supports the construction of both mobile-transparent and mobile-aware applications. The Rover toolkit combines relocatable dynamic objects and queued remote procedure calls to provide unique services for 'roving' mobile applications.
- *TSpaces [12].* This is an asynchronous messaging-based communication facility without any explicit support for context-awareness. TSpaces explores the idea of combination of tuple space and a database that is implemented in Java. Tspace targets nomadic environment where server contains tuple databases, reachable by mobile devices roaming around.
- *X-Middle [13].* This system supports disconnected operations in mobile applications. X-Middle allows mobile users to share data when they are connected, or replicate the data and perform operations on them off-line when they are disconnected. Data reconciliation takes place when the user gets reconnected.

10.1.6 Smart Spaces

The smart space environment differs from the traditional fixed service usage environment in many ways, which have been studied for some time in the research on pervasive and ubiquitous systems. Smart spaces can be seen as dynamic collections of devices collaborating to support users. A smart space is a public or private collection of data items that can be shared, synchronized, queried, and subscribed to by client systems. A smart space can offer customized services for its clients.

Let us consider two usage scenarios for smart spaces. The first scenario is a shopping mall with a number of shops and services for consumers. The second scenario is a popular public event, for example, a rock concert with a large audience. In both scenarios, a smart space can be used to improve user experience. One interesting feature is localized operation, in which offered content and user interface depends on the access location, or past history of visited locations. This localized nature is a key driver for many smart space services. A smart space may also have history and can remember clients and be able to offer them personalized service. This is expected to be useful in shopping malls, where customized product updates, discounts, and event information can be disseminated. In public events, the smart space can be used to boost the spectator experience by enabling content, such as images and video to be shared.

Requirements must also be aligned with business models to fully support product development and deployment. There must be a business incentive for both platform developers and deployers. Deployment is the key driver of the technology so we consider the possible business strategies for it. The strategies include advertisements, customer analysis and feedback, and added-value for customers. Here critical mass plays a crucial role to ensure the success of smart spaces. In order to gain popularity a smart space needs to be easy to use and require zero configuration for clients. This is our key requirement that basic browser software must alone suffice to access smart space services. Similarly, a smart space should be easy to set up and administer. Moreover, since smart spaces are often used in public spaces the clients must be able to trust the smart space providers. Hence, security plays an important part in the success of smart spaces and it should be easy for both clients and smart space providers to understand and configure proper security settings. One of the challenges is to prevent phishing attacks, in which a smart space attempts to steal user data and credentials.

10.2 Mobile Server

In this section we consider the Mobile Web Server as a example of a novel mobile service. We outline a system developed for the Symbian OS which is based on the Apache web server [14].

The Mobile Web Server makes it possible to utilize personal content in new contexts. The key notion in this system is the mobsite, which is a mobile website that provides a web-based user interface to the applications provided by the server.

Typically, mobile devices only host web clients, which use the HTTP protocol to request web pages from servers. Until recently, mobile devices have not been powerful enough to run web servers and the communications stack has not necessarily supported the required server-side Socket features.

The Mobile Web Server developed at Nokia Research Center is a port of the Apache web server for the Symbian OS. The server has been extended with a proprietary protocol for enabling efficient and proactive communications towards the mobile device.

The Mobile Web Server enables a number of interesting use cases, including the following:

- *Ad-hoc content creation.* The server supports interactivity with the mobsite owner. A web surfer can request a response from the mobsite owner, for example in interactive photography and messaging. The interactions can be linked to a certain location or time.
- *Reusing native S60 application data.* The content served by the server can be linked to native multimedia applications on the S60 platform. Therefore it is possible to re-use native S60 application data as mobsite content. For example, photos, videos, calendar entries, and contact information can be served on a mobsite.

- *Interactive photography.* The mobsite can be linked with the camera feature of the mobile device. In this use case, the mobsite owner can be requested to take a picture. This is accomplished using a 'snapshot request' feature on the mobsite. If the owner of the site accepts this request, a picture is taken with the camera and sent to the requesting web browser.
- *Locating mobsites.* The mobsite can provide information pertaining to the location of the mobile device. Typically this would involve reading GPS device or obtaining location data from the cellular network, and then making this information available. The location can then be combined with a map service to show an overlay of nearby mobile devices.
- *Remote use.* In many cases a user may want to access data stored on the mobile device remotely, for example, from a desktop machine. Remote access using a mobsite allows the owner of the site to check email, logs, calendar entries, missed calls, browse contacts, and share files. Remote use allows to initiate phone calls and send SMS messages. This feature can be used together with conventional web browsing on the desktop machine. Phone numbers and other information can then be sent to the mobile devices without complicated operations.
- *Sharing content.* The mobsite can be used to disseminate data from the mobile phone without involving user interaction with the owner. Access control information can be preconfigured to restrict access to content.
- *Browser-based user interface.* The mobsite allows content and data stored on the device to be accessed using a browser-based interface. This means that mobile content can be accessed in an ubiquitous fashion and it can be used in mashups.

The Mobile Web Server follows the remote proxy pattern outlined in Chapter 4. An outline of the system is presented in Figure 10.6. The mobile server establishes a connection with a remote gateway. The gateway then advertises the mobile device in DNS and is willing to route packets destined to the device. Some state is needed at the gateway to enable the connectivity. Now, due to this connection between the device and the gateway, the gateway is able to push web page requests to the mobile in a proactive fashion.

All Mobile Web Server implementations are associated with a gateway account and a URL that ends with *.mymobilesite.net*. These mymobilesite.net DNS mappings point to the gateway. The gateway is responsible for mapping individual browser requests to the corresponding Mobile Web Server.

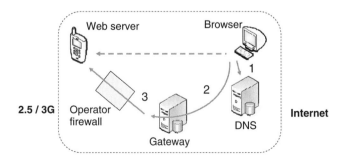

Figure 10.6   Overview of mobile server

The approach is especially useful when flat rate is used, because connection maintenance and request/response messages do not incur any cost to the user in addition to the monthly fee. Thus the mobile server can be seen as a building block for future peer-to-peer and decentralized applications.

The key software components of the system are:

- *Mobile Web Server.* A web server that is able to serve both static and dynamic content. The integrated mobsite application enables mobsite content creation.
- *Portal.* Mobsite address creation and management.
- *Gateway.* Provides Internet connectivity and additional security features. Gateway functions are transparent to the mobsite owner.

The basic connectivity scenario involves the following steps (Figure 10.7).

1. Mobile Web Server registers with the gateway. The gateway receives the Mobile Web Server IP address and it begins to route requests to the server.
2. A web surfer enters the Mobile Web Server URL resulting in a HTTP request. Before the request is sent, a DNS lookup is made by the networking stack that resolves to the gateway according to the static *mymobilesite.net* gateway IP address mapping.
3. The gateway receives the HTTP request and examines the header looking for the target Web Server (part of the domain name). Using this information, the gateway forwards the request to the proper Mobile Web Server.
4. The Mobile Web Server processes the request and sends a reply.

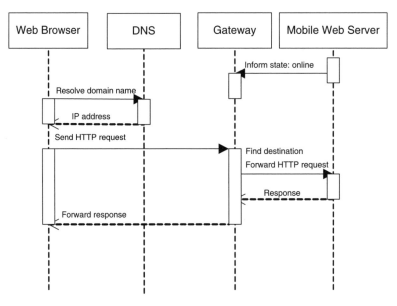

Figure 10.7 Sequence diagram of Mobile Web Server

Dynamic content can be created using the Python programming language, which has been ported to Series 60 devices (PyS60). Python allows to access S60 platform features. The life cycle of a dynamic content request consists of three steps:

1. Create a handler to manage all the tasks associated with an HTTP request. Start processing the request.
2. Execute application logic. This step typically results in content being generated.
3. Return content by using the handler.

The connector handles the creation and management of the connection between the gateway and the web server. This includes creating the initial connection to the gateway and making sure it remains functional using keep-alive packets.

The gateway supports security by using HTTPS between the gateway and a web browser accessing mobile content. The last hop communications between the gateway and Mobile Web Server is secured using TLS/SSL. This means that the gateway will be able to inspect unprotected requests and responses. Therefore gateway authentication is important to prevent man-in-the-middle attacks. On the other hand, since the gateway can inspect messages, it can also enforce security by employing various filtering techniques.

In addition to transport layer security, slow mobile connections are prone to different kinds of resource exhaustion attacks and malware. The gateway is responsible for filtering traffic and ensuring that proper requests are forwarded to the Mobile Web Server.

10.3 Mobile Advertisement

Advertising can be defined as being a non-personal presentation of ideas, products, or services to other people, in which an identifiable originator of the presented material pays compensation for the act of presentation. Currently, web advertising is the dominant form of revenue for Internet service companies, as demonstrated by the success of companies such as Google and Yahoo.

Mobile and wireless advertisements have still not proliferated; however, the expectation is that mobile and wireless advertisement will become more commonplace as the number of smart phones and smart places grows. Since smart phones typically feature rich web browsers, these devices are already taking part in the web ecosystem and consuming the more traditional Internet advertisements. The web advertisement model; however, does not take into account the unique features of the mobile and wireless environment, namely the proximity of devices, decentralized and disconnected nature of the environment, and the different contextual parameters that change over time. Hence, new advertisement solutions are needed for the mobile and wireless environment that are more context-aware and support decentralized operations, for example, in smart spaces.

Several different approaches have been proposed previously for wireless and mobile advertisement, namely utilizing multi-hop ad hoc networks (MANETs) in the distribution of advertisements. This kind of approach, in which advertisement messages are distributed hop-by-hop in the environment, are inspired by viral marketing, the messages are spread indirectly among consumers and not by the advertiser or advertisement agency. If an advertisement message leads to a purchase, those people that have contributed to the forwarding chain may receive a bonus. This of course requires that each node can be sufficiently identified.

Location-Based Service (LBS) technology is a building block for mobile advertisement. The location of a user can be used for advertisement targeting and an advertising system can suggest a route to get to advertised locations. In addition to location, other contextual parameters can also be taken into account when targeting advertisements to consumers. The three key questions in context-aware advertising are, how to represent context and match advertisement data to contextual consumer profiles, how to ensure user privacy, and how to prevent consumers from receiving unsolicited messages, so called spam.

In the MoMa system [15], end users create so called orders according to a catalogue, which is a hierarchically ordered set of possible offers described using appropriate attributes. The advertisers create offers into the system, which are also defined using the catalog. When the system detects matching orders and offers, the user is notified using some user interaction mechanism that has been agreed beforehand. In order for an end user to receive an advertisement, there must be a matching order installed by the end user. When a user issues an order and there is already a matching offer, a notification is generated immediately. This is called *pull-advertisement*. *Push-advertisement* happens when a matching offer is introduced after the order is in place.

The MoMa operator is also prevented by seeing any plain text profiles or IP addresses of end users. This is achieved by an intermediate layer that the user trusts, which performs anonymizing of data and resolving any incoming messages the proper endpoint address (IP). The system is thus based on one time pseudonyms.

Figure 10.8 illustrates the system model of MoMa. The central entities are the end user, the trusted party, the operator, and the provider. The trusted party manages end user profiles and anonymizes user profiles and other data so that other parties cannot determine user preferences. The operator is responsible for running the core system that stores orders and offers, and is able to match between them. When an order and offer match, a notification is generated towards the end user. The provider is the advertiser and responsible for the offers and providing advertisement information that can then be

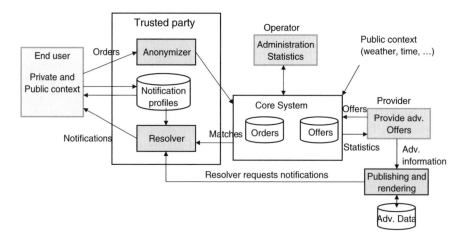

Figure 10.8   A system for mobile advertisement

delivered to end users. The trusted party hosts a component called the resolver, which receives information about matching orders and offers, and can then deliver notifications to the end user according to user notification profiles.

10.4   Mobile Push Email

Traditional email access is based on pull interactions. The *Post Office Protocol (POP3)* protocol is a popular example of a pull-based email delivery protocol. The Mail User Agent polls the Mail Delivery Agent (server) to see if there is new mail. If there is new mail, this is downloaded to a mailbox. The mail has been pushed from the sender to the final Mail Delivery Agent using the *Simple Mail Transfer Protocol (SMTP)* protocol. Push email solutions use push delivery instead of pull during the last delivery step.

Push email is more complicated because of various reachability issues in the mobile environment. The SMTP servers are advertised in DNS and they have typically public IP addresses. Mobile devices, on the other hand, are often not reachable because of NAT boxes (private addresses) and changes to the address due to mobility. The key problem is how can the mobile device and the Mail Delivery Agent be located in order to push the mail.

The *Internet Message Access Protocol (IMAP)* supports polling and notifications. The server sends a notification to a client to inform them that there is data available. This allows flexible retrieval of messages and gives the client the control of whether or not to download new message data.

In this section, we examine several well-known mobile email systems and analyze how they support message push, namely Blackberry and Windows Mobile. The mobile push solutions are based on the client-initiated communications pattern presented in Chapter 4. Using this pattern, the clients establish a communication channel, typically a TCP or TLS connection, with a server that has a public IP address. The server can then push messages to the client using this long-lived connection. This solution is appealing, because it can be made to work with firewalls and NAT boxes. Since the communication channel is initiated by the client, address translation does not prevent the server from sending packets to the client. On the other hand, the solution introduces additional state at the server and NAT devices.

Blackberry devices have become popular among business users in part because they support desktop style email usage experience with almost instant delivery of messages. Blackberry devices utilize a custom enterprise server that is connected to the traditional email system. The enterprise server monitors the email server and then can pull new messages and send them to the Blackberry device using push over the wireless network. The wireless protocol has been developed by textitResearch In Motion (RIM) and it is proprietary.

Microsoft introduced the DirectPush Technology with Windows Mobile 6. This is an additional feature for Microsoft Exchange 2003 with a new service pack that adds messaging and security features. Exchange Server utilizes the subscriber's existing wireless phone account to push messages to the user.[1]

---

[1] Microsoft DirectPush TechNet Article

Mobile devices that support DirectPush utilize a long-lived HTTPS request to the Exchange server. The Exchange server monitors activity on the user's mailbox. When a change is detected, the server sends a message to the device identifying the changes. Example changes include new or changed email messages, calendar, or contact items. The change notification message is small and indicates to the client that the device should start synchronization with the Exchange server. After receiving such a change notification, the device can issue a synchronization request to the server. After synchronization, a new long-lived HTTPS request is sent to track changes at the server.

More specifically, the DirectPush system involves the following steps:

1. A mobile device issues an HTTPS request, called a *ping*, to the server. The request informs the server to notify the device if any items of interest change in the next 15 minutes. The request is kept alive for this duration. If request cannot be kept alive the request results in a HTTP response. The 15-minute time span is called a *heartbeat interval*.
2. If there are no changes to the items in 15 minutes, the server returns a response of HTTP 200 OK. When the mobile device receives this response, it resumes activity, and sends the HTTPS request again,
3. If any items change or new items are received within the 15 minute heartbeat interval, the server sends a response that notifies the mobile device that there is new data available. The notification includes information about the change and the folder in which the new or changed item is located. The mobile device can then issue a synchronization request to the server to the folder in question. When synchronization is complete, the mobile device sends a new ping request and the process starts again.

DirectPush requires that the networking environment supports long-standing HTTPS requests. Typically in mobile access networks, such long-standing requests are not supported. In addition, some firewall and NAT configurations do not allow long-standing requests. In the case that the carrier network has a timeout value that is less than 15 minutes, the following steps are taken by DirectPush (Figure 10.9):

1. As before, a mobile device sends an HTTPS request to the server.
2. If the server does not respond after 15 minutes, the mobile device wakes up and determines that the connection to the server was timed out by the network. The device then sends a new HTTPS request, but uses a heartbeat interval of eight minutes.
3. After eight minutes, the server sends an HTTP 200 OK message. The device will then try to gain a longer connection by issuing a new HTTPS request to the server that has a heartbeat interval of 12 minutes.
4. After four minutes, a new email message is received and the server responds by sending an HTTPS request that tells the device to synchronize. The device synchronizes and reissues the HTTPS request that has a heartbeat of 12 minutes.
5. After 12 minutes, if there are no new or changed items, the server responds by sending an HTTP 200 OK message. The device wakes up and concludes that network conditions will support a heartbeat interval of 12 minutes. The device will then try to gain a longer connection by reissuing an HTTPS request that has a heartbeat interval of 16 minutes.

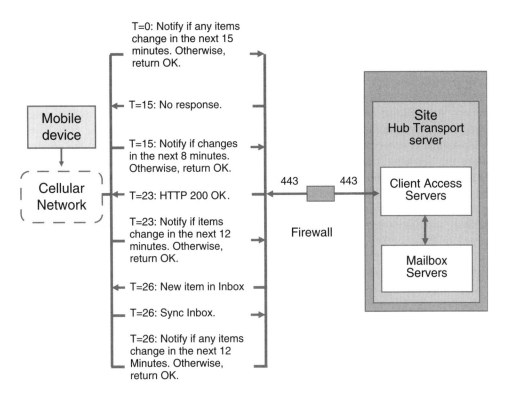

Figure 10.9   Overview of DirectPush

6. After 16 minutes, no response is received from the server. The device wakes up and concludes that network conditions cannot support a heartbeat interval of 16 minutes. The device determines that the heartbeat interval has reached its maximum value. The device then issues an HTTPS request that has a heartbeat interval of 12 minutes because this was the last successful heartbeat interval.

The basic principle of operation is that the mobile device tries to use the longest heartbeat interval that the network supports. This is motivated by the fact that unnecessary signalling wastes battery life, may increase costs, and introduces overhead to the network. Mobile carriers can specify a maximum, and initial heartbeat value in the registry settings for the mobile device.

Both Blackberry and Microsoft's DirectPush are proprietary solutions for push email. Open standards-based solutions for push email include the Push-IMAP standard and the SyncML standard. The Lemonade Profile is an IETF modification to IMAP and SMTP to make them more suited to the demands of mobile email. This solution relies on the existing IDLE (RFC 2177) command to provide instant email notification on the client device. The IDLE command is often used to signal the ability of a client to process notifications sent outside of a running command. This can be used to provide a similar user experience to push.

10.5   Mobile Video

In this section, we focus on mobile video playback and consider the two alternative technologies, namely streaming and progressive download. Streaming allows controlled playback with resource reservation using a dedicated video server that sends data to connected clients. Thus streaming allows the delivery of real-time media to the clients and supports fine-grained controlling and monitoring of the delivery. Progressive download differs from streaming, because it is based on simply downloading the media files. Due to metadata that is embedded in the media files, this technique is able to start the playback before the download has completed. This technique simply delivers the video files over standard protocols, typically HTTP. Progressive download is useful for video-on-demand, and since the media files are downloaded, they can be cached and possibly shared as well.

The *Real Time Streaming Protocol (RTSP)* was developed by the IETF (RFC 2326). RTSP is a protocol that allows the clients of a streaming media system to control the streaming. This control involves sending commands, such as play and stop, to a streaming server. The actual data transmission is not part of the protocol. Typically the RTP protocol is used in combination with RTSP. The QuickTime Streaming Server from Apple and RealNetworks streaming server support this protocol.

YouTube has become a phenomenally successful video sharing site on the web. They have recently introduced a mobile version of the site available at m.youtube.com. This site includes all the videos from the desktop version of the site. In order to access the mobile site, a mobile device is needed that supports 3GP files over RTSP. The site also includes a video upload feature. There is also a Java MIDlet available that can play the videos. YouTube uses the *Flash Video format (FLV)* and the mobile version is also based on this file format.

The two main playback options for FLV files on a mobile devices are as follows:

- Progressive download using HTTP. For example, Google Video and Youtube support progressive downloading and can seek to any part of the video before buffering is complete. Progressive download differs from multimedia streaming, because the video being watched will be cached on the client device.
- Streamed via *Real-time Messaging Protocol (RTMP)* to the Flash Player using a Flash enabled server. RTMP is a proprietary protocol from Adobe Systems for streaming audio and video over the Internet.

RTMP is based on TCP and maintains a single persistent connection with a server that is used for real-time communication. In order to guarantee smooth playback, the protocol may split video and data into fragments. The protocol supports the negotiation of the fragment size and then the multiplexing of the fragments over the connection.

The protocol defines a number of channels that are used for different kinds of traffic, such as signalling, video stream data, and audio stream data. The channels have their own quality of service requirements, such as bandwidth and latency requirements.

RTMP supports Flash Video and MP3 streams and allows clients to make RPC calls using the Action Message Format.

A variant of the protocol that uses HTTP tunneling, RTMPT, was developed to cope with firewall issues. RTMPT encapsulates RTMP data in HTTP requests and typically uses the port 80. This variant introduces some overhead due to the HTTP headers, but avoids some of the connectivity problems with RTMP and firewalls. Another variant uses the secure HTTPS protocol.

10.6   Mobile Widgets and WidSets

Widgets are lightweight web applications that can be developed using standard web technologies, such as HTML, *Cascading Style Sheets (CSS)*, RSS, Javascript, and AJAX [16, 17]. Widgets can be seen as a technology to utilize existing Web 2.0 resources and services on mobile devices in an easy way. Widgets typically use RSS feeds which offer a pull-based mechanism for obtaining updates regarding web resources. RSS feeds provide short descriptions of updates and then links to additional material. Thus this model is very suitable for the mobile environment.

Because web widgets and a widget have a reliance on web technologies, they offer much of the same functionality. However, differences exist in:

- the packaging format;
- the security model; and
- the APIs.

Web widgets are not typically packaged or downloaded as a single file. Instead, they are dynamically executed through a combination of ECMAScript (JavaScript), HTML elements, and CSS. Figure 10.10 gives an overview of the general widget environment.

A non-web-based widget can perform actions beyond the scope of web widgets stems from the usage of widget-specific APIs that are provided by some frameworks. Web widgets are part of a HTML document and they are thus bound to the security constraints of the web browser. As a consequence, a web widget cannot make cross-domain requests, and cannot access system properties or services.

Widgets and Java applets are similar in many respects. Both widgets and applets are based on a runtime engine for execution. Java applets are executed by the Java Runtime Environment, and widgets are executed by the widget engine. Both are able to do web requests and process the results.

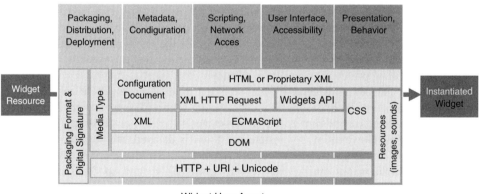

Figure 10.10   Overview of the widgets

The Series 60 platform and Symbian OS support widgets using the Web Run-Time. The Web Run-Time allows widget developers to tap into the features of the underlying S60 platform. This allows the development of more personalized and context-aware applications.

To give an example, a widget can access the user's current location by accessing built-in GPS for location, or the user's calendar information. The current location can then be used to provide weather information, nearby locations, or travel instructions. A flight tracker widget can fetch a travel itinerary from an airline's website, save it to the phone's calendar application, and set a reminder. Before departure, the widget can then automatically check the flight's status by retrieving the flight details from the calendar.

WidSets is a simple service developed by Nokia that provides mobile users with information that is normally accessed via the Internet. WidSets is based on widgets that utilize RSS feeds to retrieve current information from the web.

10.7   Airline Services

The development of new mobile services for travellers is motivated by the emergence of mobile devices with multi-mode communications, such as UMTS and WLAN, and interactive multimedia capabilities. Many of the traditional processes at airports can be automated by allowing a user to monitor and interact with information pertaining to flights and other topics of interest for the user.

Example interactions that can be provided by such a system include:

- Proactive mobile check-in.
- Notify customer of gate changes.
- Notify customers of last minute changes to their flight.
- Notify customers about disruptions or potential problems identified regarding the journey and open online web update possibility for rerouting.
- Send informative messages regarding the check-in gates, timetables, connecting flights etc.
- Schedule changes, lost baggages, waitlist acceptances etc.

An airline may deliver marketing communications to travelers based on their preferences. Travelers can define if they want to receive marketing communications and what kind of products they are interested in. The system enables the marketing communication to be marketed only to the potential customers, since the system can provide information regarding travelers' preferences and their journey.

A travel session begins when a traveler books a flight. Usually, there are no system messages before check-in in the sessions. The system asks a passenger to check-in before a flight. The request can be sent based on passenger's location or on time (fixed period before departure). However, if, for instance, a flight is delayed, the system can provide this information for all the passengers even though those who have not checked in yet.

Other system messages during the session pertain to updates to flight information, upgrade possibilities during the trip, marketing communications, and information on possible changes to the original travel plan. Typically, a request for check-in is sent automatically, whereas a proposal for substitutive travel plan is created and sent by the airline call center agent.

The travel session finishes when the trip has finished and there are no remaining system messages for the traveler. A possible post-trip message could be for instance a baggage lost message, in which the system informs the traveler about lost baggage and later about arrival time of the baggage and other information.

In such a communications system, the capability to push messages to users is crucial. This push capability can be implemented using a variety of technologies, namely SIP and HTTP. We outline a solution developed in the WeSAHMI project at Helsinki University of Technology, Tampere University of Technology, and University of Helsinki. This system is based on both SIP and HTTP. In this solution, SIP is used for control and content push messages, and HTTP for content requests. The electronic ticket is implemented using SMIL 2.0 language specified by W3C and the X-SMILES open source XML-browser. SMIL allows the design of interactive multimedia rich pages that can be updated at runtime. The SIP push message are processed by the electronic ticket SMIL code by using callbacks and an asynchronous programming model. XML content is pushed to the electronic ticket and the ticket is updated in real-time when connection with the airline service has been established. Connection can be established either using a local access point or through a wireless access network.

The XML content that is pushed to the mobile browsers can represent a whole XML document or a part of an XML document. The latter case requires a way to position the incoming content fragment with some existing document. The WeSAHMI system uses the *Remote Events for XML (REX)* format specified by W3C to do this in an inter-operable way. A REX Generator translates given data into the REX format, which is an XML-based format for representing *Document Object Model (DOM)* events. The REX message encapsulate state changes to the DOM data structure maintained by the browser.

The service access network is based on edge proxies. Edge proxies advertise service access and services using the *Service Location Protocol (SLP)* defined in RFC 2608. Mobile phones discover edge proxies and then use them to access services. All connections are initiated by the clients, which allows content to be pushed to the devices. The security domain may be federated with other domains, including SIP-based IMS systems.

Figure 10.11 illustrates how edge proxies can be used to establish long-term connections with mobile devices and then how messages can be pushed to the devices using this connection. It is proposed that security problems introduced by the wireless network environment could be solved by requiring a secure client-initiated connection between mobile devices and edge proxies. As discussed in Chapter 4, client-initiated connections can be used to alleviate problems related to device reachability due to NATs and firewalls over heterogeneous networks.

Current work at IETF is looking at client-initiated connections [18]. The key idea of the specification is to re-use the connection that was used to send the SIP REGISTER request. This connection can be a bidirectional stream of UDP datagrams, a TCP connection, or some other type of transport protocol. It is the responsibility of the User Agent (UA) to maintain connectivity. The UA may also employ multiple flows to the proxy or registrar. In addition, a keep alive mechanism is included so that failed flows can be detected.

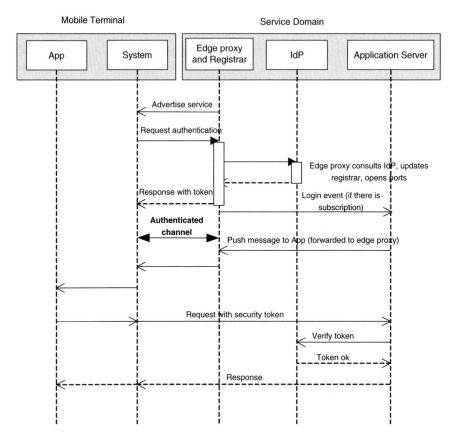

Figure 10.11   Pushing messages to mobile devices

10.8   Revisiting Mobile Patterns

In this section, we revisit the mobile design patterns presented in Chapter 4 and consider how they are used in the considered mobile applications and services. We focus our analysis on the set of mobile patterns and note that many other patterns are also employed by applications and services.

- *Remote Proxy.* Utilizing a special server that accepts incoming connections.
- *Remote facade.* Minimizing the number of remote calls in application and it provides a coarse-grained interface to one or several fine-grained objects.
- *Multiplexed Connection.* Utilizing lower layer connection for multiple higher level connections.
- *Client-initiated Connection.* Using client-initiated connections to circumvent reachability problems.
- *Rendezvous.* Using rendezvous to coordinate state updates, such as location of the mobile node.
- *Synchronization.* Synchronizing data items between hosts.

- *Caching.* Storing data locally for faster access.
- *Location Awareness.* In location-awareness the past, current, and predicted location needs to be communicated to other parties. Therefore Rendezvous and Synchronization are crucial. This can be achieved using a Remote Proxy pattern and the Connection patterns. The Remote Facade pattern is often applied to minimize the number of remote calls needed. Eager Acquisition can be used to anticipate future information needs.
- *Mobile Server.* Reachability is vital in this application and it is achieved using the Client-initiated Connection, Remote Proxy, and Rendezvous patterns. Caching can be used at the Remote Proxy to improve performance.
- *Mobile Advertisement.* This application requires a combination of patterns, namely Client-initiated connections, Rendezvous, Synchronization, Caching, Remote Proxy, and Broker. The connections ensure reachability of the mobile terminals and allow the advertisement system to synchronize advertisements and impressions with the mobile device (if they are stored on board). Rendezvous is needed to keep track of the current location of the device. Remote proxy is needed to handle the connections. The Broker is used to provide indirection between different components in the system.
- *Mobile Push Email.* Reachability is vital also in this application scenario. This is achieved using the Client-initiated Connection, Remote Proxy, and Rendezvous patterns.
- *Mobile Video.* This application can utilize the Client-initiated Connection and Multiplexed Connection for enabling continuous media delivery to the client. Video-on-demand can be Cached, and video stream buffering can be seen as a variant of the Eager Acquisition pattern.
- *Widgets.* Widgets can employ a number of patterns, typically Remote Proxy and Broker are pertinent.
- *Airline Services.* This application case is similar to Mobile Server, Location Awareness, Mobile Advertisement, and Mobile Video.

10.9 Summary

This chapter examined a number of mobile applications and services. In the first sections, we observed that current trend is towards a converged communications environment, in which web technologies and telecommunication standards are used in combination. Given the current deployed state of the art, HTTP and SIP are the basic building blocks for the services. We then considered the business models and their evolution. The open web ecosystem appears to foster much faster development cycles than is traditional for mobile services. Therefore, the adoption of web-based mobile solutions is attractive. On the other hand, the mobile device ecosystem is much more heterogeneous and fragmented than the desktop environment, which will continue to be challenging in the near future. For example, the Java ME platform itself is rather fragmented, and it is executed on an even more fragmented OS base.

We considered more sophisticated mobile applications for pervasive environments and briefly examined some research systems that have addressed challenges in these environments. In ad hoc and pervasive scenarios, issues with discovery, interoperability, security, trust, and privacy become crucial. Identity, accountability, and trust management can be seen as key enablers for pervasive services.

Location awareness is an important feature of mobile systems. Location awareness is fundamentally the knowledge of an entity of its location. A middleware system needs to be able to support various notions of location awareness. This implies also some level of context awareness. APIs are needed for obtaining location information, sharing this information, reasoning about the locations, and also ways to react to the environment. Location can be seen to be one of the most utilized context attributes.

The Mobile Web Server makes it possible to utilize personal content in new contexts. The key notion in this system is the mobsite, which is a mobile website that provides a web-based user interface to the applications provided by the server. The Mobile Web Server can host different kinds of resources, such as blogs, pictures and other media files, and preferences. The system allows arbitrary page requests to be sent to the mobile.

Mobile and wireless advertisements have still not become popular; however, the expectation is that mobile and wireless advertisement will become more commonplace as the number of smartphones and smart places grows. Since smartphones typically feature rich web browsers, these devices are already taking part in the web ecosystem and consuming the more traditional Internet advertisements. The web advertisement model, however, does not take into account the unique features of the mobile and wireless environment, namely the proximity of devices, decentralized and disconnected nature of the environment, and the different contextual parameters that change over time. Hence, new advertisement solutions are needed for the mobile and wireless environment that are more context-aware and support decentralized operations, for example, in smart spaces.

We examined several well-known mobile email systems and analyze how they support message push, namely Blackberry and Windows Mobile. The mobile push solutions are based on the client-initiated communications pattern. Using this pattern, the clients establish a communication channel, typically a TCP or TLS connection, with a server that has a public IP address. The server can then push messages to the client using this long-lived connection. This solution is appealing, because it can be made to work with firewalls and NAT boxes.

We also considered mobile video playback and examined the two alternative technologies, namely streaming and progressive download. Streaming allows controlled playback with resource reservation using a dedicated video server that sends data to connected clients. Thus streaming allows the delivery of real-time media to the clients and supports fine-grained controlling and monitoring of the delivery. Progressive download differs from streaming, because it is based on simply downloading the media files. Due to metadata that is embedded in the media files, this technique is able to start the playback before the download has completed. This technique simply delivers the video files over standard protocols, typically HTTP. Progressive download is useful for video-on-demand, and since the media files are downloaded, they can be cached and possibly shared as well.

Widgets are lightweight web applications that can be developed using standard web technologies, such as HTML, CSS, RSS, Javascript, and AJAX. Widgets can be seen as a technology to utilize existing Web 2.0 resources and services on mobile devices in an easy way. Widgets typically use RSS feeds which offer a pull-based mechanism for obtaining updates regarding web resources. RSS feeds provide short descriptions of updates and then links to additional material. Thus this model is very suitable for the mobile environment.

Bibliography

[1] *3GPP TS 33.220: Generic Authentication Architecture (GAA); Generic Bootstrapping Architecture*.

[2] Wittie MP, Stone-Gross B, Almeroth KC and Belding EM 2007 Mist: Cellular data network measurement for mobile applications *Proceedings of the 4th International Conference on Broadband Communications, Networks, and Systems (BROADNETS 2007)*, Raleigh, North Carolina, USA.

[3] Dombroviak KM and Ramnath R 2007 A taxonomy of mobile and pervasive applications *SAC '07: Proceedings of the 2007 ACM symposium on Applied computing*, pp. 1609–1615. ACM, New York, NY, USA.

[4] Zeimpekis V, Giaglis GM and Lekakos G 2003 A taxonomy of indoor and outdoor positioning techniques for mobile location services. *SIGecom Exch.* **3**(4), 19–27.

[5] Yung MJ 2003 *Enterprise J2ME: Developing Mobile Java Applications*. Prentice Hall PTR, Upper Saddle River, NJ, USA.

[6] Ahlgren B, Arkko J, Eggert L and Rajahalme J 2006 A node identity internetworking architecture *INFOCOM*. IEEE.

[7] Garlan D, Siewiorek D, Smailagic A and Steenkiste P 2002 Project Aura: Toward Distraction-Free Pervasive Computing. *IEEE Pervasive Computing* **01**(2), 22–31.

[8] Román M, Hess CK, Cerqueira R, Ranganathan A, Campbell RH and Nahrstedt K 2002 Gaia: A Middleware Infrastructure to Enable Active Spaces. *IEEE Pervasive Computing* pp. 74–83.

[9] Murphy AL, Picco GP and Roman GC 2006 Lime: A coordination model and middleware supporting mobility of hosts and agents. *ACM Trans. Softw. Eng. Methodol.* **15**(3), 279–328.

[10] Klemettinen M 2007 *Enabling Technologies for Mobile Services: The MobiLife Book*. John Wiley & Sons Ltd.

[11] Joseph AD, Tauber JA and Kaashoek MF 1997 Mobile computing with the rover toolkit. *IEEE Transactions on Computers* **46**, 337–352.

[12] Lehman TJ, Cozzi A, Xiong Y, Gottschalk J, Vasudevan V, Landis S, Davis P, Khavar B and Bowman P 2001 Hitting the distributed computing sweet spot with tspaces. *Comput. Netw.* **35**(4), 457–472.

[13] Mascolo C, Capra L, Zachariadis S and Emmerich W 2002 Xmiddle: A data-sharing middleware for mobile computing. *Wirel. Pers. Commun.* **21**(1), 77–103.

[14] Ilkka J and Vainio C 2007 *Nokia Mobile Web Server*. Nokia.

[15] Bulander R, Decker M, Schiefer G and Kolmel B 2005 Comparison of different approaches for mobile advertising *WMCS '05: Proceedings of the Second IEEE International Workshop on Mobile Commerce and Services*, pp. 174–182. IEEE Computer Society, Washington, DC, USA.

[16] Mahemoff M 2006 *Ajax Design Patterns*. O'Reilly Media, Inc.

[17] W3C 2008 *Widgets 1.0: Requirements*. [W3C Working Draft].

[18] Jennings C and Mahy R 2006 Managing Client Initiated Connections in the Session Initiation Protocol (SIP). Internet-Draft draft-ietf-sip-outbound-06, IETF. Work in progress.

11

Conclusions

In this chapter we briefly summarize the main themes of the book and consider the current and near-future trends in mobile middleware. As mentioned throughout the book, mobile platforms play a crucial role in success of mobile devices. A good platform empowers developers and fosters creativity. The current platform landscape is heterogeneous and several different operating systems, programming languages, and interfaces are used. This has traditionally resulted in rather complex mobile software development and testing processes.

In this book we have focused on platform structure and capabilities, network support, architectural and design patterns, asynchronous messaging and publish/subscribe, data synchronization, and security. A mobile platform needs to be flexible and extensible not only in the distributed environment but also in the local environment. We surveyed current and emerging platforms and it can be said that they are rather closed development environments. For example Java ME MIDP and Android, although in principle are extensible, this extension cannot be done easily. This means that it is much easier to extend and modify functionality at the server-side instead of modifying the client.

On current high-end mobile phones, Java ME MIDP and its limitations, including a very small set of supported data structures, no access to file system, can be seen to be quite restricting for programmers. As an alternative, a more powerful Java ME configuration and profile can be used; however, these are not currently well supported on phones. For the Java developer, the best option would be to have the full Java 2 API available and also to have an easy way to access platform functions. Middleware can be used to alleviate some of these issues by either providing them using Java APIs or as system functions that can be invoked using IPC or local Sockets. On more powerful devices, many functions can be provided through a web server and this was one of our example use cases. A mobile server running on the phone can serve both local and remote requests and perform various functions. This would enable a seamless integration of local resources and functions, and remote services. If mobile web servers become popular and are shipped with mobile devices, they would pave way for the integration of web and mobile technologies. On the other hand, mobile web servers require infrastructure support and care needs to be taken in their configuration so that privacy and security can be maintained.

One step towards extensibility and universality is to employ a common and interoperable message bus that supports component discovery, capability negotiation, and

Mobile Middleware   Sasu Tarkoma
© 2009 John Wiley & Sons, Ltd

communications. One possible mechanism could be a data-centric message bus or a tuple space. Indeed, either message passing or tuple spaces or both have been proposed as the key components of mobile and pervasive software. On the other hand these ideas have not yet found their way into products and standardization. HTTP is still the common denominator for communications and local sockets have turned out to be a working solution for enabling also intra-device communications.

In the introduction chapter, five key requirements were identified for mobile computing:

- Accessibility for ensuring that data is accessible by all entities.
- Reachability for ensuring that mobile devices and the service and resources that they host are reachable (and thus accessible).
- Adaptability which is needed to support good user experience and cope when the computing environment changes. Context-awareness is a key part of adaptability.
- Trustworthiness, which is needed to maintain user confidence in the services and the service ecosystem. This requires facilities for assessing and establishing trust between entities.
- Universality, which requires the use of interoperable standards-based protocols and formats. This is especially needed for loosely coupled systems that are combined at runtime.

Mobile middleware plays a crucial role in ensuring that the above requirements can be met. The middleware cannot meet these requirements alone, but needs support from the operating system and network protocol stack. Indeed, the current TCP/IP stack and the mobile and wireless environment can be seen as challenging for application development due to the use of NATs and firewalls. We presented a number of design patterns and middleware and protocol solutions that can be used to maintain accessibility and reachability even in challenging environments.

Today, the mobile and traditional IT fields of computing are converging and as a result, web technologies are becoming increasingly popular in the development and deployment of mobile applications. The motivation here is that it is not particularly difficult to implement scalable web applications, and this has led to a prospering service ecosystem exemplified by the many massively popular websites.

Indeed, many applications can be implemented based on current web technologies, such as AJAX, REST, OpenID, and other solutions. However, the web protocols do not directly work well with mobile and wireless links due to the special constraints of the environment.

Simply using current web technologies only supports applications that are natural to realize using the web's request-reply interactions style and are not particularly susceptible to variations in latency. In this book, we extensively discussed the patterns and systems for asynchronous communications and publish/subscribe. Asynchronous operation is particularly useful in mobile applications that need to react to changes in the environment.

Moreover, web technologies and the TCP/IP protocol stack have not been designed for wireless environments, which means that the performance is not optimal. A significant research effort has been put into making, for example TCP and HTTP, more efficient in wireless environments. Current standardization and development efforts include making XML and XMPP more suitable for mobile devices.

As mentioned, web technologies have their limitations and typically access to underlying system services and operating system is not possible. This has been addressed in some systems by providing specific APIs that expose some system functionality to applications, for example pertaining to location, messaging, security, and storage. Symbian and iPhone are examples of mobile platforms that give programmers the possibility to access low level functions if C++ or Objective-C are used, although it may require certification. Android is an example of a platform with a Java API for developers. Java ME is an example of a platform with a core Java API and then a number of extensions whose availability differs from platform to platform.

Many research projects and prototypes have addressed context-aware and adaptive operation. Support for adaptive operation can be seen as an important new trend in mobile applications and services. Adaptation can be realized in many ways, for example on client devices, on servers, using proxies and gateways, and through collaboration of the different entities. Indeed, the expectation is that services and software running on devices collaborate to realize a service for users.

Context-awareness brings forth several new challenges. Practical challenges with context-awareness and service personalization are context acquisition, privacy, and software testing and quality assurance. If applications and services are adaptive and context-aware, how can we test that they are working properly. This requires new kinds of testing solutions and methods for assuring that software is working properly and resulting in the desired and planned user experience.

Moreover, device and service discovery is still not available for developers in an easy to use and universal fashion. The current trend is to strive towards adaptive applications by both utilizing web technology and platform specific APIs.

Another way to approach challenges with heterogeneous device hardware and software is to utilize an interoperable hardware-based interconnect as exemplified by the NoTA system. This effectively pushes interoperability to the lower layer and supports more flexible combinations of software and hardware. The motivation behind NoTA is to be able to combine various hardware elements easily. Combined with additional middleware this paves way for hardware/software systems that can be configured managed at run-time.

Another approach would be to utilize virtualization techniques to support multiple operating systems and platforms on the same hardware, possibly at the same time. Virtualization can also be used to enhance security of a system. Current mobile devices do not yet support virtualization so this is a technology that is still maturing especially for mobile devices.

As an example, we can take any application that needs to push information to consumers, such as email or RSS. Current technological constraints require that push-based applications are implemented using polling. This is not a very efficient way of doing push. Moreover, the same constraints that make push difficult also create difficulties for VoIP calls and other media flows. The SIP protocol architecture and related NAT traversal specifications, solve most of these reachability issues; however, the resulting system is very complex. Although middleware and protocol stack extensions take care of the complexity for the developer, understanding of the environment and the protocols that are used is still vital for the system designer, developer, and tester.

In this book a number of security solutions were examined that are the basis for building trust and assessing the trustworthiness of entities in the distributed environment.

The basic solution is to utilize security tokens in combination with transport or network layer end-to-end security. Security tokens are also the building block for single-sign on (SSO) systems and domain and service federation. Given that mashups, combinations of web services, are becoming increasingly popular, this is an emerging technology domain on the web.

To summarize the current state of the art and trends in the near-future, a considerable amount of research and development has gone into developing solutions for different kinds of mobile and pervasive environments that support a wide variety of different applications. The solution landscape is still fragmented; however, the high-level XML, HTTP and HTTPS have emerged as solutions for browser-based services.

The next step here would be to support access to context information and enable more intelligent information processing on client devices. This is needed to fully realize the visions of pervasive and ubiquitous computing. On the other hand, the current distributed service trend is to build large computing clusters, server farms and data centers, and then connect these with end users. These parallel developments, namely mobile devices and data centers, can work together. Information processing can be distributed between mobile devices and data centers. This would require sophisticated rules for distributing application functionality and mechanisms for ensuring that private information is not leaked.

Given that there are about three billion mobile devices on the market today and the projections indicate that the number will approach five billion in the near future, the prospects for mobile applications, service, and middleware appear to be very promising. To be able to handle such a large amount of users with possibly widely differing device types and characteristics necessitates interoperable and high performance platforms, adaptive wireless communications, and highly scalable and available fixed infrastructure.

Index

Mobile Middleware   Sasu Tarkoma
© 2009 John Wiley & Sons, Ltd